BATTERIES

BATTERIES
The Future of Energy Storage

edited by
Subashani Maniam

JENNY STANFORD
PUBLISHING

Published by

Jenny Stanford Publishing Pte. Ltd.
101 Thomson Road
#06-01, United Square
Singapore 307591

Email: editorial@jennystanford.com
Web: www.jennystanford.com

British Library Cataloguing-in-Publication Data
A catalogue record for this book is available from the British Library.

Batteries: The Future of Energy Storage

Cover image: Created by Afzal, one of the authors of this book

ISBN 978-981-5129-38-0 (Hardcover)
ISBN 978-1-003-51288-2 (eBook)

Contents

Preface

Batteries have received outstanding attention since the conception of their principle back in the 1780s. Historically, the lead–acid battery technology is the oldest battery type but still dominates the yearly energy storage capacity despite its major drawbacks, such as limited energy density (30–40 Wh/kg) and low energy efficiency. Alkaline batteries have also gained popularity, with a preference for the metal hydride type, that can reach energy densities around 100 Wh/kg. A more recent introduction is lithium-ion batteries, which have energy densities that reach about 250 Wh/kg.

Batteries are of paramount importance for energy storage for domestic purposes and industrial use in our current and future societies. The development of batteries beyond lithium-ion has changed our understanding of the materials and architecture of batteries. The combination of lithium with other elements has produced excellent new materials for energy storage. Pushing the boundaries of materials development by incorporating polymers and organic small molecules has generated classes of compounds with unique redox properties. These organic-based batteries have huge potential to compete equally with inorganic materials. Ultimately, the design of batteries, coupled with renewable electricity sources, has evolved to hybrid-based flow batteries. Such systems are predicted to generate high energy densities with low decay rates, allowing large-scale energy storage. The chapters in this book are written by experts in the field and aim to cover the most recent advances.

Finally, I thank Prof. Hiroyuki Nishide for introducing me to battery research. I also thank my husband, Sathish, and my son, Daarshvinn, for being a constant source of motivation and inspiration.

Subashani Maniam

March 2024

Chapter 1

Advancements in Battery Technology: Beyond Lithium-Ion Batteries

Afzal,[a] Oshadie De Silva,[a] Mudasira Bhurt,[b] Shahdev Sajnani,[c] and Subashani Maniam[a]

[a]*Department of Applied Chemistry and Environment Sciences, School of Sciences, STEM College, RMIT University, GPO Box 2476, Melbourne, VIC 3001, Australia*

[b]*School of Chemical Engineering, Dalian University of Technology, 116024, Dalian, Liaoning, P.R. China*

[c]*School of Environmental Science and Engineering, Tianjin University, 300354, Tianjin, P.R. China*

subashani.maniam@rmit.edu.au

1.1 Lithium-Ion Batteries

1.1.1 Introduction

With the groundbreaking work of Alessandro Volta two centuries ago, the world's first electric battery was introduced [1]. Then the lead–acid battery was developed by Gaston Planté and lithium-ion batteries (LIBs) by John B. Goodenough, Akira Yoshino, and

Batteries: The Future of Energy Storage
Edited by Subashani Maniam

ISBN 978-981-5129-38-0 (Hardcover), 978-1-003-51288-2 (eBook)
www.jennystanford.com

M. Stanley Whittingham, followed by the introduction of the first electric battery [2–4]. These hallmark inventions revolutionized portable electronics, paving the way for mobile communication and electric transportation due to favorable energy/power densities and advancements in manufacturing.

LIBs are constantly in demand as they are one of the popular energy devices used for mobile phones, portable electric devices, electric vehicles (EVs), and aerospace. The demand for devices using LIBs has opened new ventures for researchers to explore this technology. LIB technology has both advantages and disadvantages. Despite the advantages, such as low maintenance, prolonged battery life cycle, and potential for many applications, the drawbacks of LIBs have gained much attention.

1.1.2 Fundamentals of LIBs

The basic LIB comprises a solid anode and a cathode, while the electrolyte is a liquid, as shown in Fig. 1.1. The anode (negative electrode) is made from graphite, while the cathode (positive electrode) is made of lithium metal oxide [5]. Upon charging, Li ions are deintercalated from the cathode surface and transferred across the electrolyte, and then these Li ions intercalate within graphite layers. On the other hand, the oxidation of the cathode material occurs during charging, while the lithiation occurs on the anode material. In contrast, the reverse process occurs in discharge, and electrons move through the external circuit. The cathode becomes more positive with Li ions during discharging, which facilitates the attraction of electrons toward the cathode. The battery absorbs and stores energy during the process [6, 7]. There is a separator between the anode and the cathode, which is an insulator, to avoid physical contact between the anode and cathode materials. This further facilitates the Li ions to diffuse through the electrolyte and avoid the flow of electrons inside the cell. The first discovered cathode material for LIBs is $LiCoO_2$ (LCO), and the state-of-the-art cathode materials are mainly transition metal intercalated lithium metal oxides. These include layered $LiMO_2$, spinel LiM_2O_4, olivine $LiMPO_4$, silicate Li_2MSiO_4, tavorite $LiMPO_4F$, and borate $LiMBO_3$ [6, 8]. Typically, the cell voltage of LIB is 3.7 V, and it has a capacity of 150 Ah/kg and power around 200 Wh/kg.

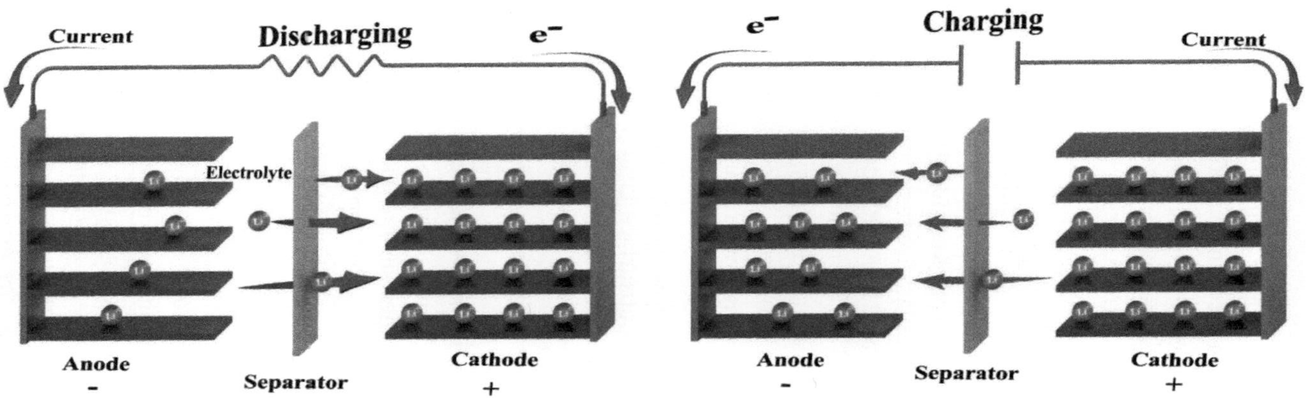

Figure 1.1 Basic working principle of a lithium-ion battery.

In prior investigations focusing on cell development and characterization, researchers utilized "harvested" electrodes to study the degradation mechanisms of 18,650 cells under different cycling and storage conditions [9]. The experiments revealed that the rise in cell resistance primarily originated from the lithium nickel cobalt aluminum (NCA) oxide cathode ($LiNi_{0.8}Co_{0.15}Al_{0.05}O_2$ or $LiNi_{0.8}Co_{0.1}Al_{0.1}O_2$), while side reactions at the graphite anode resulted in the consumption of active lithium. In a study involving pouch cells containing LCO cathodes and graphite anodes, researchers observed that the primary capacity loss occurred at the anode, and lithium deposits were identified after storage at higher temperatures [10]. Comparative tests were conducted to assess the performance of LCO and NCA cathodes in 18,650 cells, especially during storage at 45 °C. Remarkably, the LCO cathode exhibited significantly greater capacity loss compared to the NCA cathode or the graphite anode in either cell. This distinction was linked to the thickness of the surface layer with a cubic/rock salt structure on the cathode particles. Notably, after 2 years of storage, the layer was 4 nm thick on the NCA, whereas it measured 9 nm on the LCO cathode. Additionally, a 20 nm subsurface layer with a mixed rock salt/α-$NaFeO_2$ structure was detected on the LCO particles [11].

1.1.3 Limitations and Room for Improvement

LIBs dazzled the world with their high energy density, impressive cycle life, and freedom from the pesky memory effect that plagued older battery technologies such as nickel cadmium (Ni-Cd). These powerhouse performers were more than just compact; they were a quantum leap in miniaturization that spurred a rapid expansion in the realm of mobile devices. Their energy density has experienced a significant uptick, while the associated costs have seen a considerable decrease. However, the trajectory of future improvements may not follow this trend due to some of the limitations in LIBs [12].

One of the major limitations of an LIB is the degradation of its reversibility over time. This is mainly due to the subsequent growth of the passivation layer at the solid–electrolyte interface (SEI) during charging cycles [7]. This can increase the internal resistance and potentially decline the energy density of the LIB as cycles proceed due to the consumption of Li ions and solvent molecules. Therefore, generating a stable SEI in the first cycle is vital. Moreover, the formation of a suitable SEI layer is important for the safety and performance of the battery. For instance, the needle-like lithium deposits can pierce through the SEI film on the lithium substrate, especially at high current rates. This can lead to the SEI film breaking under stress, allowing the lithium to continuously grow outward, as illustrated in Fig. 1.2 [13]. Figure 1.2(a) shows deposition needles of varying lengths and diameters, while Fig. 1.2(b) demonstrates how the needle-like deposition grows through the SEI film on the lithium substrate. Incremental capacity analysis (ICA) and a mechanistic model are used to study the effects of low-temperature charging on LIBs by measuring the lithium loss indicator (LLI), lithium anode mass (LAM), and internal irreversible reaction (IIR). The model's parameters vary due to the lithium deposition on the anode. ICA revealed the loss mechanism as lithium deposition altering the anode's electrochemical behavior, aligning it with lithium metal anodes.

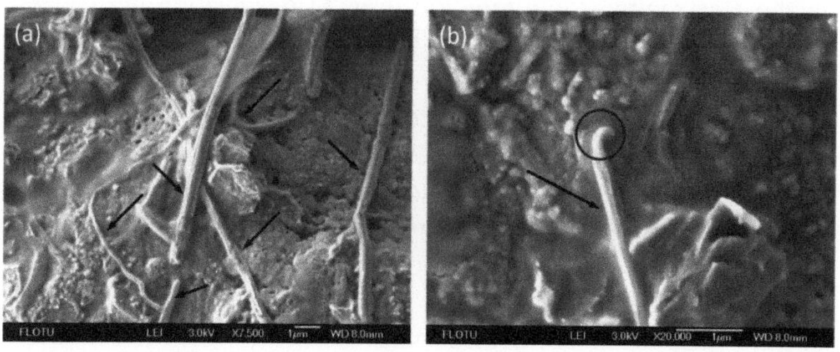

Figure 1.2 SEM images of the needle-like deposition, which might be lithium on the surface of the negative electrode cell. Reproduced from Ref. [13] with permission from Elsevier.

However, an improved SEI layer can be achieved by adding electrolyte additives such as reduction-type additives [14, 15]. Among the different types, carbonate derivatives, sulfur-containing additives, and nitrogen-containing additives are commonly identified. However, their functionality depends on their thermal stability, capacity retention, oxidation stability, impedance, and environmental hazards [14].

Another drawback of LIBs is the low capacity of the anode material/graphite (372 mAh/g), which presents a major restriction to the energy density of LIBs. However, graphite owes its popularity to its low cost, long cycle life, and relatively high energy density, but it falls short of matching the future needs of LIB technology [16]. The major bottleneck of graphite is its inability to compete with the ongoing energy demand. Alternative anode materials such as silicon and lithium have come under the scanner due to their extremely high theoretical specific capacities of 4200 mAh/g and 3860 mAh/g, respectively, far outstripping graphite's 372 mAh/g [17]. However, the severe volume change that occurs during lithium intercalation and deintercalation causes poor cyclability and serious safety issues in practical applications [18].

Moreover, low working temperatures can significantly affect the performance of LIBs. One of the practical limitations is the decrease in capacity and power when the working temperature is below 0 °C [19]. This has been observed in different applications, such as EVs, aerospace, and military devices. The main reason behind this is the reduction in ionic conductivity due to the increase in the viscosity of the electrolyte at low temperatures. Li desolvation occurs after the SEI is formed only during the first cycle, and this requires high activation energy. Further, the morphology of the electrode interfaces can be damaged due to the growth of lithium dendrites caused by the high electrochemical polarization of the electrodes and the larger Li-ion concentration polarization [20, 21]. However, this can be avoided by selecting appropriate solvents in a way to change the dielectric constant and viscosity accordingly.

On the other hand, elevated temperatures accelerate capacity decline and shorten battery lifespans. When subjected to conditions beyond their intended scope, LIBs can rapidly undergo self-heating or thermal runaway, potentially resulting in ignition. LIB abuse can be classified as thermal, electrical, or mechanical [22]. Thermal abuse involves overheating and fires [23], and electrical abuse encompasses overcharge and/or over-discharge [24], as well as short circuits. In addition, mechanical abuse includes impacts, penetration, and bending [11–14].

As mentioned earlier, one of the commonly observed drawbacks of LIBs is the thermal runaway process. Thermal runaway is evident by battery deformation or through the ability of sharp objects as a external stress to penetrate the battery. This is primarily due to the heat generated by an external heat source or internal electrochemical side reactions, leading to fires or even explosions in LIBs. The temperature increment can be caused by overcharging, over-discharging, and external circuiting [25–32]. Transportation and storage of LIBs are another important factor that needs attention. For instance, the International Civil Aviation Organization (ICAO) has restricted the cargo shipment of LIBs on passenger aircraft. LIBs transported on dedicated cargo aircraft must maintain a state of charge (SOC) below 30% [33]. Under thermal runaway conditions, batteries may release particulates, flammable gases, and toxic substances while generating jet-like flames or experiencing rupture. These incidents underscore the urgency of enhancing LIB technology safety. A deep understanding of failure mechanisms is crucial to advancing battery development that prioritizes safety alongside higher energy density.

The rate of decrease in capacity can occur during cycling (charging/discharging) and in storage. This process depends on many factors, such as temperature, electrode material, charging and discharging currents, electrolyte, and battery size. For instance, the impact of different experimental conditions on capacity degradation was examined and compared, and the summarized results are presented in Fig. 1.3.

In particular, the effects of varying cutoff voltage limits and charging current rates on capacity fade were explored. Figure 1.3(a) demonstrates that capacity degradation accelerates notably as the cutoff voltage increases, suggesting that restricting the cutoff voltage mitigates degradation, albeit at the cost of reduced charge capacity during cycling. Additionally, a reference performance test (RPT) revealed that at 50% SOC, the internal resistance was 30.49 mΩ at −10 °C and 10.23 mΩ at 25 °C, indicating that lower temperatures expedite the reaching of the cutoff voltage at the same current rate. For instance, cell 7 exhibits an average charging capacity of 8.94 Ah, equivalent to 75.54% of the nominal capacity. In Fig. 1.3(b), capacity evolution under various charging current rates is illustrated, where higher rates led to accelerated capacity degradation, likely attributed to the behavior of lithium plating at elevated rates.

On the environmental front, the production of LIBs raises serious concerns. The mining of lithium, cobalt, nickel, and other materials used in these batteries can have substantial environmental repercussions. The recycling rates for LIBs are currently low, underscoring the urgent need for efficient and cost-effective recycling technologies [34].

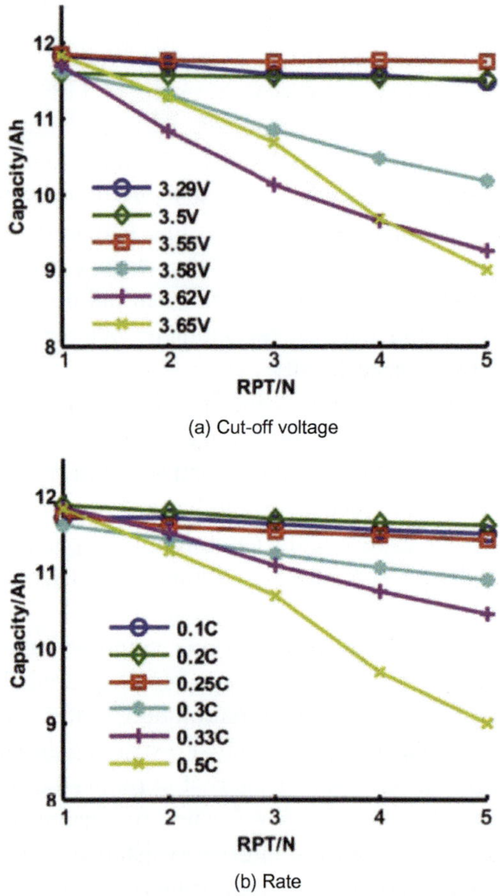

Figure 1.3 Evolution of standard test discharge capacity at intervals of 10 cycles: (a) charging cutoff voltage and (b) charge current rate. Reproduced from Ref. [13] with permission from Elsevier.

Despite these challenges, LIBs play an increasingly vital role in renewable energy storage, assisting in mitigating the intermittency issues of renewable sources such as wind and solar power. The future of LIBs lies in the effective resolution of their current limitations while continuing to enhance their energy density, lifespan, safety, and cost-effectiveness.

1.1.4 LIBs in the Future Automotive Industry

As the global automotive industry transitions toward electrification, LIBs have emerged as a cornerstone technology, offering unparalleled energy densities, rapid charging capabilities, and improved cycle lives. Their integration into modern EVs not only promises reduced greenhouse gas emissions but also ushers in a new era of efficient, sustainable, and high-performance transportation solutions. The optimal combination of energy storage devices that could reliably support engine cold cranking in terms of weight, cost, and efficiency was evaluated in some studies. For instance, Fedotov et al. [35] explored the possibilities of using LIBs in engine cold start systems. The study involved testing down-scaled specimens of LiC_6-LiNiMnCo (NMC) and LiC_6-$LiFePO_4$ batteries, along with a supercapacitor (SC) module, in a cold cranking mode. The researchers conducted various electrochemical tests to measure the resistance and performance of these batteries at high rates.

Considering the low currents and high energy capacity of LIBs, the researchers proposed combining them with SCs, which offer high power output at low temperatures. The study involved downscaling the system for physical experiments using test stations and climate chambers. The researchers selected specific LIB and SC configurations for the experiments and compared their performance to a reference lead–acid battery commonly used in starter systems. They conducted engine-cranking attempts with various currents and operating temperatures to assess the reliability and success rate of the combined energy storage system. The results showed that for a 12 V lead–acid battery with a rated capacity of 7 Ah, it was sufficient to replace it with a parallel-connected LIB and SC module with 16 V and 57 F to reliably start the engine with an overload current of 7 C.

Figure 1.4(a) displays the combined energy storage starter load currents when using an internal $LiFePO_4$ battery. Meanwhile, Fig. 1.4(b) illustrates the voltage variations in the combined energy storage during cold cranking with the $LiFePO_4$ battery incorporated. The reliable operation of the combined hybrid energy storage was confirmed at temperatures of 0 °C and higher. At −10 °C, the reliability significantly dropped, making three successful cranking attempts impossible. However, the $LiFePO_4$ battery provided one successful start attempt even at −40 °C, while the Li-NMC battery achieved its first successful start attempt at −30 °C. Overall, the research conducted estimated the optimal composition for an internal combustion engine and verified the results through experiments. The findings revealed that for stable operation at subzero temperatures, LIBs with higher discharge currents (about 0.1–0.8 C) and energy output in this temperature range are required.

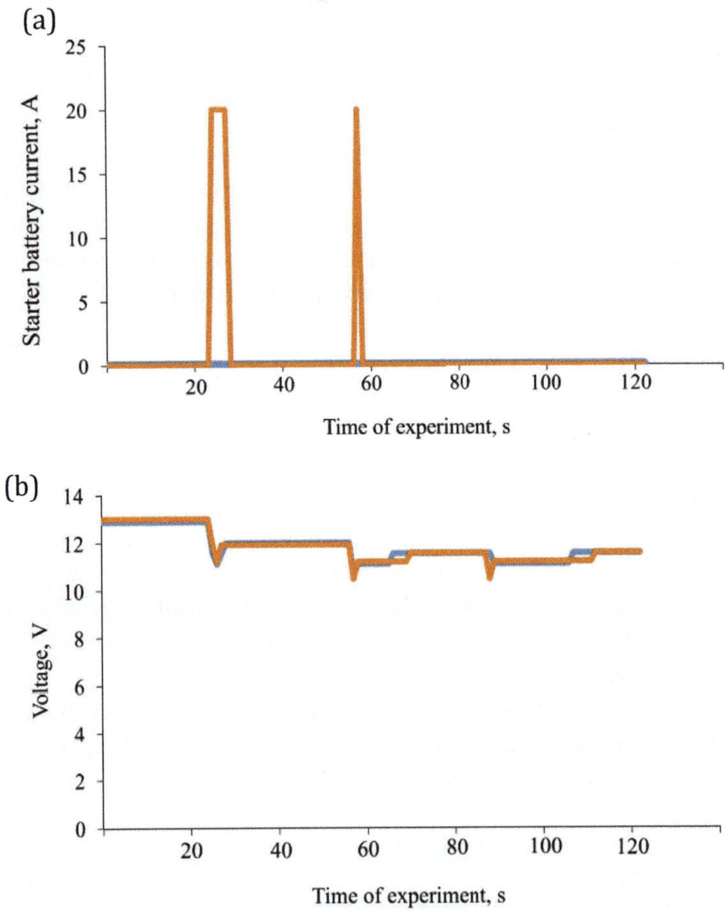

Figure 1.4 (a) Cranking current of combined energy storage with $LiFePO_4$, during a 7 C discharging at −40 °C: blue line—$LiFePO_4$ battery voltage; orange line—SC module voltage. (b) Voltage changes while cranking the combined energy storage with $LiFePO_4$, during a 7 C discharging at −40 °C: blue line—$LiFePO_4$ battery voltage; orange line—SC module voltage. Reproduced from Ref. [35] with permission from IOP Publishing.

EVs are quickly becoming a significant component of the automotive industry due to advancements in battery technology. The heart of this electric revolution is the LIB, which powers nearly every modern EV [36]. Over the past 30 years since their introduction, the energy density of LIBs has improved from around 150 Wh/kg to over 300 Wh/kg [37]. This has allowed the range of EVs to improve from around 60 miles per charge to over 300 miles in some cases [38]. Simultaneously, the cost of LIBs has fallen dramatically, from over $1100/kWh in 2010 to around $137/kWh in 2020 [39]. This is a decline of nearly 90% and has been a major factor in making EVs more affordable.

Despite their success, LIBs have their own set of challenges in automotive applications. For instance, LIBs need to meet high-performance requirements such as a calendar life of 10 years, a cycle life of 1000 cycles, a temperature range of −30 to 52 °C, and a cost of $100/kWh [40]. Current research is focused on improving the energy density of LIBs by using different anode materials, such as lithium and silicon. Silicon anodes can theoretically provide a capacity of 4200 mAh/g [41], which is 10 times greater than the capacity of traditional graphite anodes. However, silicon anodes suffer from significant volume expansion during charging, which can lead to degradation over time [42]. On the cathode side, high-nickel-content cathodes are being researched for their potential to increase energy density [43]. The most common high-nickel cathode material, NMC811 (nickel manganese cobalt oxide with a ratio of 8:1:1), has an energy density of around 20% higher than the previous generation of NMC622 material. Another area of focus is fast charging, which is a key factor in EV adoption. Currently, the best LIBs can be charged to 80% capacity in around 10 min [44]. However, Kim et al. have recently done a fabulous job by taking this trend up to a 12-min full charge [45]. But this is still significantly slower than refueling a gasoline-powered vehicle.

Looking forward, LIBs are likely to continue dominating. New research is looking at potential alternatives, including solid-state batteries, lithium-sulfur batteries, and even hydrogen fuel cells [46–48]. These technologies promise even greater energy densities and safety improvements, although they are still in the early stages of development.

1.2 Zinc-Based Batteries

1.2.1 Introduction

Zinc (Zn) is one of the most abundant elements in the earth's crust [49], and zinc and zinc oxides can be recycled using chemical, electrochemical, and thermal processes. An attractive feature of zinc is its multivalent nature, which allows the possibility of transferring multiple electrons per ion. This implies that more energy can potentially be stored in the same space [50]. Zinc is easily mined with a relatively low environmental impact compared to other metals, such as cobalt and lithium, for batteries. Zinc-based batteries are of interest because they can be manufactured at a low cost with high specific energy characteristics compared to other available types of batteries [51]. Perhaps the most appealing aspect of zinc-based batteries is their sustainability. They present an effective solution to energy storage that aligns with our growing environmental consciousness [52].

The economics of zinc further reinforce its viability. The cost of zinc is only a fraction of that of lithium. Considering the increasing demand for energy storage solutions, affordable and easily available zinc could make a significant difference in the worldwide adoption of sustainable energy technologies [53]. Longevity and cost-efficiency are pivotal factors when evaluating batteries. Conventional Zn-ion batteries (ZIBs) have been overshadowed by traditional LIBs in terms of output voltage and energy density. Generally, the energy density of a traditional LIB falls between 180 and 230 Wh/kg, while Zn batteries have shown 70–140 Wh/kg. Chao et al. [54] have spotlighted the promise of zinc-based electrochemistry in energy storage, particularly due to its high energy density (409 Wh/kg) given by the Zn-MnO_2 system.

1.2.2 Fundamentals of ZIBs

The interest in ZIBs relies on some key features, such as the high redox potential of Zn (−0.763 V versus SHE), the potential of using aqueous and nonaqueous electrolytes [55, 56], low toxicity, and the reversibility of Zn plating/stripping. The working principle of ZIBs is similar to that of LIBs. ZIBs consist of Zn metal as the anode, a cathode material with the capability to store zinc ions, and an electrolyte containing Zn ions, which allows the exchange of ions between the anode and the cathode during charging and discharging cycles.

The anode of ZIBs can be a zinc foil, zinc powder (with conductive carbon, binders, and a supportive layer), electrodeposited zinc nanoparticles (NPs), electroplating with organic additives, etc. [57]. However, these materials have their own advantages and disadvantages. For instance, the most commonly used Zn foil is readily available, which is a plus point in commercializing. However, the limited surface area has compromised the homogenous electric field distribution, and it can become an insulator once the passivation layer is formed [58]. Further, electrodeposited zinc anodes exhibit high-capacity retention (~13%) with $LiMn_2O_4$ cathode material as it has different surface texture and crystallographic orientations. Such materials are more porous and enhance the active surface area as well [59, 60]. Furthermore, electrically rechargeable Zn anodes have been investigated to eliminate different drawbacks of the Zn anode, such as dendrite growth, passivation, hydrogen evolution, and shape change [61].

In contrast, researchers have faced lots of challenges in developing a suitable Zn host cathode material, mainly due to the high polarization effect of Zn^{2+} ions. Different types of cathode materials have been investigated throughout the years, such as manganese oxide materials (α-MnO_2, β-MnO_2, γ-MnO_2, and Δ-MnO_2). The ability to insert more Zn ions, especially due to layered and tunneled structures, has been tailored to depict high electrochemical performance [62].

1.2.3 Aqueous ZIBs

Particularly promising are aqueous ZIBs, which employ water-based electrolytes. These batteries present several advantages in terms of safety, cost, and environmental impact. To add to their efficiency, aqueous electrolytes also have a higher ionic conductivity compared to nonaqueous electrolytes. Researchers have extensively studied the use of manganese dioxide and vanadium oxides as cathode materials for aqueous ZIBs. For instance, the MnO_2 cathode in Zn/MnO_2

batteries can transition from a tunnel-type to a layered structure, thus enabling the insertion or extraction of Zn^{2+} ions [63–65]. Compared to manganese-based ones, vanadium-based cathodes have several advantages. They display various oxidation states and exhibit more stability. Furthermore, the V-O coordination can adapt to different polyhedral units, changing based on the V oxidation state and offering a wide range of possibilities for Zn^{2+} (de)intercalation. Among vanadium-based cathodes, zinc vanadium oxide hydrate ($Zn_{0.25}V_2O_5{\cdot}nH_2O$) is noteworthy. The presence of water in the cathode facilitates the intercalation of Zn^{2+} due to the expanded structure. The battery performance and lifespan are enhanced as the electrochemical behavior is highly reversible. Moreover, the zinc vanadium oxide ($Zn_{0.3}V_2O_5{\cdot}1.5H_2O$) cathode, coupled with a 3 M, aq. $Zn(CF_3SO_3)_2$ (zinc trifluoromethanesulfonate or zinc triflate) electrolyte and a Zn anode, has exhibited outstanding performance [66]. As the battery cycles, the cathode undergoes a phase transition to form hierarchical cathode nanoflowers, improving the surface contact between the electrode and the electrolyte and, thereby, enabling more active Zn-ion storage sites [67–69].

In their groundbreaking work, Wan et al. [70] spotlighted the promising trajectory of aqueous ZIBs noted for their cost benefits and heightened safety features. Employing layered $VOPO_4$ cathodes in conjunction with a water-in-salt electrolyte, the team developed ZIBs that effectively harness the power of oxygen redox chemistry within a high-voltage domain. This innovation accounted for a 27% increase in capacity, pushing the average operating voltage to an impressive 1.56 V. As a direct consequence, energy density experienced a surge of approximately 36%. Highlighting the efficacy of their design, they demonstrated that Zn/$VOPO_4$ batteries could consistently achieve an average operating voltage of 1.56 V.

Furthermore, in a study conducted by Liu et al. [71], rechargeable aqueous zinc-ion batteries (RAZIB) were noted for their safety and robust capacity, with manganese oxides, such as nanorod β-MnO_2, being especially prominent due to their cost-effectiveness and nontoxicity. Using hydrothermal methods, they demonstrated that Zn//β-MnO_2, when used as a cathode, can achieve a significant capacity of 355 mAh/g at 0.1 A/g and impressively sustain 110 mAh/g after 1000 cycles at 0.2 A/g. The study, diverging from traditional mechanisms, proposes a unique proton conversion in conjunction with Mn ion dissolution/deposition. Through a series of advanced techniques, such as in-situ XRD and ex-situ SEM, the team deduced that β-MnO_2 interacts with protons, resulting in the formation of Mn-OOH, which then experiences multiple transformations during the battery life cycle. These groundbreaking findings hold potential to reshape our understanding of Zn-MnO_2 battery chemistry and might drive its broader acceptance and evolution in the industry. For instance, the previously mentioned battery with a $Zn_{0.3}V_2O_5 \cdot 1.5H_2O$ cathode demonstrated 96% capacity retention after 20,000 cycles, signifying excellent durability.

Wu et al. [72] introduced an innovative aqueous ZIB utilizing graphene scroll-coated α-MnO_2 as the cathode. The use of reduced graphene oxide (rGO) scrolls notably boosts the MnO_2 electrical conductivity and minimizes its dissolution during cycling. This cathode achieves a record energy density of 406.6 Wh/kg (362.2 mAh/g) at 0.3 A/g and showcases outstanding long-term cycling stability with a 94% capacity retention after 3000 cycles at 3 A/g. More impressively, even at a high current density of 7 A/g, it retains 87.4 mAh/g after 800 cycles.

1.2.4 Nonaqueous ZIBs

Nonaqueous zinc-ion batteries (NAZIBs) have gained considerable attention in the contemporary landscape of energy storage systems. Their emergence was spurred by the quest to harness the rich benefits of zinc, a material known for its abundance, with reserves surpassing most common battery metals and a notable volumetric capacity of 5855 mAh/cm^3 [73]. Beyond these inherent advantages, NAZIBs have showcased potential improvements over traditional battery technologies in various aspects.

At the heart of NAZIB's potential lies its diverse range of electrolyte choices. These electrolytes have the potential to augment both the stability and performance of the battery. For instance, ether-based electrolytes like 1,3-dioxolane (DOL) [74] have been recognized for their capability to improve the cycling stability of NAZIBs. Additionally, nonaqueous electrolytes offer a distinct advantage over their aqueous counterparts due to their broader operational voltage window, often exceeding 3 V. This expansive window sidesteps the limitations of proton/hydroxyl electrolytic decomposition characteristic of aqueous systems. Han et al. [75] delved into the oxidation behaviors of several salts and organic solvents. Their findings spotlighted acetonitrile's superior stability when compared to alternatives such as propylene carbonate, *N,N*-dimethylformamide, and diglyme. An electrolyte formulation consisting of 0.5 M $Zn(TFSI)_2$ in acetonitrile exhibited an extensive electrochemical stability window of up to 3.7 V against Zn/Zn^{2+}. Drawing from this insight, another nonaqueous electrolyte was crafted using a 0.3 M solution of $Zn(CF_3SO_3)_2$ in acetonitrile, which, when paired with a $Zn/ZnAlxCo_2{-}xO_4$ battery, delivered a notable voltage output nearing 1.7 V [76]. Furthermore, *N,N*-dimethylformamide also showcased its potential as a viable solvent in the context of a Zn/phenanthrenequinone macrocyclic trimer (PQ-MCT) battery [77], achieving an anodic stability close to 2.5 V against Zn/Zn^{2+}.

Rechargeable ZIBs face challenges from side reactions and dendrite growth on the zinc anode. However, work done by Li et al. [78] in which a novel lignin-containing cellulose nanofiber (LCNF)-polyacrylamide hydrogel (LPH) electrolyte has been developed addresses these issues through its unique structure and moisture retention. This results in a wider voltage window, from 0.2 V to 1.9 V. Zn/Zn symmetric cells with LPH achieve a 4000 h lifespan at 0.5 mA/cm^2 and 1 mAh/cm^2. Moreover, Zn/MgVO full batteries using LPH exhibit a capacity of 149.2 mAh/g, 91.9% retention after 4000 cycles at 5 A/g, and an impressive energy density of 255.4 Wh/kg for over 5000 cycles at 341.7 W/kg. This LPH electrolyte offers a promising direction for enhancing the performance of rechargeable ZIBs.

However, among these, bis(trifluoromethanesulfonyl)imide (TFSI)-based electrolytes have particularly risen to prominence. TFSI's adoption in NAZIBs has led to remarkable improvements in both ionic conductivity and overall stability. Recent studies have underscored the prowess of TFSI, with batteries leveraging this electrolyte showcasing more than 750 cycles [79]. Such performance metrics not only establish TFSI's centrality in NAZIB development but also hint at the future trajectory of research in this domain.

As research continues to delve into the synergistic relationship between the chosen electrolyte and cathode materials, such as vanadium-based compounds and manganese dioxide, the promise of

NAZIBs becomes more pronounced. They present an amalgamation of sustainability, courtesy of zinc's abundance, and performance, fueled by cutting-edge research on materials and electrolytes [80]. Past research on aqueous ZIBs has shown that hydrogen evolution, corrosion of Zn, and passivation are more susceptible to occurring in a protonated environment. These challenges can be mitigated by using nonaqueous electrolytes such as organic electrolytes, deep-eutectic solvents (DESs), and room-temperature ionic liquids (RTILs). Moreover, nonaqueous electrolytes are stable for use with high-voltage host materials. Han et al. [81] have developed a groundbreaking non-flammable hydrous organic electrolyte for ZIBs, combining a hydrated $Zn(BF_4)_2$ salt with an ethylene glycol solvent. This formulation effectively counters dendrite growth, corrosion, and hydrogen evolution, common challenges in aqueous ZIBs, by promoting the in-situ formation of a protective ZnF_2 layer. Key results showcase the Zn anode sustained cycling for over 4000 h at 0.5 mA/cm^2, an impressive coulombic efficiency of 99.4%, and an aerial capacity of 5 mAh/cm^2. Additionally, the electrolyte offers consistent performance across a wide temperature range, from −30 °C to 40 °C. Notably, Zn//V_2O_5 full cells using our electrolyte displayed significantly improved capacity retention compared to those with traditional aqueous $ZnSO_4$ electrolytes. This innovation paves the way for future advancements in Zn battery technology, merging safety, performance, and sustainability.

1.2.5 Challenges in Commercializing Zinc-Based Batteries

ZIBs offer many advantages, but their commercialization still faces several challenges. One of the main identified drawbacks is the formation of zinc dendrites, similar to lithium metals, tree-like structures that can cause short-circuiting and, ultimately, affect the lifetime of the battery. Moreover, Zn dendrites can easily peel off from the electrode surface and can reduce battery performance and capacity. This problem is being actively researched, and several solutions are in the works, including the addition of additives to electrolytes (polymers, organic molecules, metal ions with more positive electrode potential), the introduction of additives [82] to electrodes, the modification of electrode structure, etc. [83–86]. For instance, the addition of polymers to electrolytes can control the local current distribution near the tip, increase surface polarization, and reduce the rate of Zn reduction [87]. In contrast, organic molecules such as aldehydes, alcohols, and organic acids can block the dendrite growth sites and either prevent or slow down the dendrite growth rate [82]. However, the addition of metal ions is preventing the reduction of Zn^{2+}, and this has been explored using Bi^{3+} and Pd^{2+} [88, 89].

The deterioration of the crystal structure of electrodes is another limitation of zinc-based batteries during repetitive intercalation/deintercalation of hydrated Zn^{2+} and H^{+}. This process can eventually lead to an increase in the capacity fade rate. Further, there is a potential to dissolve Mn- and V-based cathode materials during the cycling process. However, this can be limited through the structural engineering of electrodes, such as the incorporation of oxygen vacancies and ion doping, as these can improve electronic conductivity, enhance transport channels, increase active sites, and lower ion interaction barriers [90]. A study revealed that Zn coated with N-doped graphene (NGO) and $LiMn_2O_4$ electrode systems delivered 89.1 mAh/g at 1 C over 600 cycles [91]. This is mainly due to the larger surface area given by the NGO for offering more nucleation sites, thus regulating interfacial electric field distribution and depleting dendrite formation. By addressing these shortcomings, we can potentially develop low-cost, energy-efficient, and chemically stable zinc-based batteries. This is a promising area of research, and there is still much room for improvement. However, the potential benefits are significant, and this approach could lead to the development of new batteries with better performance and a longer lifespan.

1.3 Sodium-Ion Batteries

1.3.1 Introduction

Sodium is the most common element found on earth. It is particularly plentiful in seawater, with about 10,000 parts per million of the earth's oceans consisting of sodium ions. Sodium is also found in various minerals, such as feldspar and sodalite [92]. Furthermore, while sodium is widely distributed on a global scale, the distribution of lithium is notably more concentrated, with its primary reserves located in specific regions [93].

Sodium shares close similarities with lithium in terms of properties. Its redox potential aligns closely with lithium's ($E_{Na+/Na}$ = −2.71 V versus SHE, $E_{Li+/Li}$ = −3.02 V versus SHE), and it boasts greater crustal reserves (Na: 2.8 wt%, Li: 0.0017 wt%) [94]. Notably, in the initial phases of LIB research, sodium-ion batteries (SIBs) also garnered attention. While LIBs underwent systematic investigation into storage mechanisms, material design, and performance enhancement, SIBs lagged [95]. The key distinction lies in the ion carriers, facilitating the potential for LIB manufacturers to pivot toward SIB production by leveraging existing manufacturing infrastructure [96].

On the manufacturing side, if we assume similar efficiencies as LIBs, a factory producing 10 GWh of LIBs per year might be able to produce a similar volume of SIBs. However, this would depend on a variety of factors, including the specific materials used to build the battery, the manufacturing processes required for those materials, and the yield of the manufacturing process. SIBs may be particularly well-suited for large-scale energy storage due to their potential lower cost and acceptable energy density. For grid storage, energy density is less of a concern than cost and lifespan. If SIBs can achieve similar lifespans to LIBs (typically 1000–5000 cycles) and do so at a lower cost, they could be a competitive option for stationary energy storage.

1.3.2 Mechanism of SIBs

At the core of battery design lie the cathode and anode electrodes, demarcated by a glass fiber (GF) separator [97], a safeguard against short-circuiting. The SIB system operates akin to a "rocking-chair" mechanism: During charging, Na^+ ions traverse from cathode to anode through the electrolyte, and concurrently, electrons flow through the external circuit to the anode, maintaining equilibrium. Conversely, this process reverses upon discharging. One of the challenges to be overcome for scaling up SIB production is developing materials that can withstand the larger sodium ions and still maintain good performance.

First, SIBs operate on a similar principle to LIBs: They shuttle ions from the cathode to the anode during charge and back during discharge. However, Na^+ boasts a notably larger ionic radius (1.02 Å compared to Li^+'s 0.67 Å, where 1 Å equals 10^{-10} m) and a heavier atomic weight (23 g/mol compared to Li^+'s 6.94 g/mol) [98]. This implies that the electrochemical equivalent in SIBs is over three times heavier than that in LIBs. Consequently, an increase in this magnitude is anticipated to lead to less favorable transport kinetics for Na^+, reduced phase stability due to interphase formation, lower solubility in solids, unfavorable gravimetric and volumetric energy density, and diminished Madelung energy. These factors collectively contribute to a lower operational voltage for SIBs [99]. Table 1.1 compares the properties of lithium and sodium relevant to battery technologies. It contrasts their physical and electrochemical attributes, including ionic radius and ionization energy. Notably, sodium is more abundant and cheaper than lithium, but lithium boasts higher capacities for battery performance.

Table 1.1 Comparison of key properties of lithium and sodium

Comparison	Li	Na
Atomic mass	6.941	22.99
Ionic radius (Å)	1.45	1.80
Density (kg/m^3)	534	968
SHE versus Std reduction potential (V)	−3.038	−2.712
Gravimetric capacity (mAh/g)	3829	1165
Volumetric capacity (mAh/cm^3)	2062	1131
Abundance in earth encrustation (%)	0.002	2.358
Equivalent mass abundance in earth encrustation (mol/kg)	0.00288	1.0265
Price ($/kg)	17.00	0.15
Thermal stability	Moderate	High
Electrochemical stability	High	Moderate
Conductivity (S/cm)	0.1	0.5
Toxicity	Low	Low

Prominently, ion size stands out as the decisive factor shaping crystallographic distinctions within sodium and lithium compounds.

Replacing lithium with sodium results in structure expansion, often accommodated by increased interlayer spacing. This accommodative shift leads to the stability of sodium analogues in numerous layered lithium compounds. Examples include $NaCoO_2$ and $LiCoO_2$, as well as $NaNi_{0.5}Mn_{0.5}O_2$ and $LiNi_{0.5}Mn_{0.5}O_2$. Furthermore, compounds with open frameworks readily accommodate sodium or lithium ions. The sodium superionic conductor (NASICON) structure exemplifies this versatility, as seen in both $Na_3V_2(PO_4)_3$ and $Li_3V_2(PO_4)_3$. In contrast, densely packed crystals prompt distinct structures for Na-based and Li-based compounds due to packing considerations. For instance, $NaFePO_4$ adopts the maricite structure, while $LiFePO_4$ takes on the olivine structure. It is worth noting that metastable olivine $NaFePO_4$ can also be obtained. During deintercalation, volume changes are typically more pronounced in Na-based compounds within layered structures. Notably, residual Na ions in these structures shift from an octahedral to a trigonal-prismatic coordination, while smaller Li ions prefer tetrahedral coordination.

Concurrently, to balance the positive charge introduced by incoming ions, electrons traverse an external circuit toward the cathode, facilitating ion-based battery discharge. Conversely, during battery charging, this process reverses to support electricity generation. Although SIBs and LIBs share a fundamental operational principle, SIBs present differences in energy density and reaction kinetics. These discrepancies stem from the higher redox potential of SIBs (of −2.71 V) versus the SHE for the Na^+/Na system, the larger ionic radius, and the heavier ion. This contrasts with the LIBs' Li^+/Li system, with a redox potential of −3.04 V and a lower-molecular-weight ionic radius.

1.3.3 Choice of Electrodes

Na^+ presents a 34% larger size compared to Li^+ (1.02 Å versus 0.76 Å), posing challenges in finding suitable electrode materials for both cathodes and anodes. The key lies in locating materials capable of accommodating sodium ions, facilitating their transport, and enabling swift insertion/extraction. Advancing practical SIBs necessitates suitable electrode materials and electrolyte optimization. Key considerations include developing electrode materials with crystallographic structures that boast ample interstitial spaces to host sodium ions and ensure robust electrochemical performance.

Several types of materials are being explored for this, such as hard carbon anodes [100], glassy carbon anodes [101], and cathode materials such as layered metal oxides and polyanionic compounds [102].

1.3.3.1 Cathode materials

Diverse cathode materials have been explored for SIBs, encompassing layer and tunnel-type transition metal oxides [103], transition metal sulfides and fluorides, oxyanionic compounds, Prussian blue analogues [104], and polymers [105]. Researchers have investigated five distinct crystal structures designed to accommodate sodium-ion transportation and storage. Layered transition metal oxides, such as $NaMO_2$ (M = Co, Mn, Fe, and Ni), showcase versatile configurations with pathways for sodium-ion diffusion, making them promising candidates. In addition to layered oxides, alternatives such as phosphate polyanions are also under exploration. The overarching goal in SIB research is to optimize these diverse electrode material options to achieve superior performance, efficient ion diffusion, and ultimately enable the practical implementation of SIBs.

Metal oxide cathodes such as Na_xCoO_2 stand out as a particularly noteworthy example. The exploration of cathode materials for SIBs dates back to the early 1990s [106], and metal oxides have emerged as the most promising and extensively studied cathode materials for SIBs. A reversible intercalation of Na^+ ions is exhibited in the Na_xCoO_2 phase, accompanied by a phase transition in its layered structure. Similarly, Na_xMnO_2 polymorphs have also been extensively studied as cathode materials. The α-Na_xMnO_2 phase, which has a layered structure with monoclinic distortion, has shown better stability compared to its β-phase counterpart [107–109]. Materials such as $NaFePO_4$ (sodium iron phosphate) or layered oxide structures such as $NaNi_{0.5}Mn_{0.5}O_2$ are also explored. An $NaFePO_4$ cathode can exhibit a capacity of around 100–120 mAh/g [110], while layered oxide structures have shown a capacity of up to 140–180 mAh/g. For comparison, lithium iron phosphate ($LiFePO_4$) cathodes in LIB typically show a capacity of about 170 mAh/g [111].

The development of SIB cathodes faces challenges in understanding the transformations and reaction pathways during charge and discharge processes. However, significant progress has been made in improving the long-term cyclability of cathode materials. For instance, Jo et al. aimed to enhance SIB performance through the synthesis of pure α-$NaMnO_2$ cathode materials. They used a solid-state reaction with Mn_2O_3 and Na_2CO_3 at 750 °C. They explored Na/α-$NaMnO_2$ battery performance using two electrolytes: 1 M $NaClO_4$ EC/DEC and 1 M $NaBF_4$ TEGDME. The initial discharge capacity of Na/α-$NaMnO_2$ with 1 M $NaClO_4$ EC/DEC electrolyte was 146 mAh/g at 0.2 C, while with 1 M $NaBF_4$ TEGDME, it was 136 mAh/g. After 20 cycles, they decreased to 104 mAh/g and 97 mAh/g, respectively [112]. Other cathode materials such as $NaCrO_2$, $NaFeO_2$, and multi-cation oxides have been studied, but they exhibit weaker electrochemical performance compared to $NaMnO_2$ [113–116].

In addition to metal oxides, sulfate, phosphate, and fluoride [117–120] materials have been explored as cathodes for SIBs. $Na_2Fe_2(SO_4)_3$ has shown interesting electrochemical features with a voltage of 3.8 V and good capacity retention at high rates [121, 122]. $Na_3V_2(PO_4)_3$, known as NASICON, has demonstrated high-capacity retention and diffusion channels for Na ions due to its unique structure [123]. $Na_{1.5}VPO_{4.8}F_{0.7}$, a fluorophosphate, exhibits a pseudo-layered structure and delivers high-capacity retention and energy density [124].

1.3.3.2 Anode materials

When it comes to anode materials, graphite anodes commonly used in LIBs are not suitable for SIBs due to the larger ionic radius of Na^+ ions. Hard carbon has shown promise as a host material for Na^+ ion intercalation. Hard carbon, particularly C1600, has been identified as a suitable anode material for SIBs, offering decent rate capability and cyclability [125]. Other alternatives, including pure metals, alloys, hydrides, and oxides, have been explored [126, 127], involving mechanisms such as alloying and conversion reactions.

Graphite, a typical anode material in LIBs, offers only moderate lithium storage capacity (approximately 350 mAh/g) at around 0.1 V versus Li/Li^+ [128, 129]. Recent studies have highlighted graphite's limitations in effectively intercalating sodium ions. Non-graphitic anodes, including diverse carbonaceous materials such as carbon black and pitch-based carbon fibers, enable the insertion of sodium ions. Among these, hard carbons, synthesized at elevated temperatures from carbon precursors, have undergone thorough modeling, characterization, and thermal assessments for sodium cells. Regarded as the "first-generation" anodes for SIB systems, these non-graphitic carbonaceous materials have taken precedence. Unlike lithium, sodium metal is not employed as an anode in SIBs due to challenges such as dendrite formation, heightened reactivity, and an unstable passivation layer when used with most organic electrolytes at room temperature. The reactivity of metallic sodium with organic solvents and the propensity for dendrite formation during sodium metal deposition pose more significant hurdles than in the context of lithium anodes. Moreover, sodium's lower melting point at 97.7 °C introduces safety concerns when using Na metal electrodes under ambient conditions [104]. Hence, the emphasis is on adopting a genuine sodium-ion system where sodium ions are exchanged between cathodes and anodes, following a "rocking-chair" configuration.

Carbon nanotubes (CNTs) exhibiting one-dimensional (1D) nanostructure offer substantial advantages, including a large specific surface area, exceptional conductivity, chemical stability, and graphite-like characteristics, rendering them appealing as anode materials for SIBs. Despite these benefits, CNTs encounter a limitation in sodium storage capacity in SIBs. Notable progress has been achieved in the field of CNTs through strategies such as morphological and particle size control, heteroatom doping, and defect engineering aimed at enhancing their electrochemical performance. For instance, Tong et al. [130] synthesized a notable non-graphitic CNT grown on a smaller crystal CNT, forming a distinctive CNT-CNT structure through sequential steps involving in-situ growth of ZIF-8 on CNT, followed by tannic acid (TA) etching and carbonization. This composite exhibits a bamboo-like design, yielding a remarkable specific surface area of 476.4 m^2/g. Notably, as an anode material for SIBs, CNT-CNT-600 delivers a substantial specific capacity of 177.9 mAh/g at 1 A/g and impressive cycling stability over 2000 cycles, surpassing counterparts carbonized at higher temperatures and pure CNTs. Furthermore, nitrogen and phosphorus dual-doped CNTs were prepared [130], demonstrating a remarkable reversible capacity of 172 mAh/g at 10,000 mA/g alongside excellent cycle stability (180.3 mAh/g at 5000 mA/g after 3000 cycles), owing to the synergistic impact of dual heteroatom doping.

Another advancement involved the synthesis of nitrogen-doped bamboo-shaped CNTs featuring rich lattice defects, which exhibited a high specific capacity of 278 mAh/g at 50 mA/g, surpassing that of conventional CNTs by more than two-fold. Moreover, CNTs are commonly employed as conductive networks and buffer substrates in composites, mitigating lattice stress induced by volume expansion of other active materials during cycling.

Recent advancements in SIB anode research have seen diverse electrochemical mechanisms employed for sodium-ion storage. These mechanisms encompass intercalation-type materials, conversion reactions using transition metal compounds, alloys involving elements such as Bi, Sb, Sn, Ge, Pb, and P, dual-reaction strategies combining conversion and alloying, and organics. Intercalation-type materials offer remarkable cycle stability, yet their theoretical capacity falls short for high-energy-density SIBs [131]. Conversely, anodes based on conversion and/or alloying reactions can deliver higher theoretical capacities, but they grapple with intrinsic challenges, including low ionic/electronic conductivity and substantial volume change, leading to compromised rate performance and rapid capacity deterioration.

Organic materials exhibit enhanced sodium-ion mobility due to their structural flexibility but suffer from dissolution and poor electron conductivity [132]. Diverse optimization strategies have been pursued across all anode types, including heteroatom doping, defect engineering, nanostructure design, composite formation, and phase state adjustments. Notably, the discovery of commercial-value hard carbon marked a significant milestone in SIB anode research, reigniting interest in electrode materials [133]. Advancements have spanned transition metal oxides (TMOs), sulfides, organic compounds such as $Na_2C_8H_4O_4$, Sn, Ge, and complex hybrid structures. Intercalation-type materials, while stable over long cycling periods, face limitations in specific capacity for high-energy-density applications. Strategies such as heterostructure synthesis, nanostructure design, and composite construction have yielded significant progress in this realm. Intercalation-type materials for SIB can be categorized into two main segments: carbon-based materials and titanium-based oxides [134].

The flexibility of organic anode materials makes them appealing for SIBs. Conversion, alloying, and their combined forms show promise with high capacities, yet challenges include cycle longevity and improvements in kinetics due to structural transformations. Advancements in nanostructure design and composite formulation have propelled these materials forward. In sum, these collective efforts have greatly propelled the development of SIB anode materials. Numerous review papers have outlined various anode material types, yet the need remains for a consolidated understanding of sodium storage mechanisms.

1.3.4 Choice of Electrolytes

The pursuit of a novel electrolyte for SIBs stems from concerns surrounding the practicality and safety issues associated with conventional organic liquid electrolytes. The prevalent formulations for SIB electrolytes involve utilizing $NaClO_4$ or $NaPF_6$ salts dissolved in carbonate ester solvents, primarily propylene carbonate (PC) [135]. However, these organic electrolytes pose a challenge as they trigger continuous corrosion of metallic sodium anodes rather than forming a stable SEI. Comprehensive investigations by Komaba et al. [136], employing X-ray photoelectron spectroscopy (XPS) and time-of-flight secondary ion mass spectrometry (TOF-SIMS), revealed that when $NaPF_6$-based electrolytes are utilized, the SEI film predominantly comprises inorganic salt with precipitated NaF species on the surface of the hard carbon anode. Additionally, research led by the Palacin group [137] identified that the most optimal electrolyte configuration for hard carbon anodes consists of a mixture of $NaClO_4$ and $NaPF_6$ in an ethylene carbonate: propylene carbonate solvent blend.

A shift toward aqueous electrolytes instead of conventional organic counterparts presents a significant avenue for advancements in SIBs. Encouragingly, recent experiments involving an aqueous rechargeable battery utilizing $Na_2NiFe(CN)_6$ and $NaTi_2(PO_4)$ as cathode and anode materials, respectively, demonstrated favorable rate performance, cycle life, and a theoretical energy density of 42.5 Wh/kg [138]. This intriguing outcome implies the potential to unlock enhanced energy density by judiciously selecting appropriate electrode materials.

1.4 Metal–Air Batteries

1.4.1 Introduction

Metal–air batteries have attracted significant attention due to their potential for high energy output. This promise lies in their unique design, which combines a metal anode with an oxygen (or air) cathode, allowing them to exploit the abundant oxygen in the air as one of the battery reactants. Metal–air batteries have a metal anode and an air cathode, with the metals typically being lithium, sodium, potassium, zinc, iron, aluminum, or magnesium. Metal–air batteries first saw light in the late 19th century, when the primary zinc–air battery was designed [139]. Despite being one of the older members of the battery family, they have lagged in commercial development, with their potential only recently reignited. The renewed interest is primarily driven by the urgency to improve energy storage for EVs and grid-scale stationary energy storage.

The energy density of metal–air batteries can be three to thirty times greater than conventional LIBs. For example, zinc–air batteries, one of the more promising types, have a theoretical energy density of 1353 Wh/kg, significantly higher than the 150–200 Wh/kg of LIBs [140]. This energy density is enticing, especially for EVs, where the goal is to achieve a long driving range (e.g., 500 km) while keeping the battery weight as low as possible. In practice, zinc–air batteries have an attainable energy density between 350 and 500 Wh/kg, and this drop from theory to practice is a common feature among metal–air batteries. This energy gap can be traced back to a number of technical challenges that have kept these batteries from wider adoption.

1.4.2 Oxygen Reduction Reaction

One of the main hurdles lies with the air cathode. The oxygen reduction reaction (ORR) that occurs here is a slow process [141]. Efforts to enhance this process involve the creation of active air catalysts capable of accelerating ORR kinetics. The pursuit of enhancing ORR kinetics is pivotal for the development of efficient and sustainable energy systems, necessitating the creation of active air catalysts capable of accelerating this process. Traditionally, precious metals such as platinum and silver have been the go-to candidates for this purpose; however, their application comes with significant drawbacks, including high costs and vulnerability to contamination. These challenges have catalyzed a shift in focus toward non-precious metal catalysts (NPMCs) as potential substitutes for their precious counterparts. This research trajectory involves investigating metals such as iron, cobalt, nickel, and magnesium, all of which hold promising potential as catalysts with exceptional ORR activity.

Following extensive experimental investigations, platinum (Pt) and Pt-based catalysts have emerged as the leading candidates for efficient ORR catalysts. In the realm of low-temperature proton exchange membrane fuel cells (PEMFCs), platinum NPs supported on carbon (Pt/C) or other Pt-rich materials have gained prominence as cathode catalysts [142]. Currently, enhancing atom utilization and augmenting the inherent catalytic potential of Pt involve employing strategies such as diminishing Pt nanostructure dimensions, creating alloys, and fabricating tailored nanostructures with Pt-enriched surfaces. These approaches serve to curtail the required Pt dosage [143, 144]. Numerous intricate Pt-based nanostructures have been documented to showcase notably elevated activity in ORR. The utilization of PtCo alloy catalysts in hydrogen fuel cell vehicles underscores the extensive potential of alloy catalysts within the realm of fuel cell technology. Wang et al. [145] introduced a new class of platinum–cobalt nanocatalysts, Pt(3)Co@Pt, with ordered Pt(3)Co intermetallic cores and a 2–3 atomic-layer-thick platinum shell. These nanoparticles outperformed disordered Pt(3)Co alloy nanoparticles and commercial Pt/C catalysts in both activity and stability for the ORR in fuel cells. Specifically, Pt(3)Co@Pt demonstrated a 200% increase in mass activity and a 300% increase in specific activity, with minimal activity loss after 5000 potential cycles. Electron energy loss spectroscopic (EELS) mapping confirmed the Pt-rich shell and Pt(3)Co core structure. Electrochemical testing revealed a significant positive shift in half-wave potential ($E_{1/2}$) for Pt(3)Co@Pt/C-700 compared to Pt/C, indicating marked activity improvements.

Li et al. [146], in their quest to enhance Pt mass activity for ORR, employed a strategic approach involving sequential manipulation of Pt/NiO core/shell nanowires. This method entails controlled thermal annealing to form PtNi alloy nanowires and subsequent electrochemical dealloying to yield jagged Pt nanowires (J-PtNWs). Notably, J-PtNWs exhibit an exceptional electrochemically active surface area (ECSA) of 118 m^2/gPt and a specific activity of 11.5 mA/cm^2 for ORR. This remarkable feat translates to a mass activity of 13.6 A/mg Pt, nearly twice that of previous benchmarks. Molecular dynamics simulations underscore the crucial role of highly stressed, undercoordinated, rhombus-rich surface configurations in boosting ORR activity. This innovative approach offers promising avenues for advanced energy conversion technologies through the development of highly efficient oxygen reduction catalysts.

Chen et al. [147] revealed that Pt_3Ni nanoframe catalysts demonstrate a significant enhancement in both mass and specific activity for ORR compared to the state-of-the-art platinum–carbon catalysts, with a 36-fold increase in mass activity and a 22-fold increase in specific activity. This morphological transformation of $PtNi_3$ polyhedra into Pt_3Ni nanoframes occurred at 120 °C over a 12 h period and resulted in an ECSA ratio (ECSACO/ECSAHupd) of 1.52, indicating a Pt-skin-terminated (111)-like surface structure. Moreover, the specific activity of the Pt_3Ni nanoframes at 0.95 V exhibited an improvement factor of over 16 compared to commercial Pt/C electrocatalyst, and the Tafel slope for the Pt_3Ni nanoframes was 46 mV/dec, compared to 73 mV/dec for Pt/C. Impressively, the mass activity at 0.9 V was 5.7 A/mg Pt, surpassing the U.S. Department of Energy's 2017 target of 0.44 A/mg Pt by more than an order of magnitude. Additionally, after 10,000 potential cycles between 0.6 and 1.0 V, the Pt_3Ni nanoframes showed negligible activity loss, whereas the state-of-the-art Pt/C electrocatalysts experienced a ~40% loss of specific surface area. Lastly, the ionic liquid-encapsulated Pt_3Ni nanoframes showed a 36-fold enhancement in mass activity and a 22-fold enhancement in specific activity compared to Pt/C catalysts even after prolonged (10,000) potential cycling.

Yuan et al. [148] indicated that zirconium nitride (ZrN) NPs can outperform Pt as a catalyst for ORR in alkaline environments, presenting a more stable and cost-effective alternative. Specifically, ZrN NPs and a commercial Pt-on-carbon (Pt/C) catalyst displayed the same half-wave potential ($E_{1/2}$ = 0.80 V) in 0.1 M KOH, but ZrN exhibited higher stability ($\Delta E_{1/2}$ = −3 mV) compared to Pt/C ($\Delta E_{1/2}$ = −39 mV) after 1000 ORR cycles. The synthesized ZrN NPs, with a surface area of 28 m^2/g and a thin surface layer of oxynitride or oxide, demonstrated comparable ORR performance to Pt/C but with a higher turnover frequency at 0.8 V, indicating more active surface sites for ORR. Moreover, ZrN NPs showed high selectivity for the four-electron pathway and outperformed Pt/C in terms of H_2O_2 yield and long-term durability. In a zinc–air battery, the ZrN-based battery achieved a higher power density and displayed a smaller voltage decrease after 100 h of operation compared to a Pt/C-driven battery. Computational analysis revealed that the Zr-terminated (111) surface of ZrN has a similar oxygen adsorption energy to the Pt(111) surface, explaining the high ORR activity of ZrN.

When we delve into the realm of transition metal catalysts, we find that metals such as Au and Hg predominantly undergo a two-electron reduction. This is in stark contrast to Pt, a highly esteemed catalyst, which is believed to undergo a four-electron reduction predominantly. As researchers navigate these pathways and mechanisms, they aim to identify the rate-determining step (RDS) and clarify the sequences of electron and proton transfers. Yet, the ORR kinetics on metal catalysts face three pronounced obstacles: the initial electron transfer, the hydration process of oxygen, and the release of reaction intermediates.

Concerning carbon-based materials, their ORR electrocatalytic activity is linked to the presence of defects in the graphitic layers, such as edges or point defects caused by vacancies or impurities. These defects serve as active sites for the dissociation of O_2 into atomic oxygen, which then reacts with H_2O to form hydroxide ions in redox reactions [149]. Hence, introducing defects into carbonaceous materials used as cathodes increases the active sites for electrochemical reactions, where the nucleation of reaction products takes place. This could also lead to a reduction in particle size [150]. Carbon materials can undergo various defect-engineering methods to modify their electrochemical properties. One approach is introducing "self-doping" defects into expanded graphene, which act as additional active sites for ion storage, enhancing capacitance even in ethylene glycol with a low specific surface area. Another strategy is nitrogen doping, where nitrogen is incorporated into the carbon structure, enhancing the carbon interlayer distance, improving wettability for electrolyte ions, and facilitating rapid Li-ion diffusion. Additionally, intrinsic defects, such as vacancies, edge sites, and dislocations, can be introduced during synthesis or post-treatment processes. These defects alter the electron structure of the carbon materials, potentially influencing their electrochemical performance [151]. Consequently, many metal–air batteries prepared with defective carbon-based cathodes exhibit increased discharge capacity and lower cycle overpotentials.

In a significant study by Huang et al. [150], CNT cathodes were subjected to argon plasma bombardment, introducing deliberate surface defects to investigate their impact on the performance of rechargeable lithium–air or lithium–oxygen (Li–O) batteries. These defects manifested distinct electrochemical behaviors when paired with different electrolytes, such as tetraethylene glycol dimethyl ether (TEGDME) and *N*-methyl-*N*-propylpiperidinium bis(trifluoromethansulfonyl)imide (PP13TFSI). Notably, the defects enhanced the nucleation of Li_2O_2 particles, leading to smaller particle sizes, elevated discharge capacities, and diminished cycle overpotentials. Yet, while these benefits were clear, there were diverging outcomes based on the electrolyte in use. TEGDME-based cells faced challenges as the defects spurred unwanted side reactions, predominantly forming Li_2CO_3, which adversely affected the battery's longevity. Contrastingly, in PP13TFSI-based cells, this carbonate formation was largely absent, a phenomenon likely attributed to the protective attributes of the ionic liquid, which seemed to shield the cathode's functional groups from detrimental reactions. This transformative research underscores the intricate balance between cathode surface modifications and electrolyte choice. Strategically understanding and harnessing this relationship could be the key to enhancing the efficiency and lifespan of $Li–O_2$ batteries.

Cheng et al. [152] have made significant advancements in developing a bifunctional oxygen electrocatalyst tailored for rechargeable Zn–air batteries, aiming to achieve superior power density and rapid charge/discharge dynamics. Utilizing an innovative coating-activation technique, they have synthesized defect-enriched nitrogen-doped graphitic carbon microspheres exhibiting 3D interconnected super-macropores. This distinctive structure marries the benefits of rapid electron movement with efficient mass transport, attributes credited to its continuous carbon skeletons and its super-macroporous interconnected pathways, respectively. This metal-free catalyst, termed DCM-1000, demonstrates remarkable intrinsic activity combined with an optimal porous structure, leading to a Zn–air battery power density peak of 193 mW/cm^2. Remarkably, these batteries achieved an energy density of 915.46 Wh/kgZn and showcased resilience with up to 800 cycles at 50 mA/cm^2. By leveraging unique pyrolysis and ammonia treatments, Cheng et al. offered a promising strategy to boost the potential of Zn–air batteries in both EVs and future grid energy storage applications.

Additionally, there is an increasing interest in metal-free catalysts, which have demonstrated encouraging results in ORR activities, thereby broadening the spectrum of research in this field [153]. A landmark moment in this journey occurred in 1964, when Jasinski et al. uncovered the redox activity of cobalt phthalocyanine (CoPc) under alkaline conditions. This pioneering revelation marked a crucial milestone in the evolution of NPMCs, laying a foundation for subsequent advancements in the field. The exploration of alternative catalysts, including metal-based, metal-free, and even hybrid catalysts, continues to evolve, with researchers delving into the intricacies of their composition, structure, and performance characteristics. Ultimately, these efforts aim to develop catalysts that not only exhibit superior ORR activity but also possess enhanced stability, selectivity, and tolerance to contaminants, all of which are critical parameters for the practical application of these materials in energy conversion and storage devices [154].

1.4.2.1 ORR mechanism

Thc nuances of ORR at the atomic level remain a topic of intense exploration, primarily due to the intricate kinetics involved. At its core, the complete electrochemical ORR is characterized by four synchronized proton and electron transfers to molecular oxygen at the cathode. To better understand the various aspects of this mechanism, consider the following explanations:

Direct reduction: This pathway yields H_2O in acidic environments or OH^- in alkaline contexts. It operates through a four-electron process, making it distinct from other pathways that might use fewer electrons.

Two-electron pathway: As the name suggests, this pathway involves a two-electron process, leading directly to the formation of hydrogen peroxide.

Series pathway: This pathway is a blend, incorporating both two- and four-electron reductions. It is like a combination of the previous two pathways but operates in a series.

Parallel pathway: Unique in its approach, this pathway weaves elements from the first three pathways, making it versatile in its reactions.

Interactive pathway: This is more dynamic, allowing species to move or "diffuse" from the series pathway into the direct route.

Associative mechanism: This starts with the adsorption of molecular O_2. Following this, a direct proton/electron transfer takes place, resulting in the formation of OOH. This OOH then disintegrates into O and OH.

Dissociative mechanism: Here, the O–O bond in O_2 undergoes splitting. Following this rupture, atomic O undergoes hydrogenation, producing OH and H_2O.

1.4.3 Oxygen Evolution Reaction

For rechargeable metal–air batteries, the need for the air cathode to efficiently catalyze both the ORR and the reverse reaction, the oxygen evolution reaction (OER), presents another challenge. A handful of high-performance bifunctional catalysts have been developed over the past decade [155–157], but the polarization between charge and discharge remains too high, which restricts the round-trip energy efficiency to less than 65%. Cycling stability poses a significant challenge for the air cathodes of rechargeable metal–air batteries. The changing reductive and oxidative environments during the discharge and charge cycles can lead to irreversible damage to the ORR active sites at the cathode and induce corrosion of the carbon support. Although individual ORR and OER catalysts may exhibit durability under their specific operational conditions, many struggle to withstand the cycling environment. Additionally, the cycling stability of lithium metal anodes is compromised due to the initiation and growth of lithium dendrites, which tend to enlarge the electrode surface area with each cycle and consume solvent to form new passivation layers [158–161].

1.4.4 Potential of Metal–Air Batteries

Potential solutions to these problems lie in the rational design and engineering of both cathode and anode materials. An interesting approach is to decouple the ORR and OER catalysts, loading them on two separate cathodes for discharge and charge, respectively. This bicathode configuration has demonstrated improved cycling performance, albeit at the cost of a larger battery size and lower energy density. Aqueous metal-air batteries, using metals such as zinc, iron, aluminum, and magnesium that are thermodynamically unstable in aqueous media but can be passivated under certain conditions, offer another promising path. The electrochemical reactions in these batteries can be reversed upon recharge, essentially making them a special type of fuel cell running on metal fuels.

Looking beyond the challenges, the abundance and relatively low environmental impact of the metals used in metal–air batteries are among their most attractive features. Metals such as zinc and aluminum are not only more abundant than lithium, but their extraction and processing also have a smaller environmental footprint [162–164]. And unlike fossil fuels, the oxygen used in these batteries is an inexhaustible reactant, making metal–air batteries potentially sustainable for large-scale energy storage [165, 166]. Li–O batteries, sometimes also referred to as lithium–air batteries, have been the subject of intensive research due to their theoretical energy density. This energy density is approximately 10 times higher than that of conventional LIBs, which could significantly extend the range of EVs and the life of portable electronics.

The concept of an Li–O battery is relatively simple: During discharge, oxygen from the air is reduced at the cathode and reacts with lithium ions to form lithium peroxide (Li_2O_2). During charging, the Li_2O_2 decomposes back into lithium ions and oxygen. Theoretically, this process can be repeated many times, but there are several significant challenges to be addressed. One of the primary obstacles to the development of Li–O batteries is their poor cycle life. These batteries can often only undergo a few tens of charge–discharge cycles before their performance deteriorates, far from the hundreds or thousands of cycles expected of a practical battery [167, 168]. This is due to a range of factors, including the instability of the Li_2O_2

discharge product and the electrolyte and the challenges associated with the oxygen reduction and evolution reactions. Despite these challenges, the promise of a significantly higher energy density keeps the research on Li–O batteries active. An avenue of research is the development of novel cathode materials. These materials must be highly conductive to facilitate the oxygen reduction and evolution reactions, porous to allow the passage of oxygen, and chemically stable in the battery's operating environment.

Lithium metal is often used in Li–O batteries due to its high specific capacity and low electrochemical potential. However, as mentioned earlier, a lithium metal anode can lead to the formation of lithium dendrites that can cause a short circuit and present a serious safety hazard. Various strategies have been proposed to mitigate this problem, including the use of lithium metal alloys, artificial solid-electrolyte interphase layers, and lithium-ion-conductive ceramic or glass electrolytes. The development of new electrolyte systems is also a crucial aspect of Li–O battery research. These electrolytes must be stable in the presence of reduced oxygen species and during the charge–discharge process. They should also ideally be able to solubilize or at least accommodate the Li_2O_2 discharge product. Novel solvent systems, solid electrolytes, and the addition of various additives are all being investigated in this regard [169–171].

It is worth noting that the challenges facing Li–O batteries are complex and interconnected. Solutions to one problem often have implications for another, meaning a holistic approach is needed. For example, the choice of electrolyte can affect the stability of the cathode and the efficiency of oxygen reduction and evolution reactions. While Li–O batteries present significant potential due to their high theoretical energy density, several key challenges must be addressed before they can be commercially viable. These include the instability of the Li_2O_2 discharge product and the electrolyte, the efficiency of oxygen reduction and evolution reactions, the choice of cathode and anode materials, and the inherent safety issues associated with using a lithium metal anode. Although considerable progress has been made in each of these areas, there is still much work to be done.

Energy storage is pivotal for the modern world, enabling progress in personal electronics, EVs, and grid renewable energy integration. From 2010 to 2020, the cost of LIB packs decreased from $1100/kWh to \$156/kWh, facilitating the growth of EVs and renewable energy storage. Despite this, significant challenges remain in transitioning to low- or no-carbon transportation and the electricity grid. Diverse application requirements necessitate a variety of purpose-designed batteries, each tailored to specific performance criteria. A decade ago, the Joint Center for Energy Storage Research (JCESR) set targets to increase energy density to 400 Wh/kg and reduce costs to \$100/kWh at the pack level for transportation and grid batteries. Adopting a top-down, reductionist approach led to advances in various storage platforms, including nonaqueous flow batteries, air-breathing aqueous sulfur flow batteries for the grid, and Li-S, Li-O, and Mg-ion batteries for transportation. Despite progress, considerable challenges persist, requiring ongoing research and innovation to develop purpose-built batteries for diverse applications [172].

Li-S batteries have caught the attention of scientists for their extraordinary lightness and higher energy density compared to LIBs. Despite these promising characteristics, hurdles still exist to bring down their production costs and make them a viable commercial option. Furthermore, the theoretical energy density of Li-S batteries is around 2500 Wh/kg, which is a mind-boggling figure as it is over five times the energy density goal set by the JCESR for LIBs. In a slightly different approach, some scientists are considering substituting lithium with heavier magnesium to create Mg-ion batteries, hoping to boost power density. On the other hand, Li–O batteries are theorized to hold up to 10 times the energy density of their Li-ion counterparts. However, the challenge lies in enhancing their currently short life cycles, which barely extend to 50 recharge cycles.

Among the diverse landscape of metal–air batteries, four distinct types stand out in terms of their electrochemical attributes and potential applications: Li–air, Zn–air, Al–air, and Fe–air. The lithium–air battery, with its exceptional theoretical specific energy of 5200 Wh/kg [173], holds promise for EVs, though it faces challenges with electrolyte stability, cathode instability, and achieving a practical specific energy close to its theoretical value, expected to be around ~2450 Wh/kg [174]. The zinc–air battery, operating at 1.65 V and offering a theoretical specific energy of 1086 Wh/kg, has already found commercial applications in devices such as hearing aids due to its high power density, although its rechargeable variants are limited

Table 1.2 Comparative analysis of key characteristics and performance metrics for four prominent metal–air batteries: Li–air, Zn–air, Mg–air, and Al–air

Metrics	Li–Air	Zn–Air	Mg–Air	Al–Air	Ref.
Theoretical specific energy (excluding O_2/including O_2) Wh/kg	11680/ 3500	1300/ 1086	6800/ 3910	8135/ 2800	[179]
Practical Specific Energy (Wh/kg)	~2000	~200	~700	~400	[179]
Theoretical cell voltage (versus SHE)	3.10 V	1.65 V	3.09 V	2.71 V	[179]
Electrolyte	Nonaqueous/Aqueous	Aqueous	Aqueous	Aqueous	[179]
Main Application	Electric vehicles	Hearing aids, communication devices	Military, underwater vehicles	Not specified	[179–181]
Advantages	High energy density	High power density, commercially available	High energy density, safe	Low cost	[182 – 185]
Challenges	Electrolyte stability, cathode instability, low round-trip efficiency	Limited cycle life, low round-trip efficiency	Limited cycle life, anode corrosion, high hydrogen evolution	Low round-trip efficiency, slow kinetics	[181, 186–188]
Electrochemical reaction	$2Li + O_2 \rightleftarrows Li_2O_2$	$2Zn + O_2 \rightarrow 2ZnO$	$2 Mg + O_2 + 2H_2O \rightarrow 2 Mg(OH)_2$	$4Al + 3O_2 + 6H_2O \rightarrow 4Al(OH)_3$	[179]

by their cycle life [175, 176]. Aluminum–air batteries, characterized by a voltage of 1.2 V and a substantial theoretical specific energy of 2980 Wh/kg [177], show potential for military applications and underwater vehicles, yet they grapple with anode corrosion and a propensity for high hydrogen evolution. Lastly, the iron–air battery, though less explored, presents a cost-effective option with its theoretical specific energy of 1237 Wh/kg [178], albeit facing challenges in round-trip efficiency and kinetic performance.

Table 1.2 provides a comprehensive comparison of key parameters for different metal–air batteries. The table highlights important characteristics such as voltage, theoretical specific capacity (Ah/kg), theoretical energy density (Wh/kg), and the chemical reaction involved in each battery.

1.5 Magnesium-Ion Batteries

1.5.1 Introduction

In the past decade, a growing interest has been observed in multivalent-ion batteries, particularly magnesium, due to its abundance, affordability, and relative stability. Magnesium rechargeable batteries have been discovered as an alternative energy storage device due to their low redox potential (−2.37 V versus SHE), high volumetric capacity (3833 mAh/cm^3), and high gravimetric capacity (2205 mA/h) [189, 190].

The development of magnesium battery (MB) electrolytes is rooted in the seminal work of Brenner in the 1970s, who reacted Grignards with boron Lewis acids with the scope of improving magnesium electrodeposition baths as requested by NASA for the electroforming of light-weight engineering components for satellites [191, 192]. The first significant efforts to enhance the battery applicability of Grignard electrolytes were made by Gregory et al. in the 1980s. Building on the work of Brenner, Gregory, and the team replaced boron Lewis acids with cheaper and less toxic aluminum Lewis acids, such as aluminum trichloride. Spiking alkyl Grignards with a Lewis acid, such as aluminum trichloride, enhances the electrochemical plating of magnesium metal [193]. This observation led to the birth of magnesium organohaloaluminate electrolytes with high oxidative stability obtained from the reaction of an organomagnesium reagent (such as dialklylmagnesiums or Grignards) and a Lewis acid. Gregory et al. embarked on an extensive journey, delving into the intricate world of intercalation cathodes, electrolytes, and solvent-solute-intercalation cathode combinations [194]. They deeply investigated the electrochemical properties of magnesium and unleashed its full potential as a powerful energy storage medium. A myriad of cathode materials, from sulfides to oxides and borides of transition metals, were developed.

1.5.2 Potential Anode and Cathode Materials and Effect of Electrolyte on Performance

Metallic magnesium is the most explored anode material. However, metallic Mg is not thermodynamically stable in electrolytes when it is in contact with reducible species [190]. For instance, fluoride-based electrolytes increase the resistance to the flow of Mg^{2+} ions through the electrolyte to the anode, which compacts the SEI on the anode surface. Eventually, this reduced the lifespan of the battery, leading to low coulombic efficiency [195]. Recent research has discovered that electrolytes based on fluorinated alkoxyborate or alkoxyaluminate anions can potentially develop stable MBs due to their compatibility with Mg and anodic stability. The study conducted by Mandai et al. discovered that magnesium-fluorinated alkoxyaluminate, $Mg[Al(HFIP)_4]_2$ (HFIP = hexafluoro-iso-propoxyl group), and diglyme salt–solvent combinations can increase the coulombic efficiency of the battery to 99.4% with over 250 cyclic stability [196].

On the other hand, MBs are commonly developed with inorganic cathodes, specifically inorganic oxide, as well as polyanionic and sulfide compounds. Aurbach and his team developed a battery system with Mo_6S_8 (Chevrel phase) and organohaloaluminate-based electrolytes and observed a higher energy density of 60 Wh/kg compared to Ni-Cd or lead–acid batteries [197]. Further, interest in polyanionic compounds for cathodic materials is rising as polyanions can increase the redox potential of center transition metals, which can eventually develop high-energy-dense Mg-ion cells. The study explored with the $K_2(VO)_2(HPO_4)_2(C_2O_4)\cdot 4.5H_2O$ (KVPCH) cathode with $Mg(ClO_4)_2$ in propylene carbonate showed 87% capacity retention after 1500 cycles at 5 C. The higher capacity retention is induced by the efficient Mg storage since Mg^{2+} diffusion in the lattice is pronounced as the desolvation barrier is lowered. They have proved this through density functional theory (DFT) calculations. On the other hand, researchers developed intercalation cathodes, which involve the insertion and extraction of ions into the crystal lattice of the cathode material [198, 199]. Some intercalation-type cathodes that have shown promise for MBs include materials such as vanadium pentoxide (V_2O_5), molybdenum disulfide (MoS_2), and titanium disulfide (TiS_2) [200–202]. These materials can host magnesium ions within their crystal structure, enabling reversible intercalation reactions.

1.5.3 Challenges in Development

The development of MBs has encountered challenges, primarily stemming from the need for an appropriate cathode material that can manage the distinctive double charge of magnesium ions. While lithium ions have a +1 charge, magnesium ions bear a +2 charge. Intriguingly, even with this double charge, the ionic radius of Mg^{2+} ions (0.72 Å) is closely aligned with that of Li^+ ions (0.76 Å). This similarity prompts Mg^{2+} ions to form robust covalent bonds with electrolytes, especially evident in the Mg–Cl bonds common in all-phenyl complex electrolytes. These bonds are marked by their considerable dissociation energies. Additionally, the intense electrostatic interactions between Mg^{2+} ions and their solid environments significantly impede their diffusion. As a result, Mg^{2+} ions face greater mobility challenges than Li^+ ions in comparable cathode materials [203, 204]. Therefore, reasonable modifications and research need to be done to develop stable and long-hour cyclic stability anode materials.

Considering cathodes, many studies have been focused on improving cathode activity using polyanionic compounds and different intercalations. These areas need to be thoroughly explored, as Mg storage mechanisms can vary depending on the type of system. For example, even though polyanionic compounds show high performance in MBs, their incorporation leads to poor electrical conductivity. Moreover, the poor compatibility between electrodes and electrolytes has severely affected the performance of batteries. Therefore, special attention needs to be given to optimizing the electrolyte compositions and electrodes [205, 206].

To develop high-capacity, long-life, and highly safe rechargeable MBs for large-scale energy storage, researchers must understand the magnesium-storage mechanism in electrode materials, identify and prevent failure mechanisms, rationally design the structure of electrode materials, and optimize the electrode/electrolyte system. By addressing these challenges, researchers can bring MBs closer to widespread commercialization.

1.6 Conclusion

Batteries play a crucial role in the current energy demand, and there is a growing need to advance battery technologies. In this chapter, different types of metal-based batteries have been explored, such as Li-ion, Zn-based, sodium-ion, metal–air, and magnesium-ion batteries. Each type has its own pros and cons. To meet the growing demand for batteries with superior energy density, specific energy, and power capability, researchers are pursuing a dual-pronged approach. In the commercial world, cost and energy efficiency are the most prominent factors. The development of hybrid batteries is a pioneering work that marks a pivotal milestone in energy storage technology, providing a blueprint that harnesses the distinctive attributes of disparate battery systems. Beyond its immediate impact, this research lays the foundation for energy-efficient applications, particularly within the realm of EVs.

References

1. D. Ribeiro, "Alessandro Volta," *Rev. Ciência Elem.*, **1**,1, pp. 1–2, 2013, doi: 10.24927/rce2013.038.
2. P. Kurzweil, "Gaston Planté and his invention of the lead-acid battery-The genesis of the first practical rechargeable battery," *J. Power Sources*, **195**, 14, pp. 4424–4434, 2010, doi: 10.1016/j.jpowsour.2009.12.126.
3. E. J. Lincoln, "A retrospective on abenomics," *Nat. Commun.*, **11**, 378, pp. 1–4, 2020.

4. J. Xie and Y. C. Lu, "A retrospective on lithium-ion batteries," *Nat. Commun.*, **11**(1), pp. 1–4, 2020, doi: 10.1038/s41467-020-16259-9.
5. J. L. Yang et al., "Advanced cathode for dual-ion batteries: Waste-to-wealth reuse of spent graphite from lithium-ion batteries," *eScience*, **2**, 1, pp. 95–101, 2022, doi: 10.1016/j.esci.2021.11.001.
6. B. Xu, D. Qian, Z. Wang, and Y. S. Meng, "Recent progress in cathode materials research for advanced lithium ion batteries," *Mater. Sci. Eng. R Reports*, **73**, 5–6, pp. 51–65, 2012, doi: 10.1016/j.mser.2012.05.003.
7. E. Muchuweni, E. T. Mombeshora, C. M. Muiva, and T. S. Sathiaraj, "Lithium-ion batteries: Recent progress in improving the cycling and rate performances of transition metal oxide anodes by incorporating graphene-based materials," *J. Energy Storage*, **73**, p. 109013, 2023, doi: 10.1016/j.est.2023.109013.
8. A. Manthiram, "An outlook on lithium ion battery technology," *ACS Cent. Sci.*, **3**, 10, pp. 1063–1069, 2017, doi: 10.1021/acscentsci.7b00288.
9. D. P. Abraham, J. L. Knuth, D. W. Dees, I. Bloom, and J. P. Christophersen, "Performance degradation of high-power lithium-ion cells-Electrochemistry of harvested electrodes," *J. Power Sources*, **170**, 2, pp. 465–475, 2007, doi: 10.1016/j.jpowsour.2007.03.071.
10. Q. Zhang and R. E. White, "Calendar life study of Li-ion pouch cells," *J. Power Sources*, **173**, 2, pp. 990–997, 2007, doi: 10.1016/j.jpowsour.2007.08.044.
11. S. Watanabe, M. Kinoshita, and K. Nakura, "Capacity fade of $LiNi_{(1-x-y)}Co_xAl_yO_2$ cathode for lithium-ion batteries during accelerated calendar and cycle life test. I. Comparison analysis between $LiNi_{(1-x-y)}Co_xAl_yO_2$ and $LiCoO_2$ cathodes in cylindrical lithium-ion cells during long term storage test," *J. Power Sources*, **247**, pp. 412–422, 2014, doi: 10.1016/j.jpowsour.2013.08.079.
12. M. S. Hossain Lipu et al., "Data-driven state of charge estimation of lithium-ion batteries: Algorithms, implementation factors, limitations and future trends," *J. Clean. Prod.*, **277**, p. 124110, 2020, doi: 10.1016/j.jclepro.2020.124110.
13. M. Ouyang et al., "Low temperature aging mechanism identification and lithium deposition in a large format lithium iron phosphate battery for different charge profiles," *J. Power Sources*, **286**, pp. 309–320, 2015.
14. S. H. Beheshti et al., "Development, retainment, and assessment of the graphite-electrolyte interphase in Li-ion batteries regarding the functionality of SEI-forming additives," *iScience*, **25**, 3, pp. 1–23, 2022, doi: 10.1016/j.isci.2022.103862.

15. M. Ershadi, M. Javanbakht, S. Beheshti, B. Mosallanejad, and Z. Kiaei, "A patent landscape on liquid electrolytes for lithium-ion batteries," *Anal. Bioanal. Electrochem.*, **10**, pp. 1629–1653, 2018.
16. H. Zhang, Y. Yang, D. Ren, L. Wang, and X. He, "Graphite as anode materials: Fundamental mechanism, recent progress and advances," *Energy Storage Mater.*, **36**, pp. 147–170, 2021, doi: 10.1016/j.ensm.2020.12.027.
17. B. Zhu, X. Wang, P. Yao, J. Li, and J. Zhu, "Towards high energy density lithium battery anodes: Silicon and lithium," *Chem. Sci.*, **10**, 30, pp. 7132–7148, 2019, doi: 10.1039/C9SC01201J.
18. J. Chen, G. Wang, H. Song, B. Wang, G. Wu, and J. Lei, "Stress and displacement of cylindrical lithium-ion power battery during charging and discharging," *Energies*, **15**, 21, pp. 1–22, 2022, doi: 10.3390/en15218244.
19. X. Su et al., "Liquid electrolytes for low-temperature lithium batteries: Main limitations, current advances, and future perspectives," *Energy Storage Mater.*, 56, pp. 642–663, 2023, doi: 10.1016/J.ENSM.2023.01.044.
20. Y. Guo et al., "Visualization of concentration polarization in thick electrodes," *Energy Storage Mater.*, **51**, pp. 476–485, 2022, doi: 10.1016/J.ENSM.2022.07.007.
21. R. Akolkar, "Modeling dendrite growth during lithium electrodeposition at sub-ambient temperature," *J. Power Sources*, **246**, pp. 84–89, 2014, doi: 10.1016/J.JPOWSOUR.2013.07.056.
22. F. Larsson, P. Andersson, P. Blomqvist, and B. Mellander, "Characteristics of lithium-ion batteries during fire tests," **271**, pp. 414–420, 2014.
23. Y. Fernandes, A. Bry, and S. de Persis, "Identification and quantification of gases emitted during abuse tests by overcharge of a commercial Li-ion battery," *J. Power Sources*, **389**, pp. 106–119, 2018, doi: 10.1016/j.jpowsour.2018.03.034.
24. T. H. Dubaniewicz and J. P. DuCarme, "Further study of the intrinsic safety of internally shorted lithium and lithium-ion cells within methane-air," *J. Loss Prev. Process Ind.*, **32**, pp. 165–173, 2014, doi: 10.1016/j.jlp.2014.09.002.
25. R. Spotnitz and J. Franklin, "Abuse behavior of high-power, lithium-ion cells," *J. Power Sources*, **113**, 1, pp. 81–100, 2003, doi: 10.1016/S0378-7753(02)00488-3.
26. X. Feng et al., "Characterization of penetration induced thermal runaway propagation process within a large format lithium-ion battery

module," *J. Power Sources*, 275, pp. 261–273, 2015, doi: 10.1016/j.jpowsour.2014.11.017.

27. T. D. Hatchard, S. Trussler, and J. R. Dahn, "Building a 'smart nail' for penetration tests on Li-ion cells," *J. Power Sources*, 247, pp. 821–823, 2014, doi: 10.1016/j.jpowsour.2013.09.022.
28. B. Mao, H. Chen, Z. Cui, T. Wu, and Q. Wang, "Failure mechanism of the lithium-ion battery during nail penetration," *Int. J. Heat Mass Transf.*, **122**, pp. 1103–1115, 2018, doi: 10.1016/j.ijheatmasstransfer.2018.02.036.
29. Y. Chen et al., "A review of lithium-ion battery safety concerns: The issues, strategies, and testing standards," *J. Energy Chem.*, **59**, pp. 83–99, 2021, doi: 10.1016/j.jechem.2020.10.017.
30. Z. Liao, S. Zhang, K. Li, G. Zhang, and T. G. Habetler, "A survey of methods for monitoring and detecting thermal runaway of lithium-ion batteries," *J. Power Sources*, **436**, p. 226879, 2019, doi: 10.1016/J.JPOWSOUR.2019.226879.
31. T. Yamanaka, Y. Takagishi, Y. Tozuka, and T. Yamaue, "Modeling lithium-ion battery nail penetration tests and quantitative evaluation of the degree of combustion risk," *J. Power Sources*, **416**, pp. 132–140, 2019, doi: 10.1016/J.JPOWSOUR.2019.01.055.
32. D. Ouyang, M. Chen, J. Liu, R. Wei, J. Weng, and J. Wang, "Investigation of a commercial lithium-ion battery under overcharge/over-discharge failure conditions," *RSC Adv.*, **8**, 58, pp. 33414–33424, 2018, doi: 10.1039/C8RA05564E.
33. A. Changes et al., "Lithium batteries as cargo in 2016 update III," no. February, pp. 22–23, 2016.
34. E. Mossali, N. Picone, L. Gentilini, O. Rodrìguez, J. M. Pérez, and M. Colledani, "Lithium-ion batteries towards circular economy: A literature review of opportunities and issues of recycling treatments," *J. Environ. Manage.*, **264**, p. 110500, 2020, doi: 10.1016/j.jenvman.2020.110500.
35. A. A. Fedotov, O. S. Popel, S. V. Kiseleva, and A. B. Tarasenko, "Limitations for lithium-ion batteries application in engine cold cranking," *J. Phys. Conf. Ser.*, **1787**, 1, 2021, doi: 10.1088/1742-6596/1787/1/012065.
36. B. Jones, R. J. R. Elliott, and V. Nguyen-Tien, "The EV revolution: The road ahead for critical raw materials demand," *Appl. Energy*, **280**, p. 115072, 2020, doi: 10.1016/j.apenergy.2020.115072.
37. Y. Liu, R. Zhang, J. Wang, and Y. Wang, "Current and future lithium-ion battery manufacturing," *iScience*, **24**, 4, p. 102332, 2021, doi: 10.1016/j.isci.2021.102332.

38. Y. L. Ding, Z. P. Cano, A. P. Yu, and Z. W. Chen, "Automotive Li-ion batteries: Current status and future perspectives," *Pap. Knowl. Towar. Media Hist. Doc.*, pp. 1–82, 2014.

39. "Battery Pack Prices Cited Below $100/kWh for the First Time in 2020, While Market Average Sits at $137/kWh | BloombergNEF." https://about.bnef.com/blog/battery-pack-prices-cited-below-100-kwh-for-the-first-time-in-2020-while-market-average-sits-at-137-kwh/ (accessed Aug. 12, 2023).

40. A. Masias, J. Marcicki, and W. A. Paxton, "Opportunities and challenges of lithium-ion batteries in automotive applications," *ACS Energy Lett.*, **6**, 2, pp. 621–630, 2021, doi: 10.1021/acsenergylett.0c02584.

41. Z. Li et al., "Research progress of SiO_x-based anode materials for lithium-ion batteries," *Chem. Eng. J.*, **473**, p. 145294, 2023, doi: 10.1016/j.cej.2023.145294.

42. H. Shen, Q. Wang, Z. Chen, C. Rong, and D. Chao, "Application and development of silicon anode binders for lithium-ion batteries," *Materials (Basel).*, **16**, 12, p. 4266, 2023, doi: 10.3390/ma16124266.

43. Y. Hong Luo et al., "Nickel-rich and cobalt-free layered oxide cathode materials for lithium-ion batteries," *Energy Storage Mater.*, **50**, pp. 274–307, 2022, doi: 10.1016/j.ensm.2022.05.019.

44. M. Li, M. Feng, D. Luo, and Z. Chen, "Fast charging Li-ion batteries for a new era of electric vehicles," *Cell Reports Phys. Sci.*, **1**, 10, p. 100212, 2020, doi: 10.1016/j.xcrp.2020.100212.

45. U. H. Kim, S. B. Lee, N. Y. Park, S. J. Kim, C. S. Yoon, and Y. K. Sun, "High-energy-density Li-ion battery reaching full charge in 12 min," *ACS Energy Lett.*, **7**, 11, pp. 3880–3888, 2022, doi: 10.1021/acsenergylett.2c02032.

46. C. Li et al., "An advance review of solid-state battery: Challenges, progress and prospects," *Sustain. Mater. Technol.*, **29**, p. e00297, 2021, doi: 10.1016/j.susmat.2021.e00297.

47. M. Zhao, B. Q. Li, X. Q. Zhang, J. Q. Huang, and Q. Zhang, "A perspective toward practical lithium-sulfur batteries," *ACS Cent. Sci.*, **6**, 7, pp. 1095–1104, 2020, doi: 10.1021/acscentsci.0c00449.

48. M. K. Singla, P. Nijhawan, and A. S. Oberoi, "Hydrogen fuel and fuel cell technology for cleaner future: A review," *Environ. Sci. Pollut. Res.*, **28**, 13, pp. 15607–15626, 2021, doi: 10.1007/s11356-020-12231-8.

49. J. Wolf, H. H. Sandstead, and L. Rink, "Zinc," *Handb. Toxicol. Met. Fifth Ed.*, **2**, pp. 963–984, 2022, doi: 10.1016/B978-0-12-822946-0.00034-9.

50. J. Xie and Q. Zhang, "Recent progress in multivalent metal (Mg, Zn, Ca, and Al) and metal-ion rechargeable batteries with organic materials as promising electrodes," *Small*, **15**, 15, 2019, doi: 10.1002/SMLL.201805061.
51. N. Borchers, S. Clark, B. Horstmann, K. Jayasayee, M. Juel, and P. Stevens, "Innovative zinc-based batteries," *J. Power Sources*, **484**, p. 229309, 2021, doi: 10.1016/j.jpowsour.2020.229309.
52. L. E. Blanc, D. Kundu, and L. F. Nazar, "Scientific challenges for the implementation of Zn-ion batteries," *Joule*, **4**, 4, pp. 771–799, 2020, doi: 10.1016/j.joule.2020.03.002.
53. R. F. Service, "Zinc aims to beat lithium batteries at storing energy," *Science*, **372**, 6545, pp. 890–891, 2021, doi: 10.1126/SCIENCE.372.6545.890/ASSET/1A0CF7E4-A457-40C7-B702-0CE08D1968F8/ASSETS/GRAPHIC/372_890_F2.JPEG.
54. D. Chao et al., "An electrolytic Zn–MnO_2 battery for high-voltage and scalable energy storage," *Angew. Chemie Int. Ed.*, **58**, 23, pp. 7823–7828, 2019, doi: 10.1002/anie.201904174.
55. J. A. Dean and N. A. Lange, "Lange's chemistry handbook," *McGraw-Hill, Inc*, **4**, p. 1424, 1999.
56. J. Ming, J. Guo, C. Xia, W. Wang, and H. N. Alshareef, "Zinc-ion batteries: Materials, mechanisms, and applications," *Mater. Sci. Eng. R Reports*, **135**, pp. 58–84, 2019, doi: 10.1016/J.MSER.2018.10.002.
57. J. Yu et al., "Modifying the Zn anode with nano-silica: A strategy to realize dendrite-free zinc-ion hybrid supercapacitors," *J. Electrochem. Soc.*, **170**, 2, p. 020508, 2023, doi: 10.1149/1945-7111/ACB614.
58. W. Du, E. H. Ang, Y. Yang, Y. Zhang, M. Ye, and C. C. Li, "Challenges in the material and structural design of zinc anode towards high-performance aqueous zinc-ion batteries," *Energy Environ. Sci.*, **13**, 10, pp. 3330–3360, 2020, doi: 10.1039/D0EE02079F.
59. W. Xiong et al., "Controlling the sustainability and shape change of the zinc anode in rechargeable aqueous Zn/$LiMn_2O_4$ battery," *Energy Storage Mater.*, **15**, pp. 131–138, 2018, doi: 10.1016/J.ENSM.2018.03.023.
60. K. E. K. Sun, T. K. A. Hoang, T. N. L. Doan, Y. Yu, and P. Chen, "Highly sustainable zinc anodes for a rechargeable hybrid aqueous battery," *Chem. A Eur. J.*, **24**, 7, pp. 1667–1673, 2018, doi: 10.1002/CHEM.201704440.
61. T. H. Wu, Y. Zhang, Z. D. Althouse, and N. Liu, "Nanoscale design of zinc anodes for high-energy aqueous rechargeable batteries," *Mater. Today Nano*, **6**, p. 100032, 2019, doi: 10.1016/J.MTNANO.2019.100032.

62. B. Tang, L. Shan, S. Liang, and J. Zhou, "Issues and opportunities facing aqueous zinc-ion batteries," *Energy Environ. Sci.*, **12**, 11, pp. 3288–3304, 2019, doi: 10.1039/C9EE02526J.

63. Y. Jiang, D. Ba, Y. Li, and J. Liu, "Noninterference revealing of 'layered to layered' zinc storage mechanism of Δ-MnO_2 toward neutral Zn–Mn batteries with superior performance," *Adv. Sci.*, **7**, 6, 2020, doi: 10.1002/advs.201902795.

64. X. Guo, J. Zhou, C. Bai, X. Li, G. Fang, and S. Liang, "Zn/MnO_2 battery chemistry with dissolution-deposition mechanism," *Mater. Today Energy*, **16**, p. 100396, 2020, doi: 10.1016/j.mtener.2020.100396.

65. X. Liu, J. Yi, K. Wu, Y. Jiang, Y. Liu, and B. Zhao, "Rechargeable Zn–MnO_2 batteries: Advances , challenges and perspectives," *Nanotechnology*, **31**, 12, p. 122001, 2020, doi: 10.1088/1361-6528/ab5b38.

66. L. Wang, K. W. Huang, J. Chen, and J. Zheng, "Ultralong cycle stability of aqueous zinc-ion batteries with zinc vanadium oxide cathodes," *Sci. Adv.*, **5**, 10, pp. 1–10, 2019, doi: 10.1126/sciadv.aax4279.

67. X. Meng, Z. Cheng, and L. Li, "The promotion of research progress of zinc manganate cathode materials for zinc-ion batteries by characterization and analysis technology," *Molecules*, **28**, 11, p. 4459, 2023, doi: 10.3390/molecules28114459.

68. L. Wang and J. Zheng, "Recent advances in cathode materials of rechargeable aqueous zinc-ion batteries," *Mater. Today Adv.*, **7**, p. 100078, 2020, doi: 10.1016/j.mtadv.2020.100078.

69. Y. Liu, G. He, H. Jiang, I. P. Parkin, P. R. Shearing, and D. J. L. Brett, "Cathode design for aqueous rechargeable multivalent ion batteries: Challenges and opportunities," *Adv. Funct. Mater.*, **31**, 13, 2021, doi: 10.1002/adfm.202010445.

70. F. Wan et al., "Reversible oxygen redox chemistry in aqueous zinc-ion batteries," *Angew. Chemie Int. Ed.*, **58**, 21, pp. 7062–7067, 2019, doi: 10.1002/anie.201902679.

71. W. Liu et al., "β-MnO_2 with proton conversion mechanism in rechargeable zinc ion battery," *J. Energy Chem.*, **56**, pp. 365–373, 2021, doi: 10.1016/j.jechem.2020.07.027.

72. B. Wu et al., "Graphene scroll-coated α-MnO_2 nanowires as high-performance cathode materials for aqueous Zn-ion battery," *Small*, **14**, 13, p. 1703850, 2018, doi: 10.1002/smll.201703850.

73. H. Li et al., "Advanced rechargeable zinc-based batteries: Recent progress and future perspectives," *Nano Energy*, **62**, pp. 550–587, 2019, doi: 10.1016/j.nanoen.2019.05.059.

74. L. Ma, S. Chen, X. Li, A. Chen, B. Dong, and C. Zhi, "Liquid-free all-solid-state zinc batteries and encapsulation-free flexible batteries enabled by in situ constructed polymer electrolyte," *Angew. Chemie Int. Ed.*, **59**, 52, pp. 23836–23844, 2020, doi: 10.1002/ANIE.202011788.

75. S. D. Han et al., "Origin of electrochemical, structural, and transport properties in nonaqueous zinc electrolytes," *ACS Appl. Mater. Interfaces*, **8**, 5, pp. 3021–3031, 2016, doi: 10.1021/ACSAMI.5B10024/SUPPL_FILE/AM5B10024_SI_001.PDF.

76. C. Pan, R. G. Nuzzo, and A. A. Gewirth, "$ZnAl_xCo_{2-x}O_4$ spinels as cathode materials for non-aqueous Zn batteries with an open circuit voltage of ≤2 V," *Chem. Mater.*, **29**, 21. pp. 9351–9359, 2017, doi: 10.1021/acs.chemmater.7b03340.

77. N. Wang et al., "Zinc–organic battery with a wide operation-temperature window from −70 to 150 °C," *Angew. Chemie Int. Ed.*, **59**, 34, pp. 14577–14583, 2020, doi: 10.1002/ANIE.202005603.

78. Z. Li et al., "High-energy and dendrite-free solid-state zinc metal battery enabled by a dual-network and water-confinement hydrogel electrolyte," *Chem. Eng. J.*, **472**, p. 144992, 2023, doi: 10.1016/J.CEJ.2023.144992.

79. F. Ilyas, M. Ishaq, M. Jabeen, M. Saeed, A. Ihsan, and M. Ahmed, "Recent trends in the benign-by-design electrolytes for zinc batteries," *J. Mol. Liq.*, **343**, p. 117606, 2021, doi: 10.1016/J.MOLLIQ.2021.117606.

80. Y. Wang et al., "Electrolyte engineering enables high performance zinc-ion batteries," *Small*, **18**, 43, 2022, doi: 10.1002/smll.202107033.

81. D. Han et al., "A non-flammable hydrous organic electrolyte for sustainable zinc batteries," *Nature Sustainability*, **5**, pp. 205–213, 2022, doi: 10.1038/s41893-021-00800-9.

82. W. Lu, C. Xie, H. Zhang, and X. Li, "Inhibition of zinc dendrite growth in zinc-based batteries," *ChemSusChem*, **11**, 23, pp. 3996–4006, 2018, doi: 10.1002/CSSC.201801657.

83. K. Wang et al., "Dendrite growth in the recharging process of zinc-air batteries," *J. Mater. Chem. A*, **3**, 45, pp. 22648–22655, 2015, doi: 10.1039/c5ta06366c.

84. Y. Zuo et al., "Zinc dendrite growth and inhibition strategies," *Mater. Today Energy*, **20**, p. 100692, 2021, doi: 10.1016/j.mtener.2021.100692.

85. L. Guo, H. Guo, H. Huang, S. Tao, and Y. Cheng, "Inhibition of zinc dendrites in zinc-based flow batteries," *Front. Chem.*, **8**, pp. 1–8, 2020, doi: 10.3389/fchem.2020.00557.

86. K. E. K. Sun et al., "Suppression of dendrite formation and corrosion on zinc anode of secondary aqueous batteries," *ACS Appl. Mater. Interfaces*, **9**, 11, pp. 9681–9687, 2017, doi: 10.1021/acsami.6b16560.
87. J. Kan, H. Xue, and S. Mu, "Effect of inhibitors on Zn-dendrite formation for zinc-polyaniline secondary battery," *J. Power Sources*, **74**, 1, pp. 113–116, 1998, doi: 10.1016/S0378-7753(98)00040-8.
88. J. M. Wang, L. Zhang, C. Zhang, and J. Q. Zhang, "Effects of bismuth ion and tetrabutylammonium bromide on the dendritic growth of zinc in alkaline zincate solutions," *J. Power Sources*, **102**, 1–2, pp. 139–143, 2001, doi: 10.1016/S0378-7753(01)00789-3.
89. F. Mansfeld and S. Gilman, "The effect of lead ions on the dissolution and deposition characteristics of a zinc single crystal in 6N KOH," *J. Electrochem. Soc.*, **117**, 5, p. 588, 1970, doi: 10.1149/1.2407584/XML.
90. Q. Song, S. Zhou, S. Wang, S. Li, L. Xu, and J. Qiu, "Insights into the oxygen vacancies in transition metal oxides for aqueous Zinc-Ion batteries," *Chem. Eng. J.*, **461**, p. 142033, 2023, doi: 10.1016/J.CEJ.2023.142033.
91. Y. Hao et al., "Artificial N-doped Graphene protective layer enables stable Zn anode for aqueous Zn-ion batteries," *ACS Appl. Energy Mater.*, **4**, 6, pp. 6364–6373, 2021, doi: 10.1021/ACSAEM.1C01306/SUPPL_FILE/AE1C01306_SI_001.PDF.
92. B. Oral, B. Tekin, D. Eroglu, and R. Yildirim, "Performance analysis of Na-ion batteries by machine learning," *J. Power Sources*, **549**, p. 232126, 2022, doi: 10.1016/j.jpowsour.2022.232126.
93. K. M. Abraham, "How comparable are sodium-ion batteries to lithium-ion counterparts," *ACS Energy Lett.*, **5**, no 11, pp. 3544–3547, 2020, doi: 10.1021/acsenergylett.0c02181.
94. C. Yang, S. Xin, L. Mai, and Y. You, "Materials design for high-safety sodium-ion battery," *Adv. Energy Mater.*, **11**, 2, p. 2000974, 2021, doi: 10.1002/AENM.202000974.
95. A. Yoshino, "The birth of the lithium-ion battery," *Angew. Chemie Int. Ed.*, **51**, 24, pp. 5798–5800, 2012, doi: 10.1002/ANIE.201105006.
96. Y. Huang et al., "Electrode materials of sodium-ion batteries toward practical application," *ACS Energy Lett.*, **3**, 7, pp. 1604–1612, 2018, doi: 10.1021/ACSENERGYLETT.8B00609/ASSET/IMAGES/LARGE/NZ-2018-00609A_0004.JPEG.
97. D. M. D. Babiker, Z. R. Usha, C. Wan, M. M. El. Hassaan, X. Chen, and L. Li, "Recent progress of composite polyethylene separators for lithium/sodium batteries," *J. Power Sources*, **564**, p. 232853, 2023, doi: 10.1016/J.JPOWSOUR.2023.232853.

98. H. Yu, K. Seomoon, J. Kim, and J. Kim, "Low-cost and highly safe solid-phase sodium ion battery with a Sn–C nanocomposite anode," *J. Ind. Eng. Chem.*, **100**, pp. 112–118, 2021, doi: 10.1016/j.jiec.2021.05.033.

99. S. Chen et al., "Challenges and perspectives for NASICON-type electrode materials for advanced sodium-ion batteries," *Adv. Mater.*, **29**, 48, p. 1700431, 2017, doi: 10.1002/ADMA.201700431.

100. B. Xiao, T. Rojo, and X. Li, "Hard carbon as sodium-ion battery anodes: Progress and challenges," *ChemSusChem*, **12**, 1, pp. 133–144, 2019, doi: 10.1002/CSSC.201801879.

101. F. Legrain, J. Sottmann, K. Kotsis, S. Gorantla, S. Sartori, and S. Manzhos, "Amorphous (glassy) carbon, a promising material for sodium ion battery anodes: A combined first-principles and experimental study," *J. Phys. Chem. C*, **119**, 24, pp. 13496–13501, 2015, doi: 10.1021/acs.jpcc.5b03407.

102. S. Mukherjee, S. Bin Mujib, D. Soares, and G. Singh, "Electrode materials for high-performance sodium-ion batteries," *Materials*, **12**, 12, p. 1952, 2019, doi: 10.3390/ma12121952.

103. R. M. Gao, Z. J. Zheng, P. F. Wang, C. Y. Wang, H. Ye, and F. F. Cao, "Recent advances and prospects of layered transition metal oxide cathodes for sodium-ion batteries," *Energy Storage Mater.*, **30**, pp. 9–26, 2020, doi: 10.1016/J.ENSM.2020.04.040.

104. J. Peng et al., "Prussian blue analogues for sodium-ion batteries: Past, present, and future," *Adv. Mater.*, **34**, 15, p. 2108384, 2022, doi: 10.1002/ADMA.202108384.

105. Z. Lv et al., "Vanadium-based polyanionic compounds as cathode materials for sodium-ion batteries: Toward high-energy and high-power applications," *J. Energy Chem.*, **55**, pp. 361–390, 2021, doi: 10.1016/J.JECHEM.2020.07.008.

106. V. R. R. Boddu, D. Puthusseri, P. M. Shirage, P. Mathur, and V. G. Pol, "Layered Na_xCoO_2-based cathodes for advanced Na-ion batteries: Review on challenges and advancements," *Ionics*, **27**, 11, pp. 4549–4572, 2021, doi: 10.1007/S11581-021-04265-W.

107. R. J. Clément, P. G. Bruce, and C. P. Grey, "Review: Manganese-based P2-type transition metal oxides as sodium-ion battery cathode materials," *J. ELectrochem. Soc.*, **162**, 14, p. A2589, 2015, doi: 10.1149/2.0201514jes.

108. K. Sada, B. Senthilkumar, and P. Barpanda, "Potassium-ion intercalation mechanism in layered $Na_2Mn_3O_7$," *ACS Appl. Energy Mater.*, **1**, 10, pp. 5410–5416, 2018, doi: 10.1021/acsaem.8b01016.

109. A. El Kharbachi, O. Zavorotynska, M. Latroche, F. Cuevas, V. Yartys, and M. Fichtner, "Exploits, advances and challenges benefiting beyond Li-ion battery technologies," *J. Alloys Compd.*, **817**, p. 153261, 2020, doi: 10.1016/j.jallcom.2019.153261.

110. X. Xiang, K. Zhang, and J. Chen, "Recent advances and prospects of cathode materials for sodium-ion batteries," pp. 5343–5364, 2015, doi: 10.1002/adma.201501527.

111. C. Huang, T. Kuo, S. Yougbaré, and L. Lin, "Journal of colloid and interface science design of $LiFePO_4$ and porous carbon composites with excellent high-rate charging performance for lithium-ion secondary battery," *J. Colloid Interface Sci.*, **607**, pp. 1457–1465, 2022, doi: 10.1016/j.jcis.2021.09.118.

112. I.-H. Jo et al., "The effect of electrolyte on the electrochemical properties of Na/α-$NaMnO_2$ batteries," *Mater. Res. Bull.*, **58**, pp. 74–77, 2014, doi: 10.1016/j.materresbull.2014.02.024.

113. Y. Zhao, L. Li, Y. Wu, Y. Fang, and H. Xie, "Progress of the elements doped $NaFeO_2$ cathode materials for high performance sodium-ion batteries," *ChemistrySelect*, **6**, 36, pp. 9701–9708, 2021, doi: 10.1002/SLCT.202102304.

114. K. Kubota et al., "New insight into structural evolution in layered $NaCrO_2$ during electrochemical sodium extraction," *J. Phys. Chem. C*, **119**, 1, pp. 166–175, 2015, doi: 10.1021/JP5105888/SUPPL_FILE/JP5105888_SI_001.PDF.

115. C. Y. Yu et al., "$NaCrO_2$ cathode for high-rate sodium-ion batteries," *Energy Environ. Sci.*, **8**, 7, pp. 2019–2026, 2015, doi: 10.1039/C5EE00695C.

116. X. Xia and J. R. Dahn, "$NaCrO_2$ is a fundamentally safe positive electrode material for sodium-ion batteries with liquid electrolytes," *Electrochem. Solid-State Lett.*, **15**, 1, p. A1, 2012, doi: 10.1149/2.002201ESL/XML.

117. Y. Gao et al., "Low-cost polyanion-type sulfate cathode for sodium-ion battery," *Adv. Energy. Mater.*, 2101751, pp. 1–18, 2021, doi: 10.1002/aenm.202101751.

118. W. Li, C. Han, W. Wang, F. Gebert, S. Chou, and H. Liu, "Commercial prospects of existing cathode materials for sodium ion storage," *Adv. Energy. Mater.*, 1700274, pp. 1–10, 2017, doi: 10.1002/aenm.201700274.

119. C. Sun, N. Wang, and Q. Zhang, "Ni- and/or Mn-based layered transition metal oxides as cathode materials for sodium ion batteries: Status, challenges and countermeasures," *J. Mater. Chem. A.*, **7**, 17, 2019, doi: 10.1039/c8ta12441h.

120. M. Jiang, X. Wang, S. Wei, Y. Shen, and H. Hu, "An ionic-liquid-assisted approach to synthesize a reduced graphene oxide loading iron-based fluoride as a cathode material for sodium-ion batteries," *J. Alloys Compd.*, **670**, pp. 362–368, 2016, doi: 10.1016/j.jallcom.2016.01.248.

121. J. Kang et al., "Highly stable Fe^{2+}/Ti^{3+}-based fluoride cathode enabling low-cost and high-performance Na-ion batteries," *Adv. Funct. Mater.*, **32**, p. 2201816, 2022, doi: 10.1002/adfm.202201816.

122. A. Plewa, A. Kulka, D. Baster, and J. Molenda, "Abundant cathode materials for Na-ion batteries," *Solid State Ionics*, **335**, pp. 15–22, 2019, doi: 10.1016/j.ssi.2019.02.007.

123. M. Chen et al., "A novel graphene oxide wrapped $Na_2Fe_2(SO_4)_3$/C cathode composite for long life and high energy density sodium-ion batteries" *Adv. Energy. Mater.*, **8**, p. 1800944, 2018, doi: 10.1002/aenm.201800944.

124. B. Zhang et al., "Recent advances of NASICON-$Na_3V_2(PO_4)_3$ as cathode for sodium-ion batteries: Synthesis, modifications, and perspectives," *J. Alloys Compd.*, **867**, p. 159060, 2021, doi: 10.1016/j.jallcom.2021.159060.

125. Y. Park et al., "A new high-energy cathode for a Na-ion battery with ultrahigh stability," *J. Am. Chem. Soc.*, **135**, 37, pp 13870–13878, 2013, doi: 10.1021/ja406016j.

126. J. Zhao, L. Zhao, K. Chihara, S. Okada, and J. Yamaki, "Electrochemical and thermal properties of hard carbon-type anodes for Na-ion batteries," *J. Power Sources*, **244**, pp. 752–757, 2013, doi: 10.1016/j.jpowsour.2013.06.109.

127. W. Pantoja and J. A. Perez-taborda, "Tug-of-war in the selection of materials for battery technologies," *Batteries*, **8**, 9, p. 105, 2022.

128. D. P. Divincenzo and E. J. Mele, "Cohesion and structure in stage-1 graphite intercalation compounds," *Phys. Rev. B*, **32**, 4, p. 2538, 1985, doi: 10.1103/PhysRevB.32.2538.

129. R. Alcántara, J. M. Jiménez Mateos, and J. L. Tirado, "Negative electrodes for lithium- and sodium-ion batteries obtained by heat-treatment of petroleum cokes below 1000 °C," *J. Electrochem. Soc.*, **149**, 2, p. A201, 2002, doi: 10.1149/1.1431963.

130. Z. W. Tong, Y. F. Yuan, S. M. Yin, B. X. Wang, S. Y. Guo, and C. L. Mo, "Bamboo-like carbon nanotube-carbon nanotube for high-performance sodium-ion batteries," *Mater. Lett.*, **311**, p. 131587, 2022, doi: 10.1016/J.MATLET.2021.131587.

131. M. Jäckle and A. Groß, "Microscopic properties of lithium, sodium, and magnesium battery anode materials related to possible dendrite growth," *J. Chem. Phys.*, **141**, 17, 2014, doi: 10.1063/1.4901055/313588.

132. S. Wu et al., "Highly durable organic electrode for sodium-ion batteries via a stabilized α-C radical intermediate," *Nat. Commun.*, **71**, 7, pp. 1–11, 2016, doi: 10.1038/ncomms13318.

133. L.-F. Zhao et al., "Hard carbon anodes: Fundamental understanding and commercial perspectives for Na-ion batteries beyond Li-ion and K-ion counterparts," *Adv. Energy Mater.*, **11**, 1, p. 2002704, 2021, doi: 10.1002/AENM.202002704.

134. Y. Y. Zhu, Y. H. Wang, Y. T. Wang, T. J. Xu, and P. Chang, "Research progress on carbon materials as negative electrodes in sodium- and potassium-ion batteries," *Carbon Energy*, **4**, 6, pp. 1182–1213, 2022, doi: 10.1002/CEY2.221.

135. H. Che et al., "Electrolyte design strategies and research progress for room-temperature sodium-ion batteries," *Energy Environ. Sci.*, **10**, 5, pp. 1075–1101, 2017, doi: 10.1039/C7EE00524E.

136. S. Komaba et al., "Electrochemical Na insertion and solid electrolyte interphase for hard-carbon electrodes and application to Na-ion batteries," *Adv. Funct. Mater.*, **21**, 20, pp. 3859–3867, 2011, doi: 10.1002/ADFM.201100854.

137. A. Ponrouch, E. Marchante, M. Courty, J. M. Tarascon, and M. R. Palacín, "In search of an optimized electrolyte for Na-ion batteries," *Energy Environ. Sci.*, **5**, 9, pp. 8572–8583, 2012, doi: 10.1039/C2EE22258B.

138. X. Wu, Y. Cao, X. Ai, J. Qian, and H. Yang, "A low-cost and environmentally benign aqueous rechargeable sodium-ion battery based on $NaTi_2(PO_4)_3$–$Na_2NiFe(CN)_6$ intercalation chemistry," *Electrochem. Commun.*, **31**, pp. 145–148, 2013, doi: /10.1016/j.elecom.2013.03.013.

139. Y. Li and J. Lu, "Metal-air batteries: Future electrochemical energy storage of choice?" *ACS Energy Lett.*, **2**, 6, pp. 1370–1377, 2017, doi: 10.1021/acsenergylett.7b00119.

140. M. Salado and E. Lizundia, "Advances, challenges, and environmental impacts in metal–air battery electrolytes," *Mater. Today Energy*, **28**, p. 101064, 2022, doi: 10.1016/j.mtener.2022.101064.

141. X. Li et al., "Magneli phase Ti_4O_7 electrode for oxygen reduction reaction and its implication for zinc-air rechargeable batteries," *Electrochim. Acta*, **55**, 20, pp. 5891–5898, 2010, doi: 10.1016/j.electacta.2010.05.041.

142. M. Shao, Q. Chang, J. P. Dodelet, and R. Chenitz, "Recent advances in electrocatalysts for oxygen reduction reaction," *Chem. Rev.*, **116**, 6, pp. 3594–3657, 2016, doi: 10.1021/acs.chemrev.5b00462/asset/images/medium/cr-2015-00462u_0039.gif.
143. W. Yu, M. D. Porosoff, and J. G. Chen, "Review of Pt-based bimetallic catalysis: From model surfaces to supported catalysts," *Chem. Rev.*, **112**, 11, pp. 5780–5817, 2012, doi: 10.1021/CR300096B.
144. Y. Nie, L. Li, and Z. Wei, "Recent advancements in Pt and Pt-free catalysts for oxygen reduction reaction," *Chem. Soc. Rev.*, **44**, 8, pp. 2168–2201, 2015, doi: 10.1039/C4CS00484A.
145. D. Wang et al., "Structurally ordered intermetallic platinum-cobalt core-shell nanoparticles with enhanced activity and stability as oxygen reduction electrocatalysts," *Nat. Mater.*, **12**, 1, pp. 81–87, 2013, doi: 10.1038/nmat3458.
146. M. Li et al., "Ultrafine jagged platinum nanowires enable ultrahigh mass activity for the oxygen reduction reaction," *Science*, **354**, 6318, pp. 1414–1419, 2016, doi: 10.1126/science.aaf9050.
147. C. Chen et al., "Highly crystalline multimetallic nanoframes with three-dimensional electrocatalytic surfaces," *Science*, **343**, 6177, pp. 1339–1343, 2014, doi: 10.1126/SCIENCE.1249061.
148. Y. Yuan et al., "Zirconium nitride catalysts surpass platinum for oxygen reduction," *Nat. Mater.*, **19**, 3, pp. 282–286, 2020, doi: 10.1038/s41563-019-0535-9.
149. Z. Fang, L. Li, D. A. Dixon, R. R. Fushimi, and E. J. Dufek, "Nature of oxygen adsorption on defective carbonaceous materials," *J. Phys. Chem. C*, **125**, 37, pp. 20686–20696, 2021, doi: 10.1021/acs.jpcc.1c06741/asset/images/large/jp1c06741_0009.jpeg.
150. S. Huang, W. Fan, X. Guo, F. Meng, and X. Liu, "Positive role of surface defects on carbon nanotube cathodes in overpotential and capacity retention of rechargeable lithium-oxygen batteries," *ACS Appl. Mater. Interfaces*, **6**, 23, pp. 21567–21575, 2014, doi: 10.1021/am506564n/suppl_file/am506564n_si_001.pdf.
151. X. Luo et al., "Defect engineering of carbons for energy conversion and storage applications," *Energy Environ. Mater.*, **6**, 3, pp. 1–22, 2023, doi: 10.1002/eem2.12402.
152. D. Cheng, Z. Wang, C. Chen, and K. Zhou, "Defect-enriched, nitrogen-doped graphitic carbon microspheres within 3D interconnected super-macropores as efficient oxygen electrocatalysts for breathing Zn-Air battery," *Carbon*, **145**, pp. 38–46, 2019, doi: 10.1016/j.carbon.2019.01.010.

153. L. Zhang, Q. Shao, and J. Zhang, "An overview of non-noble metal electrocatalysts and their associated air cathodes for Mg-air batteries," *Mater. Reports Energy*, **1**, 1, p. 100002, 2021, doi: 10.1016/j.matre.2020.11.001.

154. R. Jasinski, "A new fuel cell cathode catalyst," *Nature*, **201**, pp. 1212–1213, 1964.

155. A. Karczmarska, M. Adamek, S. El Houbbadi, P. Kowalczyk, and M. Laskowska, "Carbon-Supported noble-metal nanoparticles for catalytic applications: A review," *Crystals*, **12**, 5, p. 584, 2022.

156. Y. Wang et al., "Graphene-based metal/acid bifunctional catalyst for the conversion of levulinic acid to γ-valerolactone," *ACS Sustainable Chem. Eng.*, **5**, 2, pp. 1538–1548, 2017, doi: 10.1021/acssuschemeng.6b02244.

157. K. Karuppasamy, K. Prasanna, V. R. Jothi, and D. Vikraman, "Recent advances in nanostructured transition metal carbide- and nitride-based cathode electrocatalysts for Li–O_2 batteries (LOBs): A brief review," *Nanomaterials*, **10**, 11, p. 2106, 2020.

158. P. Du et al., "Compatibility of lithium salts with solvent of the non-aqueous electrolyte in Li–O_2 batteries," *Phys. Chem. Chem. Phys.*, **15**, 15, pp. 5572–5581, 2013, doi: 10.1039/c3cp50500f.

159. D. J. Lee et al., "Composite protective layer for Li metal anode in high-performance lithium-oxygen batteries," *Electrochem. Commun.*, **40**, pp. 45–48, 2014, doi: 10.1016/j.elecom.2013.12.022.

160. R. Borah, F. R. Hughson, J. Johnston, and T. Nann, "On battery materials and methods," *Mater. Today Adv.*, **6**, p. 100046, 2020, doi: 10.1016/j.mtadv.2019.100046.

161. S. Ferrari et al., "Solid-state post Li metal ion batteries: A sustainable forthcoming reality?" *Adv. Energy Mater.*, **11**, 43, p. 2100785, 2021, doi: 10.1002/aenm.202100785.

162. T. Leisegang et al., "The aluminum-ion battery: A sustainable and seminal concept?" *Front. Chem.*, **7**, pp. 1–21, 2019, doi: 10.3389/fchem.2019.00268.

163. R. C. Mass, C. Liu, Y. Li, L. Mai, and G. Cao, "Energy storage through intercalation reactions: Electrodes for rechargeable batteries," *Natl. Sci. Rev.*, **4**, 1, pp. 26–53, 2017, doi: 10.1093/nsr/nww093.

164. M. Wu, G. Zhang, M. Wu, J. Prakash, and S. Sun, "Rational design of multifunctional air electrodes for rechargeable Zn–Air batteries: Recent progress and future perspectives," *Energy Storage Mater.*, **21**, pp. 253–286, 2019, doi: 10.1016/j.ensm.2019.05.018.

165. C. Lv et al., “Precipitation-free aluminum-air batteries with high capacity and durable service life,” *Chem. Eng. J.*, **462**, 2022, p. 142182, 2023, doi: 10.1016/j.cej.2023.142182.

166. L. Wang et al., “Insights for understanding multiscale degradation of $LiFePO_4$ cathodes,” *eScience*, **2**, 2, pp. 125–137, 2022, doi: 10.1016/j.esci.2022.03.006.

167. J. Chen and F. Ciucci, “Charging processes in lithium-oxygen batteries unraveled through the lens of the distribution of relaxation times,” *Chem*, **9**, 8, pp. 2267–2281, 2023, doi: 10.1016/j.chempr.2023.04.022.

168. C. Hermanu, B. Apribowo, S. Sarjiya, S. P. Hadi, and F. D. Wijaya, “Optimal planning of battery energy storage systems by considering battery degradation due to ambient temperature: A review, challenges, and new perspective,” *Batteries*, **8**, 12, p. 290, 2022.

169. B. D. Mccloskey, D. S. Bethune, R. M. Shelby, G. Girishkumar, and A. C. Luntz, “Solvents’ critical role in nonaqueous lithium–oxygen battery,” *J. Phys. Chem. Lett.*, **2**, 10, pp. 1161–1166, 2011, doi: 10.1021/jz200352v.

170. F. Li and J. Chen, “Mechanistic evolution of aprotic lithium–oxygen batteries,” *Adv. Energy Mater.*, **7**, **24**, p. 1602934, 2017, doi: 10.1002/aenm.201602934.

171. B. D. Adams et al., “Towards a stable organic electrolyte for the lithium oxygen battery,” *Adv. Energy Mater.*, **5**, 1, p. 1400867, 2015, doi: 10.1002/aenm.201400867.

172. L. Trahey et al., “Energy storage emerging: A perspective from the joint center for energy storage research,” *Proc. Natl. Acad. Sci. U.S.A.*, **117**, 23, pp. 12550–12557, 2020, doi: 10.1073/pnas.1821672117.

173. J. Lu, L. Li, J. B. Park, Y. K. Sun, F. Wu, and K. Amine, “Aprotic and aqueous Li-O_2 batteries,” *Chem. Rev.*, **114**, 11, pp. 5611–5640, 2014, doi: 10.1021/cr400573b/asset/images/large/cr-2013-00573b_0007.jpeg.

174. T. Zhang, N. Imanishi, Y. Takeda, and O. Yamamoto, “Aqueous lithium/air rechargeable batteries,” *Chem. Lett.*, **40**, 7, pp. 668–673, 2011, doi: 10.1246/cl.2011.668.

175. Y. Li and H. Dai, “Recent advances in zinc-air batteries,” *Chem. Soc. Rev.*, **43**, 15, pp. 5257–5275, 2014, doi: 10.1039/c4cs00015c.

176. P. Pei, K. Wang, and Z. Ma, “Technologies for extending zinc–air battery’s cyclelife: A review,” *Appl. Energy*, **128**, pp. 315–324, 2014, doi: 10.1016/j.apenergy.2014.04.095.

177. M. Wei et al., “An enhanced-performance Al-air battery optimizing the alkaline electrolyte with a strong Lewis acid $ZnCl_2$,” *Appl. Energy*, **324**, p. 119690, 2022, doi: 10.1016/J.APENERGY.2022.119690.

178. X. Zhao, Y. Gong, X. Li, N. Xu, and K. Huang, "Performance of solid oxide iron-air battery operated at 550 °C," *J. Electrochem. Soc.*, **160**, 8, pp. A1241–A1247, 2013, doi: 10.1149/2.085308jes.

179. T. Li, M. Huang, X. Bai, and Y. X. Wang, "Metal–air batteries: A review on current status and future applications," *Prog. Nat. Sci. Mater. Int.*, **33**, pp. 151–171, 2023, doi: 10.1016/j.pnsc.2023.05.007.

180. K. G. Gallagher et al., "Quantifying the promise of lithium–air batteries for electric vehicles," *Energy Environ. Sci.*, **7**, 5, pp. 1555–1563, 2014, doi: 10.1039/C3EE43870H.

181. J. Zhang, Q. Zhou, Y. Tang, L. Zhang, and Y. Li, "Zinc–air batteries: Are they ready for prime time?" *Chem. Sci.*, **10**, 39, pp. 8924–8929, 2019, doi: 10.1039/C9SC04221K.

182. Y. Liu, Q. Sun, W. Li, K. R. Adair, J. Li, and X. Sun, "A comprehensive review on recent progress in aluminum–air batteries," *Green Energy Environ.*, **2**, 3, pp. 246–277, 2017, doi: 10.1016/J.GEE.2017.06.006.

183. N. Imanishi and O. Yamamoto, "Rechargeable lithium–air batteries: Characteristics and prospects," *Mater. Today*, **17**, 1, pp. 24–30, 2014, doi: 10.1016/J.MATTOD.2013.12.004.

184. S. Zhao et al., "High-power-density and high-energy-efficiency zinc-air flow battery system for long-duration energy storage," *Chem. Eng. J.*, **470**, p. 144091, 2023, doi: 10.1016/J.CEJ.2023.144091.

185. M. Liu et al., "An ultrahigh energy density Mg–air battery with organic acid–solid anolyte biphasic electrolytes," *Sustain. Energy Fuels*, **7**, 14, pp. 3244–3249, 2023, doi: 10.1039/D3SE00509G.

186. P. Meng et al., "A low-cost and air-stable rechargeable aluminum-ion battery," *Adv. Mater.*, **34**, 8, p. 2106511, 2022, doi: 10.1002/ADMA.202106511.

187. P. Tan et al., "Advances and challenges in lithium-air batteries," *Appl. Energy*, **204**, pp. 780–806, 2017, doi: 10.1016/J.APENERGY.2017.07.054.

188. Z. Song et al., "Study on the limiting factors of the cycle life of alkaline poly(vinyl alcohol)-based zinc–air batteries," *J. Energy Storage*, **57**, p. 106202, 2023, doi: 10.1016/J.EST.2022.106202.

189. M. A. Deyab and M. M. Al-Qhatani, "Controlling magnesium self-corrosion in Mg-air batteries with the conductive nanocomposite PANI@3D-FCNT," *ACS Omega*, **6**, 40, pp. 26640–26645, 2021, doi: 10.1021/acsomega.1c03993/asset/images/large/ao1c03993_0008.jpeg.

190. R. Dominko, J. Bitenc, R. Berthelot, M. Gauthier, G. Pagot, and V. Di Noto, "Magnesium batteries: Current picture and missing pieces of the puzzle," *J. Power Sources*, **478**, p. 229027, 2020, doi: 10.1016/J.JPOWSOUR.2020.229027.
191. R. Deivanayagam, B. J. Ingram, and R. Shahbazian-Yassar, "Progress in development of electrolytes for magnesium batteries," *Energy Storage Mater.*, **21**, pp. 136–153, 2019, doi: 10.1016/J.ENSM.2019.05.028.
192. A. Brenner and J. L. Sligh, "Electrodeposition of magnesium and beryllium from organic baths," NASA CR-111850, 1971.
193. C. B. Bucur, T. Gregory, A. G. Oliver, and J. Muldoon, "Confession of a magnesium battery," *J. Phys. Chem. Lett.*, **6**, 18, pp. 3578–3591, 2015, doi: 10.1021/acs.jpclett.5b01219.
194. J. Muldoon, C. B. Bucur, and T. Gregory, "Fervent hype behind magnesium batteries: An open call to synthetic chemists—Electrolytes and cathodes needed," *Angew. Chemie Int. Ed.*, **56**, 40, pp. 12064–12084, 2017, doi: 10.1002/anie.201700673.
195. T. Mandai et al., "Modifications in coordination structure of $Mg[TFSA]_2$-based supporting salts for high-voltage magnesium rechargeable batteries," *Phys. Chem. Chem. Phys.*, **21**, 23, pp. 12100–12111, 2019, doi: 10.1039/C9CP01400D.
196. T. Mandai, Y. Youn, and Y. Tateyama, "Remarkable electrochemical and ion-transport characteristics of magnesium-fluorinated alkoxyaluminate–diglyme electrolytes for magnesium batteries," *Mater. Adv.*, **2**, 19, pp. 6283–6296, 2021, doi: 10.1039/D1MA00448D.
197. D. Aurbach et al., "Prototype systems for rechargeable magnesium batteries," *Nature*, **407**, 6805, pp. 724–727, 2000, doi: 10.1038/35037553.
198. M. M. Huie, D. C. Bock, E. S. Takeuchi, A. C. Marschilok, and K. J. Takeuchi, "Cathode materials for magnesium and magnesium-ion based batteries," *Coord. Chem. Rev.*, **287**, pp. 15–27, 2015, doi: 10.1016/j.ccr.2014.11.005.
199. C. Liu, Z. G. Neale, and G. Cao, "Understanding electrochemical potentials of cathode materials in rechargeable batteries," *Mater. Today*, **19**, 2, pp. 109–123, 2016, doi: 10.1016/j.mattod.2015.10.009.
200. X. Sun, P. Bonnick, and L. F. Nazar, "Layered TiS_2 positive electrode for Mg batteries," *ACS Energy Lett.*, **1**, 1, pp. 297–301, 2016, doi: 10.1021/acsenergylett.6b00145.
201. X. Zhang et al., "Vanadium-based cathode materials for rechargeable magnesium batteries," *Mater. Today Energy*, **32**, 2023, doi: 10.1016/j.mtener.2022.101232.

202. L. F. Wan, B. R. Perdue, C. A. Apblett, and D. Prendergast, "Mg desolvation and intercalation mechanism at the Mo6S8 chevrel phase surface," *Chem. Mater.*, **27**, 17, pp. 5932–5940, 2015, doi: 10.1021/ACS.CHEMMATER.5B01907.

203. E. Levi, Y. Gofer, and D. Aurbach, "On the way to rechargeable Mg batteries: The challenge of new cathode materials," *Chem. Mater.*, **22**, 3, pp. 860–868, 2010, doi: 10.1021/cm9016497.

204. P. Canepa et al., "Odyssey of multivalent cathode materials: Open questions and future challenges," *Chem. Rev.*, **117**, 5, pp. 4287–4341, 2017, doi: 10.1021/acs.chemrev.6b00614/asset/images/medium/cr-2016-00614e_0045.gif.

205. F. Liu, T. Wang, X. Liu, and L. Z. Fan, "Challenges and recent progress on key materials for rechargeable magnesium batteries," *Adv. Energy Mater.*, **11**, 2, p. 2000787, 2021, doi: 10.1002/AENM.202000787.

206. T. Mori et al., "Anti-site mixing governs the electrochemical performances of olivine-type $MgMnSiO_4$ cathodes for rechargeable magnesium batteries," *Phys. Chem. Chem. Phys.*, **18**, 19, pp. 13524–13529, 2016, doi: 10.1039/C6CP00448B.

Chapter 2

Cathode Materials for Lithium–Sulfur Batteries: Fundamentals, Challenges, and Solutions

Saad Zafar, Sangeeta Sahu,
Soumyasri Nikhilesh Mahapatra, and Bimlesh Lochab
Materials Chemistry Laboratory, Department of Chemistry,
School of Natural Sciences, Shiv Nadar University, Gautam Buddha Nagar,
Uttar Pradesh 201314, India
bimlesh.lochab@snu.edu.in

2.1 Introduction

The affordability of energy without harming the environment is one of the greatest challenges faced by society. The increasing consumption and limited availability of fossil fuels and the adverse environmental effects of fossil fuel combustion have led to the development of alternative sustainable renewable energy technologies. Renewable sources of energy, such as wind and solar power, are attractive; however, the dependence on intermittent energy sources demands the exploration of electrical energy storage (EES) systems that are

Batteries: The Future of Energy Storage
Edited by Subashani Maniam

ISBN 978-981-5129-38-0 (Hardcover), 978-1-003-51288-2 (eBook)
www.jennystanford.com

efficient and economical. For EES, rechargeable batteries are the most viable option. Lithium-ion batteries (LIBs) are the most widely used rechargeable battery technology and have come a long way since they were first introduced.

The current LIBs face drawbacks such as short lifetimes, overheating, and supply chain challenges for certain raw materials. Conventional LIBs with insertion-compound cathodes (such as $LiCoO_2$, $LiMn_2O_4$, and $LiFePO_4$) and anodes (graphite) have certain limitations [1–3]. The conventional cathode materials, which are based on lithiated transition metal oxide and phosphate, offer low energy densities (theoretical: 350–400 Wh/kg, experimental: 100–220 Wh/kg) and are expensive to supplement large-scale energy demands [4] and influence the cost of electric vehicles (EVs) for regular consumers [5]. Additionally, it is challenging to raise the cathode's operating voltage above 4.3 V due to the constrained electrochemical stability window of the currently available liquid electrolytes. Furthermore, the capacities of insertion-oxide cathodes are limited to 250 mAh/g. On the other hand, the graphite anode's capability is similarly restricted to 370 mAh/g. Therefore, to meet the rising demands of EVs and smart grids, the next generation beyond LIB needs to be explored. To meet the needs for sustainable energy, it is, therefore, necessary to investigate novel rechargeable systems based on new chemical principles, such as lithium–sulfur (Li-S), lithium–air, and sodium–sulfur batteries, or new electrode materials with high energy densities, involving multielectron reaction and offering high-voltage operating windows.

Lithium–sulfur batteries (LSBs) are one of the prospective energy storage systems that have received significant attention due to their superior energy density in comparison to currently used LIBs [6–11]. The primary difference between these two types of batteries resides in their energy storage mechanisms. LIBs work on the intercalation of lithium ions into layered electrode materials, such as graphite anodes and lithium metal oxide cathodes [12–15]. Since lithium ions can only be intercalated topotactically into specific locations, the theoretical energy density of LIBs is typically capped at ~420 Wh/kg or 1400 Wh/L (Fig. 2.1) [6, 16]. LSBs, on the other hand, operate based on metal deposition and stripping on the lithium anode side and a chemical reaction on the sulfur cathode side [7, 9–11]. Due to the non-topotactic nature of these reactions, lithium anodes and sulfur cathodes have theoretical specific capacities of 3860 and 1673 mAh/g, respectively. With an average cell voltage of 2.15 V, LSBs have a high theoretical energy density of ~2500 Wh/kg or 2800 Wh/L.

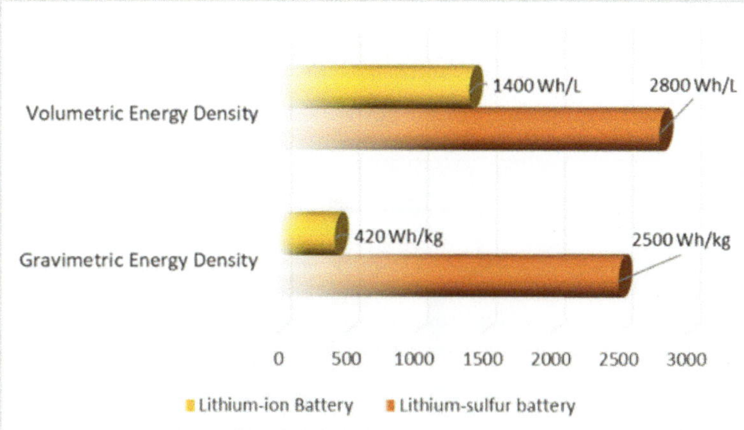

Figure 2.1 Energy density plots of lithium–sulfur versus lithium-ion batteries (based on graphite anodes and $LiNi_{1/3}Mn_{1/3}Co_{1/3}O_2$ cathodes). Values obtained from Ref. [6, 16].

2.2 Fundamentals and Challenges of LSBs

2.2.1 Mechanism for Charge–Discharge in LSBs

In general, an LSB consists of a sulfur cathode, an organic electrolyte, and a lithium anode. The redox process between sulfur and lithium completes the energy storage and conversion of LSBs, which is described as

$$16Li + S_8 \rightarrow 8Li_2S.$$

Lithium metal is oxidized during the discharge reaction, releasing lithium ions and electrons. A current is generated when lithium ions pass from the negative electrode to the positive electrode through the electrolyte while electrons circulate through the external electrical circuit. By accepting lithium ions and electrons, sulfur at the positive electrode is reduced to lithium sulfide (Li_2S) [7]. The opposite reactions will occur during the charging procedure (Fig. 2.2). The discharge reactions can be described as follows:

Negative electrode:

$$2Li \rightarrow 2Li^+ + 2e^-$$

Positive electrode:

$$S + 2Li^+ + 2e^- \rightarrow Li_2S$$

Overall discharge reaction:

$$2Li + S \rightarrow Li_2S$$

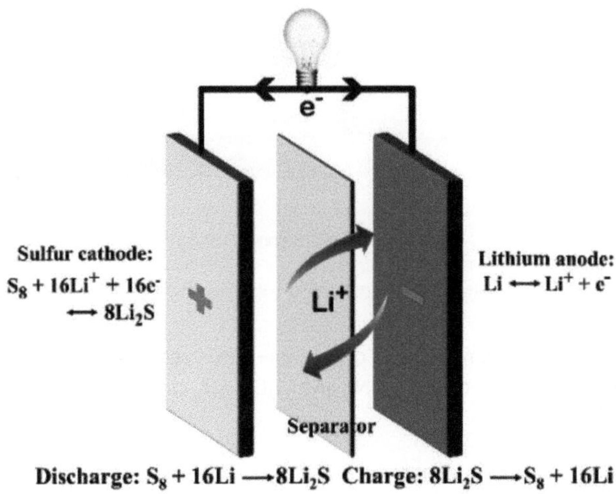

Figure 2.2 Schematic of the electrochemistry charge/discharge process involved in a rechargeable LSB consisting of lithium metal anode and sulfur cathode.

The intercalation method of LIBs is different from the multistep electrochemical mechanism followed by LSBs. As shown in Fig. 2.3, the first discharge curve contains two voltage plateaus. The higher voltage plateau at approximately 2.4 V represents the conversion of cyclic-octasulfur, S_8, associated with the ring-opening reduction reaction of cyclic-octasulfur to various long-chain (high-order) soluble lithium polysulfides (LPSs) (Li_2S_n, $6 > n > 8$) via multiple step reactions [6, 17, 18].

$$S_8(s) + 2e^- \rightarrow S_8^{2-}(aq)$$

$$3S_8^{2-}(s) + 2e^- \rightarrow 4S_6^{2-}(aq)$$

$$2S_6^{2-}(aq) + 2e^- \rightarrow 3S_4^{2-}(aq)$$

The subsequent plateau at approximately 2.4–2.1 V corresponds to the further reduction of these polysulfides into shorter chains (Li_2S_n, $n < 6$) and then at 2.1 V represents the transformation into solid lithium sulfide (Li_2S_2). Finally, at 2.1–1.7 V, it represents the formation of solid Li_2S_2 to Li_2S and then from Li_2S to S_8 [19–21]. Typically, the two charge voltage plateaus are superimposed [7].

$$S_4^{2-}(aq) + 4Li^+(aq) + 2e^- \rightarrow 2Li_2S_2(s)$$

$$Li_2S_2(s) + 2Li^+(aq) + 2e^- \rightarrow 2Li_2S(s)$$

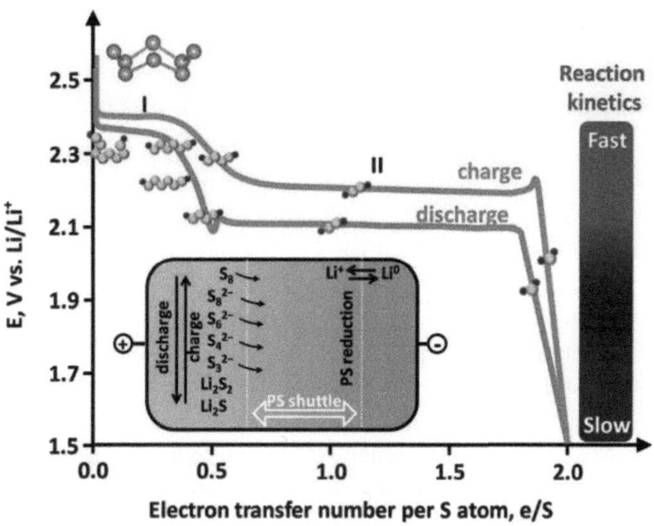

Figure 2.3 Electrochemistry of sulfur showing an ideal charge–discharge profile. PS: polysulfide. Reproduced from Ref. [19] with permission from the Royal Society of Chemistry.

2.2.2 Advantages of LSBs

In LSBs, sulfur and sulfur-based materials are used as the cathode, which are cheap, nontoxic, and abundant. Hence, in contrast to LIBs, which use heavy transition metals for cathodes, LSBs are more sustainable and economical. An LSB can be drained to a 0% charge state and does not need to be later recharged for storage, which is an additional benefit [22]. This indicates that an LSB can be stored in the discharged state for a long time without experiencing significant capacity deterioration. This is crucial because irreversible capacity loss in LIBs must be prevented by charging them prior to long-term storage. The average energy density of LSBs (400–600 Wh/kg) is two to three times higher than that of the current LIBs, which allows achieving the target of a 500 km range [11].

2.2.3 Sulfur Cathode and Its Challenges

One of the main barriers to the commercialization of LSBs is the fact that, despite having an extraordinarily high theoretical energy density, LSBs have a significantly low practical energy density value. The high insulating nature (5×10^{-30} S/cm) of sulfur accounts for a poor electrochemical performance. To mitigate the low electrical and ionic conductivity of sulfur cathodes, conductive fillers are generally blended in a significant amount [21]. In addition, the final step of converting Li_2S_2 to Li_2S is also challenging and hindered by the slow solid-state diffusion and low electrical conductivity of Li_2S (3.6×10^{-7} S/cm) [6, 23]. Consequently, active material utilization is minimal because of incomplete discharge.

Furthermore, during the discharge process, the viscosity of the electrolyte changes as the lengths of the soluble polysulfide (PS) chains shorten. Consequently, the volume expansion within the cathode is as high as 80% when sulfur is completely converted to Li_2S [19]. This change in volume results in severe cathode pulverization and consequently leads to rapid capacity decay. The PS species (S_8^{2-}, S_6^{2-}, and S_4^{2-}) produced during the charge–discharge process are readily soluble in the electrolyte and diffuse freely from the cathode to the anode under the influence of a concentration gradient (Fig. 2.3, inset). Upon reaction with the lithium anode, the lengths of the polysulfide chains are reduced to form short chains. Finally, they transformed to insoluble Li_2S_2/Li_2S deposits on the lithium anode, which can react further with the polysulfide ions, producing soluble medium-chain ions that can diffuse back to the cathode and re-oxidize [24]. Therefore, the shuttling of polysulfides between the cathode and the anode extends the charging time of the cell to infinity. This polysulfide-shuttle phenomenon is responsible for the low coulombic efficiency and self-discharge of cells [25].

Additionally, these soluble polysulfide species can be reduced to form an Li_2S insulating layer at the lithium anode surface, thus passivating the anode and causing an increase in impedance and material loss. Furthermore, at the cathode surface, the insoluble and electrically insulating Li_2S_2/Li_2S can form and precipitate. Once a thin Li_2S layer covers the entire cathode, further lithiation will be significantly impeded, resulting in low active material utilization and reduced coulombic efficiency. As the cycling continues, the dissolution and precipitation process alters the morphology of the cathode with each cycle, resulting in a significant loss of capacity and a shortened cycle life. The following characteristics are required for high-performance sulfur cathodes: (a) a closed structure to encapsulate polysulfides efficiently, improving sulfur–electrolyte contact, (b) attainment of good conductivities, usually achieved by blending conductive materials in the cathode, and (c) a flexible framework to provide sufficient space to accommodate sulfur volumetric expansion and retain the morphology of electrodes during the transportation of electrons and Li ions.

Numerous strategies have been explored to advance LSB technology to previously unexplored heights. In this chapter, different nanosized and nanostructured sulfur-based cathodes based on morphologies and compositions of the electrode for LSBs are presented, including (a) carbon-based cathode materials for sulfur; (b) pristine and/or composite of MXene; and (c) organosulfur materials. Although several new materials have been developed, which have boosted remarkable progress in the electrochemical property of cathodes for LSBs, it is still more important to design and synthesize special and innovative cathode materials to promote cycling stability, as well as improve excellent capacity density and rate capability through facile and environmentally friendly processes. A systematic discussion is provided in the subsequent sections for a better and comprehensive understanding of this emerging area of research.

2.3 Carbon-Based Cathode Materials

In 2009, Nazar et al. made a significant advancement by developing a highly ordered, mesoporous carbon–sulfur cathode for LSBs [26].

As a proof of concept, the mesoporous carbon (CMK-3) framework encapsulates elemental sulfur at 155 °C, where sulfur has the lowest viscosity (Fig. 2.4). CMK-3 offered channels with a uniform and thin, narrow mesopore diameter of 3.33 nm, and the existence of a substantial pore volume of 2.1 cm^3/g enabled the hosting of elemental sulfur in appreciable loading and provided good conductivity (0.20 S/cm). CMK-3 assisted in the capture of LPSs, and a high specific capacity of 1005 mAh/g with stable cycling over 20 cycles at 0.1 C was achieved. To further restrict the diffusion of polysulfides out of the cathode, the CMK-3/S composite was coated with a thin layer of hydrophilic poly(ethylene glycol) (PEG) polymer. The PEG@ CMK-3/S composite showed an initial discharge capacity as high as 1320 mAh/g at 0.1 C. This work led to a significant development in the field of cathode materials for metal–sulfur batteries, in particular the use of highly conductive carbon materials to physically encapsulate sulfur.

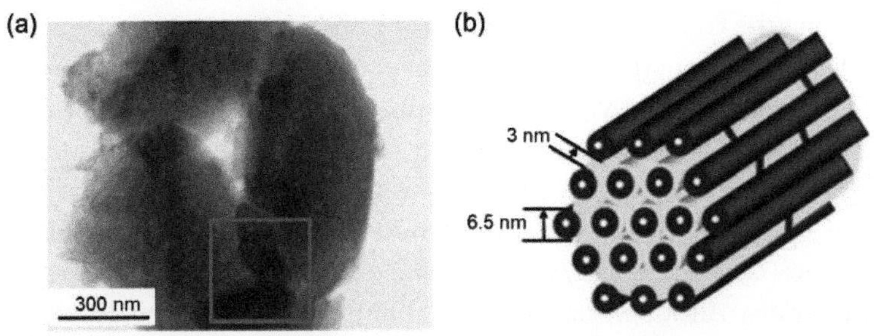

Figure 2.4 (a) TEM image and (b) schematic representation of a CMK-3/S composite. A highly ordered mesoporous carbon (CMK-3) was used to infiltrate sulfur by melt-diffusion at 155 °C. Reproduced from Ref. [26] with permission from the Royal Society of Chemistry.

2.3.1 One-Dimensional Materials

2.3.1.1 Carbon nanotubes

The arrangement of sp^2-hybridized carbon atoms in a very high aspect ratio as in carbon nanotubes (CNTs) makes them particularly desirable for metal–sulfur batteries. Due to the support framework and interconnected structure of CNTs, long-range conductivity is achieved in sulfur cathode materials. Lee et al. [27] conducted one of the earliest investigations into the use of CNTs as additives in LSBs. By physically blending multiwalled carbon nanotubes (20 wt%) with sulfur, they observed an increase in specific capacity from ~400 mAh/g to 485 mAh/g with a capacity retention from 25% (without additive) to 62% (with additive) after 50 cycles. However, inadequate contact between sulfur and CNTs hampered the capacity.

Therefore, to allow maximum contact between the two, a capillary effect to seep in molten sulfur into CNTs was performed at 155 °C [28]. Even greater contact between the two materials was noticed where the outer surface of CNTs was also covered with sulfur. Consequently, sulfur–carbon nanotube cathodes prepared at 155 °C showed reduced charge transfer resistance and improved cycling performance compared to cathodes prepared by a simple mixing of the two at room temperature. Particularly, the former case achieved a high discharge capacity of ~900 mAh/g with 74% retained capacity after 60 cycles at 0.06 C.

The effect of variation on the morphologies of CNTs was explored. For instance, vertically aligned CNTs were synthesized on a nickel foil using ethene as a precursor and iron, molybdenum, and cobalt–ethyl hexanoate based catalyst system (Fig. 2.5) [29]. The vertical arrangement of CNTs on the substrate permits rapid ion and electron diffusion throughout the entire electrode cross section. The structure was then treated with a solution of sulfur in toluene to assist sulfur impregnation at 120 °C. LSBs based on the as-produced CNT–sulfur composite showed stable discharge capacities of ~800 mAh/g over 40 cycles at 0.08 C.

It has been observed that in unmodified CNTs, the interaction between sulfur and carbon atoms happens mainly via weak van der Waals forces, which may lead to weak adsorption of LPSs. The interaction and capture of sulfur and related species are expectedly enhanced by introducing functional groups, polymers, metal oxides, heteroatoms, and other redox mediators, overall improving the electrochemical dynamics. Guo et al. [30] pioneered the flexible CNT-based electrode fabrication process by employing the template method to create a binder-free, highly conductive, and flexible LSB cathode. The sulfur-impregnated disordered CNTs were synthesized by carbonizing the polyacrylonitrile (PAN) membrane attached to the surface and interior of the anodic aluminum oxide (AAO) template, followed by etching the AAO and liquid sulfur impregnation. This study offers insight into the next-generation CNT-based self-supporting electrodes.

Figure 2.5 Low- and high-magnification SEM images of vertically aligned carbon nanotubes (a) before and (b) after sulfur infiltration. Reproduced from Ref. [29] with permission from the Royal Society of Chemistry.

2.3.1.2 Carbon nanofibers

The morphological and physical properties of carbon nanofibers (CNFs) are comparable to those of CNTs. However, CNFs possess a graphitic structure, which empowers a high conductivity and offers a narrow hollow channel. CNFs contribute to improving the electronic contact between sulfur and the current collector due to their strong mechanical strength and good electrical conductivity, which improves the utilization of active materials during the electrochemical reaction [31]. As a result, CNFs are regarded as viable options for host matrices for sulfur. However, due to the extremely small diameters and impermeable walls of CNFs, it is difficult to infiltrate sulfur within their hollow spaces. In most of the studies involving CNTs, sulfur was found to coat only the outermost layer, leaving sulfur/LPSs accessible to the electrolyte. Utilizing hollow carbon tubes with significantly larger diameters would be advantageous for encapsulating and confining sulfur/LPSs inside the tubes.

Zheng et al. [32] prepared hollow CNFs by carbonizing a polymer in an AAO template and then infusing sulfur into the tubes at 155 °C using the melt-diffusion strategy. After sulfur infusion, the template was eliminated by acid dissolution, leaving sulfur on the interior surface of the CNFs but not at the outer surface. At 0.2 C and 0.5 C, high specific capacities of ~1400 and ~1300 mAh/g were attained. However, the initial capacity of the electrode dropped significantly to 730 mAh/g after 150 cycles at 0.2 C. The reason behind the observed significant capacity loss in sulfur-impregnated porous CNFs was unknown.

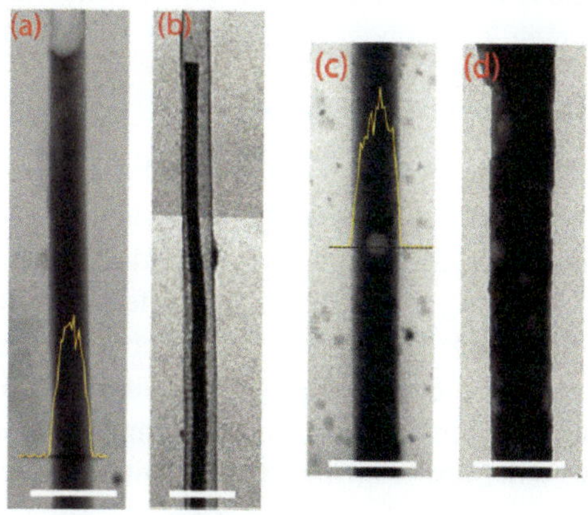

Figure 2.6 TEM images of sulfur cathodes encapsulated in unmodified hollow carbon tubes (a) before and (b) after discharge to 1.7 V versus Li^+/Li. TEM images of sulfur cathodes encapsulated in PVP-modified hollow carbon tubes (c) before and (d) after discharge under identical conditions. The yellow lines in (a) and (c) show the EDX line scan for sulfur. Scale bars in (a–d) are 500 nm. Reprinted with permission from Ref. [32]. Copyright 2013 American Chemical Society.

A subsequent study determined that the capacity degradation was primarily caused by the peeling of the discharge product, lithium sulfide, from the inner walls of the hollow carbon tubes, resulting in a loss of electrical contact (Fig. 2.6a,b) [33]. This separation was attributed to the weak contact between the highly polar lithium sulfide and the nonpolar carbon walls. The nonpolar carbon surface weakly binds to highly polar LPSs, thereby reducing the polysulfide trapping effect. An amphiphilic poly(vinylpyrrolidone) (PVP) was used to modify the surface of the interior carbon walls to address this issue. PVP has both a hydrophobic carbon backbone and hydrophilic amide groups that can anchor firmly to nonpolar carbon surfaces and bind strongly to lithium sulfide and LPSs, respectively. The ab initio calculations demonstrated that the amide groups of PVP are capable of forming robust Li–O interactions with LPSs/sulfide species (LiS and Li_2S), with binding energies of 1.30 and 1.10 eV, respectively [34]. Due to the amphiphilic nature, the detachment of lithium sulfide was prevented from the polymer-modified carbon surface, as can be seen in TEM images (Fig. 2.6c,d). Consequently, a high discharge capacity of 838 mAh/g was attained with a capacity retention of 80% after 300 cycles at 0.5 C, which is a significant improvement over a system without polymer modification.

2.3.1.3 Heteroatom modification

The modification of the carbon surface with heteroatoms, such as nitrogen, is a promising strategy for enhancing the performance of LSBs, as it can result in a stronger binding with LPS species, consequently lowering their solubility in the electrolyte. Using ab initio calculations, the possibility of using nitrogen-doped CNTs in LSBs was first predicted theoretically [35]. Nitrogen doping allowed easy penetration of lithium atoms into the nanotube walls (energy barrier decreased from ~9.0 to 1.0 eV), but no effect on lithium diffusion along the nanotubes was observed. In contrast, the energy barrier for sulfur atoms to diffuse through nitrogen-doped walls remained extremely high at 9.3 to 12.0 eV due to the presence of a strong S–N interaction, indicating that sulfur species are confined and bound effectively.

The Zhang group explored nitrogen-doped carbon nanotubes as conductive hosts in LSBs to accomplish stable cycling [36]. The conductivity of nitrogen-doped CNTs (7.98 S/cm) is relatively higher than that of pristine nanotubes (11.85 S/cm). Theoretical calculations also revealed a strong Li–N interaction between pyridinic nitrogen atoms and different LPS/sulfide species (Li_2S_4, Li_2S_3, Li_2S_2, and Li_2S), with binding energies ranging from 1.04 to 1.42 eV. Based on these results, the as-prepared nitrogen-doped CNT–sulfur composite demonstrated a discharge capacity of 937 mAh/g with a capacity retention of 70% after 200 cycles at 1 C, which is significantly superior to pristine CNTs.

2.3.2 Two-Dimensional Materials

2.3.2.1 Graphene and graphene oxide

Graphene is a single layer of two-dimensional (2D) carbon nanosheet synthesized from graphite. Graphene has a high surface area (2600 m^2/g) and excellent electrical conductivity (~10^6 S/cm), is lightweight and flexible, and displays superior mechanical strength and a chemically inert nature [37]. Additionally, the ease of functionalization and the large surface area of graphene have further encouraged its use in EES devices [38].

Wang et al. [39] were the first researchers to encapsulate sulfur particles in graphene and explore them as sulfur cathodes. Sub-micrometer sulfur particles covered with PEG-containing surfactants and graphene sheets were synthesized to form the composites. Besides trapping the soluble polysulfide intermediates formed during the electrochemical reaction, graphene significantly increased the overall electronic conductivity. The volumetric change in the sulfur particles during the electrochemical reaction was accommodated by PEG coating. A battery based on a graphene and sulfur composite demonstrated a high and consistent specific capacity of ~750 mAh/g with 70% capacity retention after 100 cycles at 0.2 C.

As an alternative to graphene, oxygen-rich graphene oxide was utilized as a host material for sulfur cathodes [40]. The composite was prepared by the acidification of sodium polysulfides to precipitate sulfur in the presence of graphene oxide sheets ($S_x^{2-} + 2H^+ \rightarrow (x\text{-}1)S + H_2S$). The S–O and S–C interactions between sulfur and the oxygen functional groups (epoxide and hydroxyl groups) in graphene oxide were predicted by ab initio calculations. In a subsequent publication, the presence of these interactions was experimentally confirmed by X-ray photoelectron spectroscopy (XPS) and X-ray absorption near the edge structure [41]. These favorable interactions between the immobilized sulfur species and graphene oxide resulted in a high specific capacity of ~1000 mAh/g with 95% capacity retention after 50 cycles at 0.1 C.

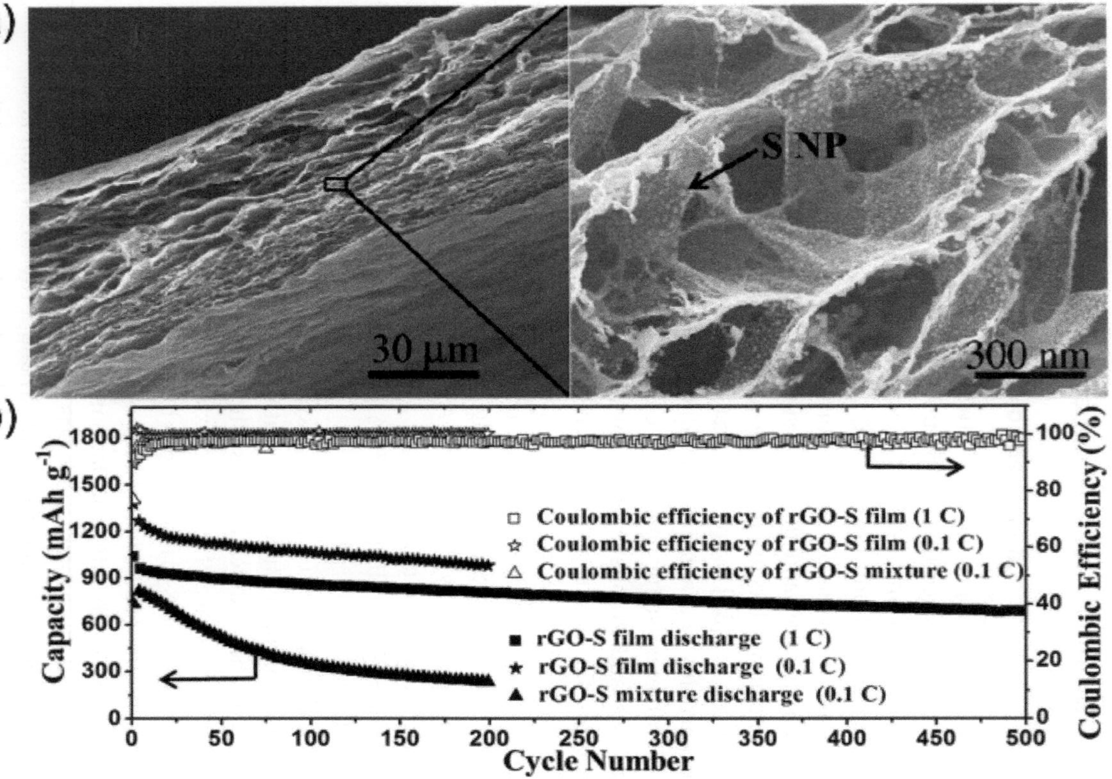

Figure 2.7 (a) Cross-sectional scanning electron micrograph of an rGO-S composite film at low and high magnifications. (b) Cycling performance and coulombic efficiency of the rGO-S mixture and rGO-S composite film electrodes at 0.1 C for 200 cycles and long-term cycle stability of the rGO-S composite film electrode at 1 C for 500 cycles. Reprinted with permission from Ref. [44]. Copyright 2016, Wiley-VCH.

The interaction between graphene oxide and LPSs was elucidated through additional computational studies [42, 43]. Li_2S_8 can effectively bind to oxygen on the graphene basal plane (in the form of epoxides) via a strong Li–O interaction. The adsorption energy of Li_2S_8 was calculated to be 1.1 eV, which is greater than the energy of binding with graphene in its pristine state (0.8 eV). The adsorption energy can be increased to 1.5 eV if oxygen attaches to a single vacancy site. Additionally, the surface hydroxyl groups can bind to Li_2S_8 with a binding energy of 1.4 eV, elucidating the role of polar oxygen-rich functionalities in increasing the cycling stability.

Cao et al. [44] reported a freestanding nanostructured reduced graphene oxide-S (rGO-S) composite film (Fig. 2.7a), prepared by simultaneous reduction and development of GO sheets with S nanoparticles on a metal surface. The composite film was expected to provide a physical barrier for capturing polysulfides, effectively limit the diffusion of LPSs in organic electrolytes, and offer quick electron and ion transfer. However, the inactive components in the electrode, such as the conductive agent, binder, and current collector (e.g., Al foil) in the electrode, negate the benefits expected for high-energy LSBs. To overcome these drawbacks, a freestanding rGO-S film electrode was fabricated without the use of a binder. A high initial discharge capacity of 1302 mAh/g and a sustained stable cycling performance for 200 charge/discharge cycles at 0.1 C were recorded (Fig. 2.7b). The excellent conductive and mechanical strength of the rGO-S composite film allowed the preparation of flexible LSBs.

Another work reported attainment of capacity of rGO-sulfur composite near to the theoretical value of sulfur when the deposited sulfur particle size on rGO is reduced to ultrasmall nano-dimension [45]. This was achieved by initially reacting sulfur and ethylenediamine to form a sulfur–amine complex ($2(R–NH_2) + S_8 \rightarrow (R–NH_3^+)(RNH–S_8^-)$). Then, this complex was added to rGO in the presence of hydrochloric acid, and as a result, the complex decomposed to nucleate sulfur on the surface of the graphene sheets ($(R–NH_3^+)(RNH–S_8^-) + 2H^+ \rightarrow 2(R–NH_3^+) + S_8$). When the pH and reaction duration were adjusted, sulfur nanoparticles with tunable diameters (5 to 150 nm) could be produced (Fig. 2.8). The ultrasmall size of sulfur nanoparticles enabled a rapid charge transfer kinetics, which compensated for the low electric and ionic conductivity of bare, large-sized sulfur particulates. At 0.1 C, the theoretical specific capacity of 1672 mAh/g was attained with sulfur nanoparticles of 5 nm diameter. Long-term cycling at 0.5 C and 1 C revealed high specific capacities of 1661 and 1574 mAh/g, with capacity retentions of ~61% after 500 cycles.

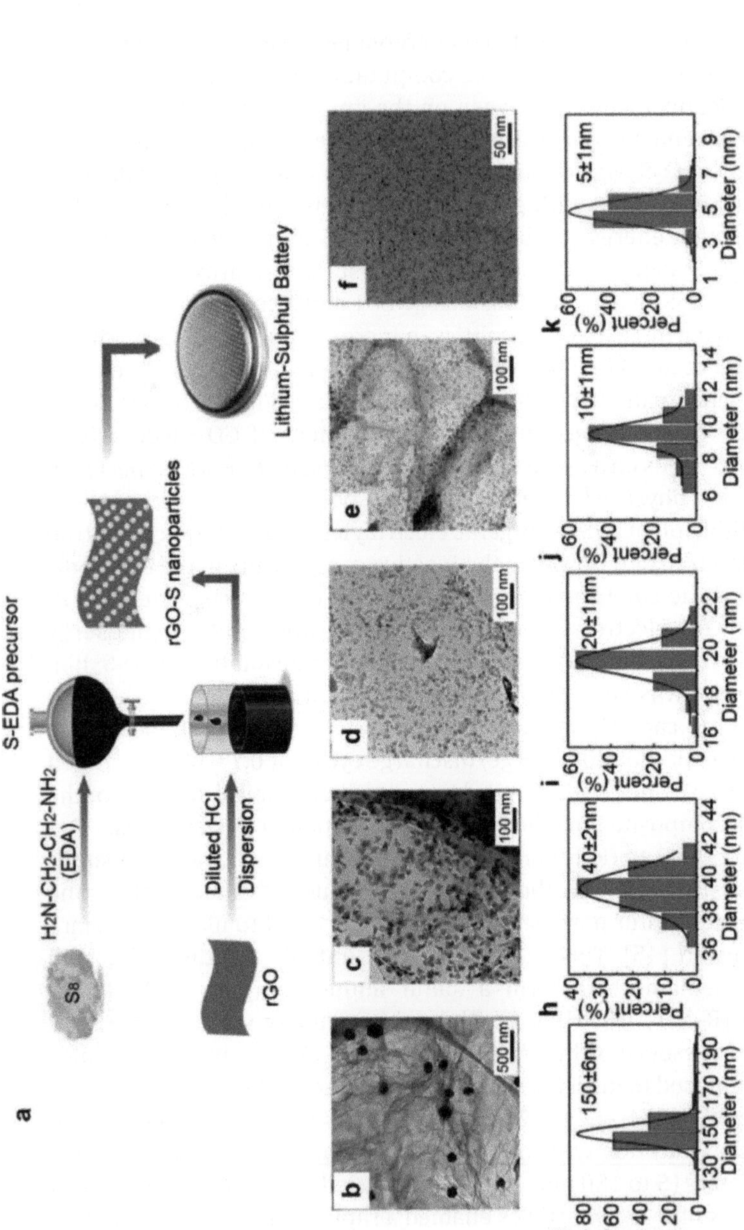

Figure 2.8 (a) Schematic of the synthesis of ultrasmall sulfur nanoparticles supported on rGO via sulfur–amine chemistry. (b–f) TEM images of sulfur nanoparticles. Reprinted with permission from Ref. [40]. Copyright 2015 American Chemical Society.

2.3.2.2 Heteroatom modification

In heteroatom-doped graphene, carbon atoms are replaced by heteroatoms (such as nitrogen and boron) in the graphene sheet. Heteroatom-doped graphene often maintains superior conductivity compared to functionalized graphene. The local positive and negative charge distributions are unequal due to the distinct electronegativity between heteroatoms and carbon atoms, leading to polar electroactive sites that can efficiently adsorb polar polysulfides.

The different nitrogen-doped forms of graphene can be categorized according to the bonding modes of the nearby carbon atoms. The three main categories of nitrogen doping are pyridinic, pyrrolic, and quaternary nitrogen atoms in graphene. In comparison to raw graphene, nitrogen-doped graphene materials have a higher degree of surface polarity, which helps them better reduce the shuttle effect of LPSs through stronger chemical adsorption as well as enhance LSB cycle stability.

Huang et al. synthesized nitrogen-doped graphene as the host material for the sulfur cathode in LSBs [46]. They achieved 1.05 wt% nitrogen doping by heating graphene oxide with an ammonia solution at 200 °C. The electronic conductivity of nitrogen-doped graphene was 1.02 S/cm, which was greater than that of undoped graphene. Due to the reaction of sodium thiosulfate and hydrochloric acid, sulfur was subsequently deposited on the nitrogen-doped graphene sheets. The as-synthesized composites demonstrated a discharge capacity of ~854 mAh/g and stable cycling performance (capacity retention of 93%) after 145 cycles at 0.4 C, which is substantially superior to that of undoped graphene.

The difference in polysulfide binding strength between pyrrolic, pyridinic, and quaternary nitrogen atoms in graphene was elucidated by computational studies [47]. In comparison with quaternary nitrogen, pyrrolic nitrogen and pyridinic nitrogen bind more strongly to LPS/sulfide species (Li_2S_8, Li_2S_4, and Li_2S). This is because, unlike quaternary nitrogen, pyrrolic and pyridinic nitrogen atoms possess a lone pair of electrons, resulting in a stronger coordinate-like interaction with the electropositive lithium in LPSs. The calculation results also indicated that pyrrolic nitrogen has a slightly higher binding energy with LPSs than pyridinic nitrogen, which was ascribed to a higher concentration of charge density and a larger available volume. This theoretical work provides a useful guideline that can optimize the various types of nitrogen atoms in graphene and further improve the polysulfide adsorption energy along with corresponding cycling performance.

2.4 Metal Chalcogenide Composites as Cathode Materials

2.4.1 Metal Oxides

Transition metal oxides (TMOs), like many other 2D materials, have also been showcased as an excellent platform in which sulfur can be hosted for exploration as cathodes in LSBs. An exfoliated 2D lepidocrocite titanium (LT) oxide nanosheet has been reported as a host for a high loading of sulfur (80%), and the interaction occurs due to the Lewis acidic nature of Ti because of the vacant d-orbitals [48]. Exfoliated 2D nanosheets and nanomaterials can provide many binding sites (Fig. 2.9) that favor the formation of enriched LT nanocomposites with high sulfur content. The LT sulfur nanocomposites also demonstrated a capacity retention (after 300 cycles at 1000 mA/g) of 82.3% and a specific capacity of 1024 mAh/g at 50 mA/g in electrochemical testing.

One of the major reasons for an abrupt decline in capacity is the shuttling effect, which poses significant difficulties in the practical application of LSBs. In general, carbon materials are attractive as a host substrate for sulfur and found to partially restrict hydrophilic LPS intermediates, and slowly these species diffuse out of the carbon host over long-term cycling. On the contrary, inorganic compounds, such as metal oxides, being highly polar, can be considered one of the most effective host matrices that bind polysulfides in LSBs. Polar metal oxides have been thoroughly investigated for their potential inclusion in LSB cathodes [49]. Metal oxides provide several benefits, including great adsorption capacity, conductive nature, and a large specific surface area. Thus, the dissolution of the LPS in the electrolyte can be inhibited due to the large contact area between the electrode and the electrolyte. LPSs can be efficiently trapped by metal oxides, which typically contain an anion of oxygen in the oxidation state of O^{2-}, thus increasing the overall efficiency with better material utilization. Metal oxides are usually insoluble in most organic solvents because of the strong bond that exists between the oxygen and the metal. Using a metal oxide and a sulfur composite cathode increases the volumetric energy density of LSBs and significantly accelerates the reaction between polysulfide intermediates and Li_2S_2/Li_2S.

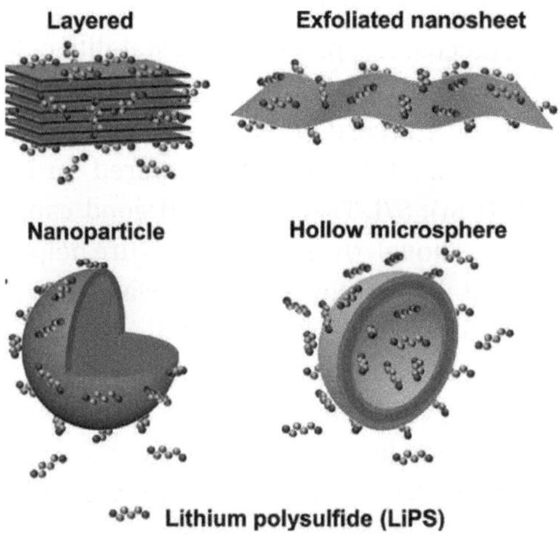

Figure 2.9 Schematic illustrations of LPSs adsorption on the surfaces of exfoliated nanosheets. Reproduced with permission from Ref. [48]. Copyright 2018, American Chemical Society.

To trap LPSs and enhance LSB electrochemical processes, many studies have explored composite materials such as MnO_2 [50], WO_{3-x} [51], V_2O_5 [52], $LaMnO_{3-x}$ [53], SnO_2 [54], CeO_2 [55], ZrO_2 [56], and TiO_2 [57]. Due to the adsorption of LPS by TiO_2, the S-TiO_2 composite demonstrated superior discharge capacity retention than pure sulfur (58% after 50 cycles) and a discharge capacity of 530 mAh/g after 50 cycles [57].

In most electrochemical reactions, an oxygen-deficient modified cathode based on a metal oxide exhibits increased electrocatalytic activity. Composites based on oxygen-deficient WO_3 with sulfur have been explored as cathodes, and a capacity decay of only 0.13% per cycle at a 0.5 C rate has been found. Additionally, the accumulation of LPSs on the cathode surface was suppressed by the interaction between WO_3 and LPS [51]. A composite material made of CeO_2

and nitrogen-doped carbon effectively reduced LPS dissolution and accelerated LPS conversion processes. After 200 cycles, the CeO_2-modified electrode had a reversible capacity of 1064 mAh/g at a 0.2 C rate [55].

2.4.2 MXenes

2.4.2.1 Pure MXene as sulfur hosts

The outstanding mechanical property and stability of MXene (Ti_3C_2) inspired Zhao et al. [58] to be the first to use an HF-etched layered MXene (L-Ti_3C_2) with accordion-like morphology as the host for sulfur in LSBs (Fig. 2.10a). Due to the layered structure and low resistivity (0.05 Ω m), S/L-Ti_3C_2 displayed good capacity and high cycle performance. Though the layered structure helped in obtaining high capacity and stability, a low specific surface area (SSA) resulted in a sulfur loading of only 58%. The initial discharge capacity of S/L-Ti_3C_2 was found to be 1291 mAh/g at a current density of 200 mA/g and had 970 mAh/g remaining after 100 cycles. Nazar et al. [59] explored the interaction between LPSs and various types of MXenes (Ti_2C, Ti_3C_2, and Ti_3CN) as sulfur hosts (Fig. 2.10b). XPS studies confirmed the presence of a Ti–S bond due to the strong Lewis acid–base interaction. Also, the polar–polar interaction between MXene and LPSs was studied. The prepared Ti_2C showed an SSA of 67.9 m^2/g, and 70% sulfur loading was achieved. The cathode showed promising cycling performance with a specific capacity of nearly 1200 mAh/g at a 5 h charge/discharge (C/5) current rate, and a remaining discharge capacity of 723 mAh/g after 650 cycles.

Due to its 2D layered morphology and excellent mechanical integrity, $Ti_3C_2T_x$ MXene can be molded into a freestanding film without the use of any binders or conductive agents, which reduces the overall cost of processing. MXene flakes are self-stacked, forming a thin film that shows excellent electrical conductivity. Wang et al. [60] prepared an MXene foam to increase the SSA from 26.3 m^2/g (a freestanding film) to 182.9 m^2/g. They prepared the foam by soaking a freestanding MXene film in hydrazine monohydrate through the hydrothermal treatment process. This enhancement in surface area resulted in an increased loading of sulfur in the foam to 5.1 mg/cm^2. The cathode showed a high initial discharge capacity (1226 mAh/g), good rate performance (711 mAh/g at 5 C), along with long-term cycling stability (690 mAh/g at 1 C) after 1000 cycles. The gravimetric energy density of the device was also determined as very high (1297.8 Wh/kg).

Taking advantage of the multilayered structure of MXenes, Zhang et al. [61] prepared Ti_3C_2 with uniform doping of sulfur nanoparticles. The sulfur nanoparticles were distributed uniformly on the MXene surface and anchored between the layers as well, and the sulfur loading was found to be 4 mg/cm^2. The loading of sulfur enhanced the capacity of MXene, and the specific capacity was obtained to be 1477 mAh/g, with a very small capacitive loss of 0.18% per cycle (100 cycles at 0.2 C).

In an effort to modify the floral-like structure of MXenes, Xiao et al. [62] prepared a porous flower-like $Ti_3C_2T_x$ (FLPT) material, without the addition of carbon hosts or conductive additives. The flower-like structures provide an advantage over other structures by favoring the electron transfer because of the quick electron/ion transfer and high loading of sulfur. Further, the redox reaction of sulfur on the surface of FLPT is propelled by the catalytic conversion of LPSs due to the anchoring behavior of the abundant active sites in the structure. The sulfur-loaded FLPT (FLPT-S) showed an excellent areal capacity of ~10 mAh/cm^2.

2.4.2.2 Modified MXenes as sulfur hosts

Chemically modifying the surface of the host and doping with elements such as N and O are believed to increase interaction with sulfur and inhibit the shuttling effect of LPSs. Similar modifications on the surface of MXene have been studied. The surface terminations of MXene have been modified to allow increased polar interactions between MXene and LPSs.

The most frequent element to be doped is N; the pyridinic nitrogen shows good interaction with LPSs. Wang et al. [63] prepared crumpled nitrogen-doped $Ti_3C_2T_x$ nanosheets (N-$Ti_3C_2T_x$) (Fig. 2.10c) using melamine as the nitrogen source via a one-step thermal treatment method. The physical encapsulation of sulfur allowed a high sulfur loading in the material (5.1 mg/cm^2) due to the high SSA (385 m^2/g) of the crumpled structure MXene. The N-$Ti_3C_2T_x$/S material showed very good electrochemical performance with a reversible capacity of 1144 mAh/g and retained a specific capacity of 610 mAh/g after 1000 cycles at the 2 C rate. In a similar structural design, using the melamine-formaldehyde (MF) template, Sun et al. [64] prepared porous N-doped Ti_3C_2 (P-NTC). The sulfur nanoparticles were seen to occupy the MF spheres after thermal annealing. The S/P-NTC composite showed excellent electrochemical performance and achieved a low capacitive decay rate of 0.033% at 2 C after 1200 cycles.

Besides non-metal doping, even metal-doped substrates are attractive as they influence the kinetics of redox reaction. Metals such as Fe and Ni are introduced into the graphene skeleton to promote the solid-state deposition of Li_2S_2 and Li_2S. Similarly, in the MXene skeleton, metals are doped through the Lewis molten salt replacement method. The Al atoms in Ti_3AlC_2 are displaced by the Zn atom from the molten $ZnCl_2$, producing single-atom zinc-implanted MXene (SA-Zn-MXene) [65]. Due to the electronegativity of Zn, SA-Zn-MXene showed higher binding energy toward the LPSs compared to pristine MXene as supported by the nucleation profile analysis. The modified sulfur host revealed a high specific capacity of 706 mAh/g and high cyclic stability (80% retention in capacity after 200 cycles at 4 C), which is higher than the pristine MXene.

MXenes already have active functional groups on the surface. To improve the adsorption and catalytic conversion of MXene, additional heteroatom doping can be facilitated through various techniques to achieve high-performing electrochemical activities in LSBs.

2.4.2.3 MXene-based composites as sulfur hosts

The chemisorption of LPSs on the surface of MXene is favorable, but the van der Waals interactions existing between the layers of MXene cause the sheets to restack, which significantly reduces the SSA and hinders the utilization of the surface terminations. To avoid the problem of restacking, MXenes are compounded with other materials. Materials such as conducting carbon or graphene are often used as they provide large SSA and high sulfur loading with good mechanical stability.

Wang et al. [66] attempted to fabricate a composite of graphene, a 2D carbon-based material, with MXene. They treated the etched $Ti_3C_2T_x$ with graphene oxide (GO) solution to obtain $Ti_3C_2T_x$/GO. The obtained $Ti_3C_2T_x$/GO was further treated with hydrazine hydrate under sonication to reduce GO and obtain the $Ti_3C_2T_x$/rGO composite, which was then treated with carbon disulfide (CS_2) to get the hybrid composite loaded with sulfur. The obtained hybrid had a high initial capacity of 1144 mAh/g at 0.5 C and good cyclic stability with a capacity retention of 878 mAh/g after 300 cycles. In another attempt at using graphene, an rGO aerogel with a 3D porous structure was prepared. MX/G (rGO) aerogel was prepared with the hydrothermal reduction method followed by freeze drying (Fig. 2.10d). The prepared rGO aerogel acted as an excellent host for sulfur when compounded with MXene ($Ti_3C_2T_x$). A $Ti_3C_2T_x$ MXene/G hybrid aerogel was used as a freestanding polysulfide reservoir to maximize the loading of sulfur [67]. It provided fast Li^+/electron transport and enhanced the redox kinetics by strongly anchoring MXene to LPSs. The hybrid showed a high capacity of 1270 mAh/g at 0.1 C and a minute capacitive decay of 0.07% after 500 cycles.

Apart from 2D layered materials, conductive carbon materials such as CNTs and CNFs are often used to prevent the restacking of MXene. Using one-step pyrolysis and spray drying, Wang et al. [68] prepared a micro-spherical MXene/CNT host material, where CNTs were grown uniformly on the surface of MXene and further interspersed in the MXene nanosheets. To further enhance the chemisorption of LPSs, N-doping was achieved using melamine as the nitrogen source. The N-CNT and N-Ti_3C_2 were internally connected to produce a 3D porous conductive structure (N-Ti_3C_2@CNT), which enhanced the SSA and the electron transfer rate. The initial specific capacity of the material was found to be 927 mAh/g with high cyclic stability (0.016% fading rate after 1000 cycles). Using MXene as a scaffold, Zhao et al. [69] designed CNT@MXene precursors by growing a zeolitic imidazolate framework (ZIF) on the N-MXene material. The sulfur hosts were prepared as hollow cobalt-doped CNTs and MXene ($Ti_3C_2T_x$) composites. The Co-doped CNTs had shown excellent adsorption toward LPSs along with electrochemical performance. At a lower loading of sulfur (2.5 mg/cm^2), it showed up to 840 stable cycles with 900 mAh/g capacity at 1 C. At an increased loading of sulfur (6 mg/cm^2), 86% capacitive retention was observed after 170 cycles at 0.2 C. The enhanced performance of LSBs resulted from a combined effect of heteroatom doping, microsphere-like morphology, and resolution of the restacking issue.

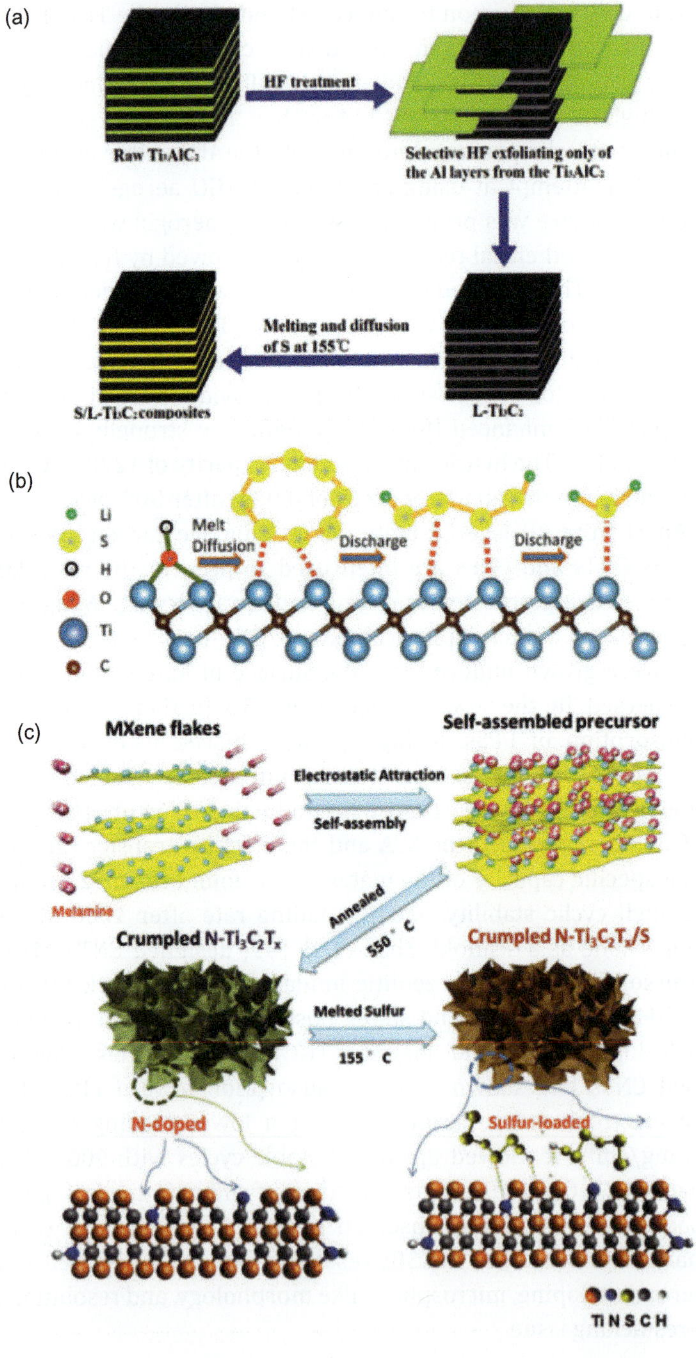
(a)
HF treatment
Raw Ti3AlC2
Selective HF exfoliating only of the Al layers from the Ti3AlC2
Melting and diffusion of S at 155℃
S/L-Ti3C2composites
L-Ti3C2
(b)
Li
S
H
O
Ti
C
Melt Diffusion
Discharge
Discharge
(c)
MXene flakes
Self-assembled precursor
Electrostatic Attraction
Self-assembly
Melamine
Annealed
550 °C
Crumpled N-Ti3C2Tx
Crumpled N-Ti3C2Tx/S
Melted Sulfur
155 °C
N-doped
Sulfur-loaded
TiNSCH

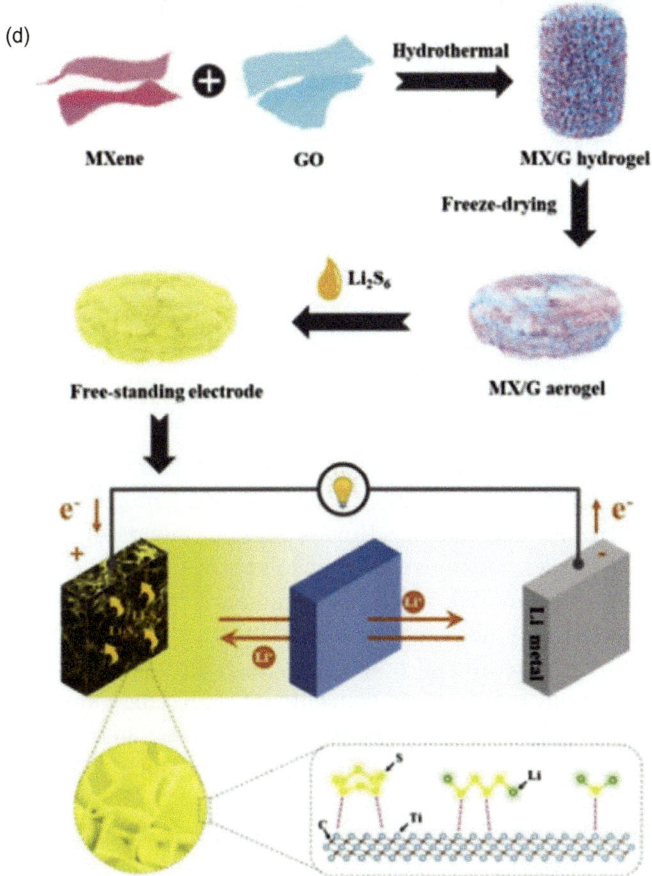

Figure 2.10 Schematic illustration of the (a) exfoliation of Ti_3AlC_2 and the subsequent synthesis of S/L-Ti_3C_2; reproduced with permission from Ref. [70], Copyright 2012, Royal Society of Chemistry. (b) Replacement of the Ti–OH bond on the MXene surface with an S–Ti–C bond on heat treatment or by contact with polysulfides; reproduced with permission from Ref. [59], Copyright 2015, Wiley-VCH. (c) Crumpled N-$Ti_3C_2T_x$/S composites synthesis; reprinted with permission from Ref. [63], Copyright 2018, Wiley-VCH. (d) MX/G aerogel electrode preparation and the schematic of the Li-S cell assembly; reproduced with permission from Ref. [70], Copyright 2012, Royal Society of Chemistry.

To further improve the binding and chemisorption of LPSs at the MXene surface, metal sulfides have also been incorporated into the framework of MXene. Guo et al. [71] compounded MXene with nano-flowers of molybdenum disulfide (MoS_2). The defects in MoS_2 provided additional active sites on the surface, including those present at MXene, preventing diffusion of LPSs. The MXene-MoS_2 based LSBs showed an excellent initial discharge capacity of 1041 mAh/g with stable cycling stability (799 mAh/g capacity after 300 cycles). In another effort to enhance the efficiency of MXene as a host for LSBs, Wu et al. [72] explored TiO_2 to inhibit polysulfide shuttling. The MXene/TiO_2 heterostructure was prepared by the in situ oxidation of $Ti_3C_2T_x$ to form TiO_2. Sulfur was uniformly loaded on MXene/TiO_2 by the melt-diffusion method. The MXene/TiO_2/S cathode showed excellent multiplicity performance by maintaining a capacity of 775 mAh/g at 2 C and recovering 1175 mAh/g at 0.1 C.

2.5 Organosulfur-Based Cathode Materials

Organic electrode materials have been developed to ensure safety, environmental friendliness, sustainability, molecular design flexibility, and affordability. Unlike physical blends of sulfur with other materials, organosulfides involve covalent bonds between matrix and sulfur [73, 74]. Organosulfides appeared as intriguing cathodes as they demonstrated benefits such as large theoretical capacity and operational potential [75]. Additionally, they offered significant advantages over conventional LSBs, including accurate redox processes, novel redox mechanisms, and exceptional functional tunability. Organosulfides can be majorly categorized as organosulfur small molecules and polymers.

Thiol groups, S–S bonds, –Sn– bonds, and thioether are the primary active sites for organosulfur compounds [76], which can be found predominantly in animals and plants, as well as in bioactive natural products and pharmaceuticals, serving as essential scaffolds [77]. Typical organosulfide electrodes follow a reversible redox reaction of the disulfide bonds as shown below; the reversible cleavage and formation of S–S bonds during discharge and charging processes account for capacity.

$$\text{R-SS-R} + 2\text{Li}^+ + 2e^- \leftrightarrow 2\text{R-SLi}$$

Figure 2.11 illustrates the development of diverse small organosulfur compounds bearing different functionalities, primarily intended for use as cathode materials in rechargeable LSBs [78]. These compounds leverage the reversible formation and breakage of S–S bonds, while also addressing the issue of reduction in polysulfide production. Figure 2.12 outlines several strategies employed to overcome existing challenges within the LSB field [79].

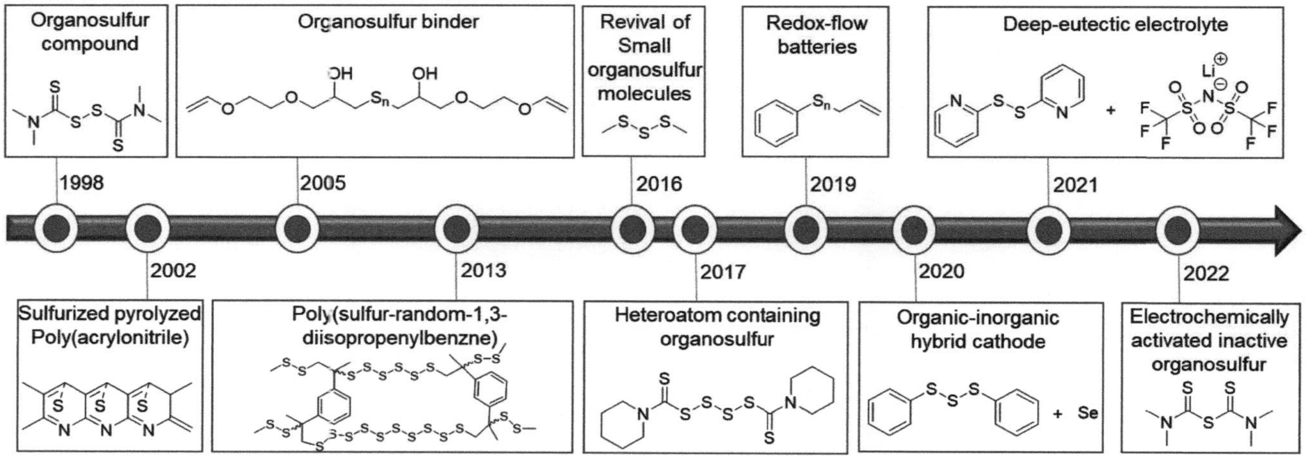

Figure 2.11 Progress of organcsulfur materials in the field of LSBs [78].

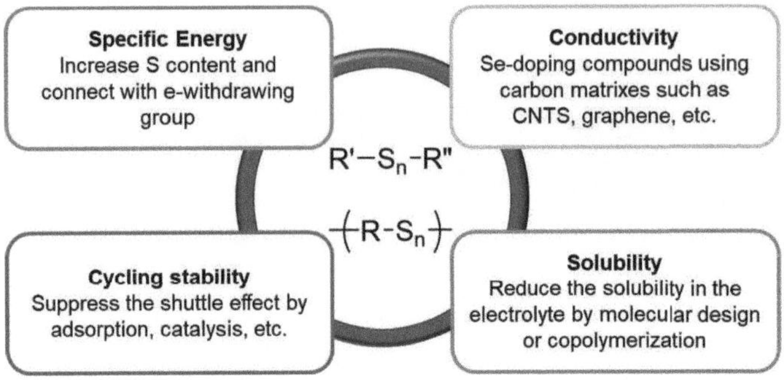

Figure 2.12 Strategies to improve the performance of organosulfide cathode materials [79].

2.5.1 Organosulfur Small Molecules

In 1988, Visco et al. [80] conducted a study to investigate the electrochemical behavior of the first organosulfur molecule, tetraethylthiuram disulfide (TETD), with the aim of promoting the development of high-energy-density batteries and replacing conventional inorganic salt electrodes. The experimental cell exhibited an energy density of 82 Wh/kg at current densities of 16 mA/cm^2. It demonstrated stable cyclic performance over 135 cycles, although it experienced a large polarization at 105 °C. Building upon this research, Inamasu et al. [81] further extended the study by exploring the application of various organodisulfides in lithium batteries. Examples of these compounds include diphenyl disulfide (DPDS), dibenzo[c,e][1,2]dithiin (DBDT), naphtho[1,8-cd][1,2]dithiol (NDT), and naptho[1,8-cd:4,5-c′d′]bis[1,2]-dithiol (NDBT). Among them, NDBT exhibited the highest theoretical specific capacity of 425 mAh/g, based on the storage of four electrons ($4e^-$) and four lithium ions ($4Li^+$) per molecule. The reversible nature and kinetics of the DPDS electrode were further improved by the formation of a composite with CNTs, DPDS@CNT, which displayed a high reversible capacity of 200 mAh/g in the ether-based electrolyte as well as 70% capacity retention after 150 cycles at the 1 C rate [82]. It is also possible to introduce additional active sites on the benzene ring, for example lithium benzenedithiolate (LBDT), which achieves a specific capacity of 349 mAh/g. Moreover, the presence of SLi on the ortho position imparted a higher capacity of 286 mAh/g than meta- and para-substituted benzene dithiolate [83].

An intelligent molecular design offers a potential solution to address the challenges associated with polysulfide dissolution in organosulfur small molecules. In this regard, the use of conjugated molecules enables increased intermolecular interactions, simultaneously reducing solubility and enhancing conductivity [84]. Notably, molecules such as hexathiapentacene (HTT) and tetrathiotetracene (TTT) exhibit stronger intermolecular interactions, resulting in theoretical specific capacities of 317 mAh/g and 265 mAh/g, respectively, at a rate of 0.1 C. The higher specific capacity of HTT can be attributed to its crystal form, which demonstrates significantly stronger intermolecular interactions compared to TTT [85].

The introduction of heteroatoms such as N, P, O, and Se into organosulfur molecules enhanced their electrochemical performance by reducing the presence of soluble polysulfides in the battery cell [86]. This improvement in performance has been observed in cost-effective N-atom-containing thiuram polysulfides such as dipentamethylenethiuram tetrasulfide (PMTT) and dipentamethylenethiuram hexasulfide (PMTH) [87], which exhibit theoretical specific capacities of 418 mAh/g and 597 mAh/g, respectively. Additionally, other N-containing heterocyclic pyridyl compounds, such as dipyridyl disulfide (2,2'-DpyDS), 4,4'-dipyridyl disulfide (4,4'-DpyDS), and 2,2'-dipyridyl disulfide-*N,N'*-dioxide (DpyDSDO), have also been investigated as cathode materials (Fig. 2.13a–c). Among them, the DpyDSDO molecule, which possesses strong electron-withdrawing properties [70], demonstrated the highest discharge voltage.

Dimethyl trisulfide (DMTS) stands out among organotrifulfides with its highest theoretical specific capacity of 849 mAh/g, as depicted in Fig. 2.13d [88]. The compound 1,4-bis(diphenylphosphanyl) tetrasulfide (BDPPTS), which contains phosphorus, demonstrates an initial discharge capacity of 312 mAh/g with a capacity retention of 57% of its initial capacity after 200 cycles [90]. Oxygen-containing organosulfur compounds offer improved reaction kinetics and specific capacity. An innovative carbonyl molecule called 2,3,4,6,8,9,10,12-octathia biscyclopenta[b,c]-5,11-anthraquinone–1,7-dithione (TPQD) serves as a cathode material [91], exhibiting insolubility in ether-based electrolytes and a theoretical

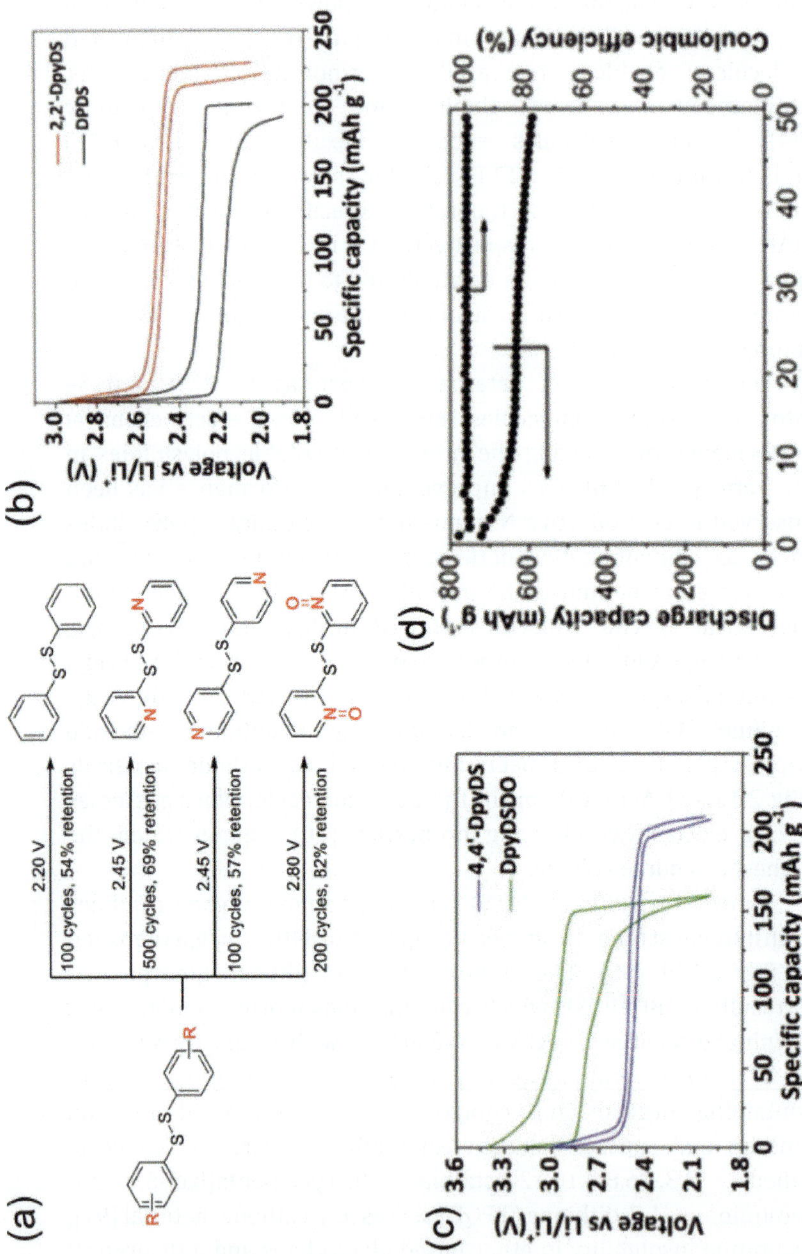
(a)
2.20 V
100 cycles, 54% retention
2.45 V
500 cycles, 69% retention
2.45 V
100 cycles, 57% retention
2.80 V
200 cycles, 82% retention
(b)
2,2'-DpyDS
DPDS
Voltage vs Li/Li+ (V)
Specific capacity (mAh g-1)
(c)
4,4'-DpyDS
DpyDSDO
Voltage vs Li/Li+(V)
Specific capacity (mAh g-1)
(d)
Discharge capacity (mAh g-1)
Coulombic efficiency (%)
Cycle number

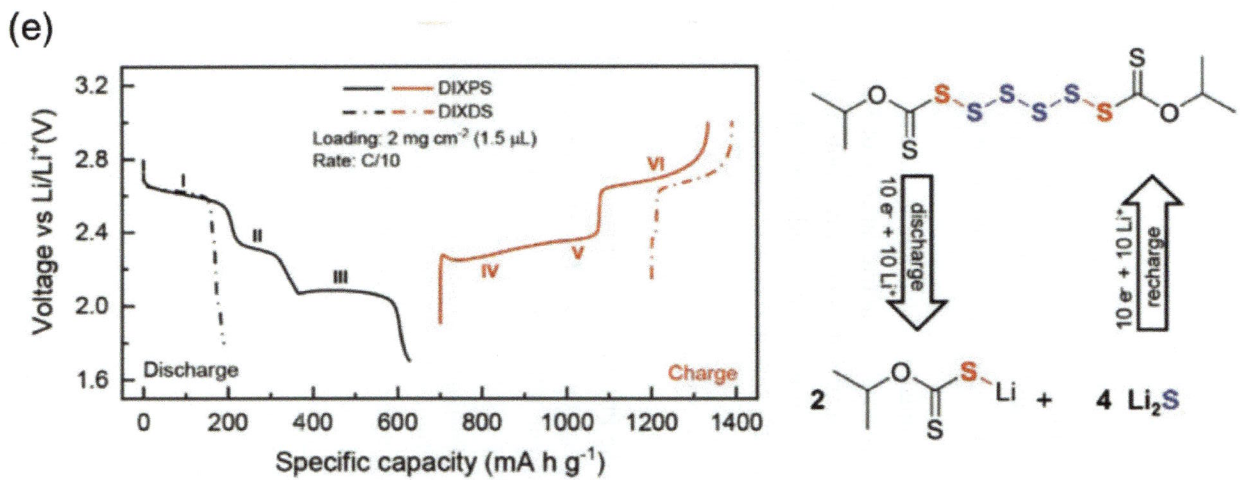

Figure 2.13 (a) Summary of the electrochemical behavior of DPDS, 2,2'-DpyDS, 4,4'-DpyDS, and DpyDSDO. (b) Voltage-capacity profile of Li/2,2'-DpyDS and Li/DPDS cells during the first cycle at a C/10 rate. (c) Voltage-capacity profile of Li/4,4'-DpyDS and Li/DpyDSDO cells during the first cycle at a C/10 rate. Reproduced with permission from Ref. [70]. Copyright 2019, Royal Society of Chemistry. (d) Cycling performance and coulombic efficiency of a cell cycled between 1.7 and 2.7 V at a C/10 rate with DMTS loading of 6.7 mg/cm^2. Reproduced with permission from Ref. [88]. Copyright 2016, Wiley-VCH. (e) Voltage profiles of diisopropyl xanthogen polysulfide (DIXPS) and diisopropyl xanthogen disulfide (DIXDS) when cycled at a C/10 rate, along with a simplified scheme of the reactions occurring during battery operation. Reproduced with permission from Ref. [89]. Copyright 2020, Wiley-VCH.

specific capacity of 324 mAh/g. The addition of functional groups such as $-OCH_3$ and $-CF_3$ to phenyl tetrasulfide to create *p*-methoxyphenyl tetrasulfide (CH_3OPTS) and *p*-trifluoromethylphenyl tetrasulfide (CF_3PTS), respectively [92], enhances cell performance. While the CF_3 group boosts the discharge voltage, the OCH_3 group reduces overpotential and improves cycling stability. Xanthogen polysulfide (DIXPS) demonstrates an impressive theoretical specific capacity of 973 mAh/g in the first cycle and retains 74% of its initial capacity after 1000 cycles, surpassing most organosulfur cathodes, as shown in Fig. 2.13e [89].

Selenium (Se) holds promise for high-energy-density batteries due to its impressive theoretical volumetric capacity of 3253 mAh/cm^3. Adding selenium (Se) to the molecules can potentially open up a promising pathway for organosulfur cathodes[93]. When an Li/diphenyl selenosulfide (Ph-SeS-Ph) cell was compared to an Li/DPDS cell, it was observed that the former exhibited significantly improved cycling stability and low overpotential. Additionally, the incorporation of phenyl selenosulfide (PhSeSSe-Ph) and phenyl selenodisulfide (Ph-SeSSSe-Ph) further enhanced the theoretical specific capacity and cycling stability [94, 95]. Another recent study by Chen et al. [96] focused on tellurium insertion via the Te–S bond in the phenyl tellurosulfide (Ph-STe-Ph) molecule. During the initial discharge process, the investigation unveiled the presence of a dynamic covalent bond network. This network played a crucial role in achieving a stable cycling performance, allowing the system to maintain an impressive capacity retention rate of 88.5% over a span of 400 cycles in a dilute electrolyte.

The structures of small organosulfur molecules are displayed, and their corresponding theoretical capacities are presented in Table 2.1.

Table 2.1 Various organosulfur small-molecule cathodes and their electrochemical performance

Materials	Structure of Organosulfur Molecule	Theoretical Capacity(mAh/g)	Initial Capacity (mAh/g) (Current Density)	Capacity Retention (%), Cycles	Ref.
DPDS	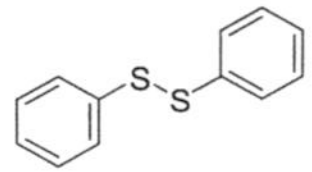	245.5	218 (1 C)	70, 150	[82]
1,2-LBDT	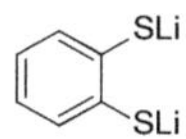	347.8	336.8 (2nd, 0.1 C)	82.2, 50	[83]
TTT	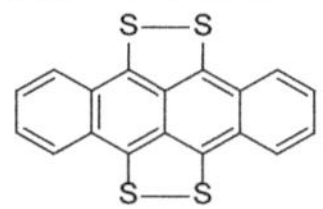	281	257 (0.1 C)	20, 100	[85]
HTP	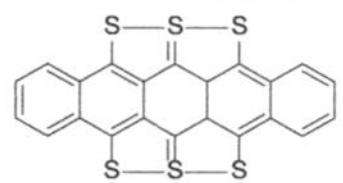	317	265 (0.1 C)	72, 100	[85]
PMTT	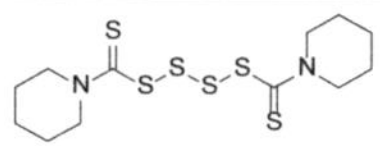	418	406 (1 C)	87, 100	[87]
PMTH	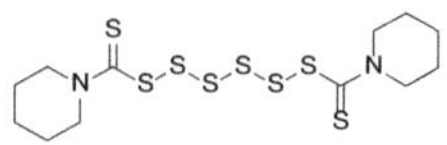	597	539.4 (0.2 C)	85, 100	[87]

(*Continued*)

Table 2.1 (*Continued*)

Materials	Structure of Organosulfur Molecule	Theoretical Capacity(mAh/g)	Initial Capacity (mAh/g) (Current Density)	Capacity Retention (%), Cycles	Ref.
2,2'-DpyDS	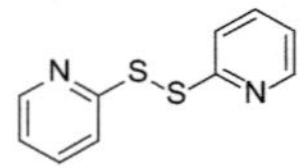	243	237 (1 C)	69, 500	[70]
DpyDSDO	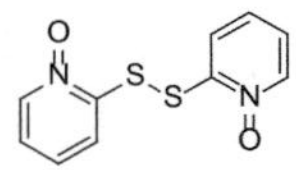	212	160 (1 C)	82, 200	[70]
DMTS	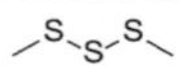	849	720 (0.1 C)	82, 50	[88]
DIXPS	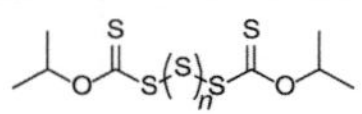	672	467 (4 C)	74, 1000	[89]
BDPPTS	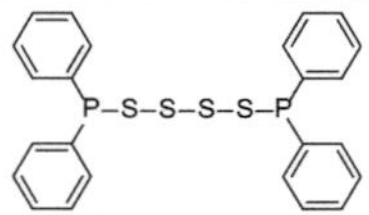	322.8	311.6 (0.2 C)	57, 200	[90]
TPQD	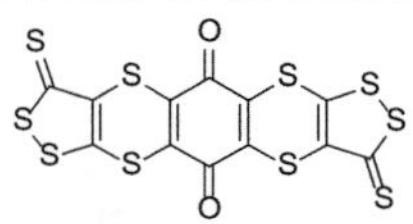	324	~230 (2nd, 0.1 C)	~87, 50	[91]

Materials	Structure of Organosulfur Molecule	Theoretical Capacity(mAh/g)	Initial Capacity (mAh/g) (Current Density)	Capacity Retention (%), Cycles	Ref.
PTS	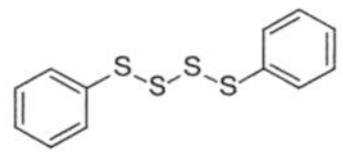	569.3	486 (0.1 C)	74, 100	[92]
CH_3OPTS	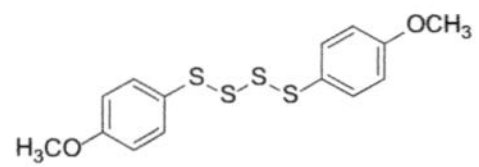	469.5	324 (0.1 C)	81, 100	[92]
CF_3PTS	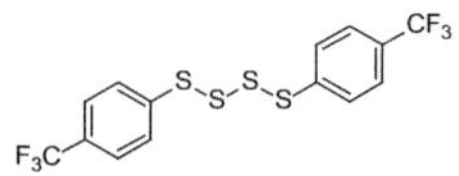	348.3	272 (0.1 C)	43, 100	[92]
PhSeSPh	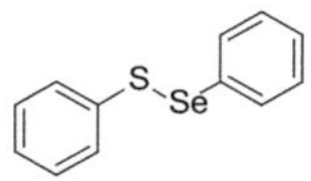	202.1	152 (0.2 C)	52, 200	[94]
PDSe-S	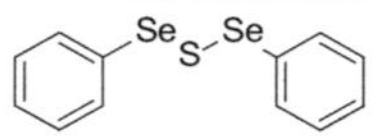	311.4	252 (0.2 C)	77, 200	[95]
PDSe-S_2	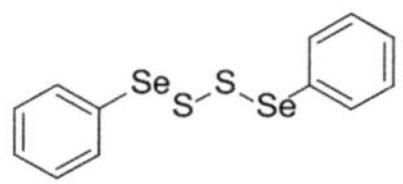	427.4	330 (0.2 C)	73, 200	[95]

2.5.2 Synthetic and Natural Polymer-Sulfur Composite Cathodes

In addition to small organosulfur-containing molecules, the utilization of polymers as cathode materials in LSBs has shown promising progress in the past decade. Copolymerization of monomers and polymers with sulfur formed chemically stable materials, which helped reduce the dissolution of long-chain polysulfides in the electrolyte, thereby addressing the shuttle effect to a significant extent [97]. One of the early works in exploring sulfur polymer electrodes for solid-state thin-film batteries was conducted by Liu et al. [98] in 1991. Since then, research efforts have continued to investigate and develop sulfur copolymer cathodes. Sulfur copolymer cathodes offer the advantage of providing relatively high specific capacities based on the redox mechanism (-S-R-S-)n + 2nLi+ + 2ne$^-$ ↔ nLiS-R-SLi. Furthermore, the presence of heteroatoms such as N, O, P, and Se in the sulfur copolymer structures can further enhance the electrochemical performance of LSBs.

In 2013, Pyun et al. [99] introduced a process called inverse vulcanization, which involves the reaction of sulfur with unsaturated monomers to form a random copolymer with a high sulfur content (90%). One such copolymer, Li/poly(S-r-DIB), demonstrated an initial discharge capacity of 1100 mAh/g and retained 75% of its capacity after 100 cycles at the 0.1 C rate (Fig. 2.14a,b). Subsequent studies focused on developing sulfur cathode materials based on alkenes and alkynes, as well as composites of poly(S-r-DIB) with substances such as MoS_2, CNT, graphene, activated porous carbon, rGO, and aluminum oxide, aiming to enhance the electrochemical performance of the cathode [100, 101].

Among the alkenes and alkynes investigated, certain structures such as allyl-terminated poly(3-hexylthiophene-2,5-diyl) (P3HT) [102], eugenol allyl ether (EAE) [103], poly(sulfur-r-4-vinyl-1,2-epoxycyclohexane) (SVE) [104], and S-triallyl isocyanurate (STI) [105] exhibited improved electrochemical activity of sulfur and minimized the dissolution of polysulfides into the electrolyte. Electrolytes with low ionic conductivity typically limit the practical use of sulfur cathodes, but the structural modifications in sulfur copolymers offer the potential for the solid–solid conversion of the sulfur cathode in conventional electrolytes, thereby reducing the shuttle effect caused by polysulfides [106]. In the quest for enhanced performance, aminophenol-formaldehyde resin/CNT (SAC) mesoporous hollow sulfurized hybrid spheres demonstrated similar reversible capacities in ether and ester electrolytes over 500 cycles at a 1 C rate, with a minimal capacity decay of 0.022% per cycle at a 2 C rate after 1000 cycles [107]. Alkyne-based copolymers and composite cathodes, such as the semi-interpenetrated network (semi-IPN) of 1,3-diethylbenzene (DEB) with elemental sulfur and its carbon black composite, exhibited improved conductivity [108]. The use of long cylindrical porous multiwalled CNTs (MWCNTs) instead of carbon black further increased conductivity and adsorption capacity [109]. The introduction of polar thiophenes into copolymers, such as poly(sulfur-co-1,4-diphenylbutadiyne) or poly(S-co-DiPhDY), effectively inhibited the shuttle effect of polysulfides, enabling the Li/poly(S-co-DiPhDY) cell to achieve a long lifetime of 850 cycles at a 0.2 C rate [110]. A copolymer of sulfur and thiourea aldehyde resin (cp(S-TAR)), which possessed higher electrical conductivity and abundant mesopores, facilitated efficient Li^+ diffusion while suppressing the shuttle effect of polysulfides [111].

Copolymers containing carbonyl groups (>C=O), such as sulfur-polyvinylpyrrolidone (PVPS), exhibited promising performance as cathode materials. PVPS delivered an initial discharge capacity of 1079 mAh/g at a 0.1 C rate and experienced a capacity fading of 0.093% per cycle over 500 cycles at a 1 C rate [112]. Another noteworthy material, organic polysulfane nanosheets (OPNS), featured a structure of polycyclic sulfur grafted onto a carbon chain. OPNS demonstrated a high specific capacity of 891 mAh/g, with a capacity retention of over 91% after 620 cycles at a 1 C rate [113]. Poly(diallyl tetrasulfide) (PDATtS) and Se-doped PDATtSSe, synthesized via the free radical addition reaction with the presence of azodiisobutyronitrile (AIBN), exhibited stable cycling performance. These copolymers showed no soluble polysulfide intermediates during the discharge process, which contributed to their improved performance [114, 115].

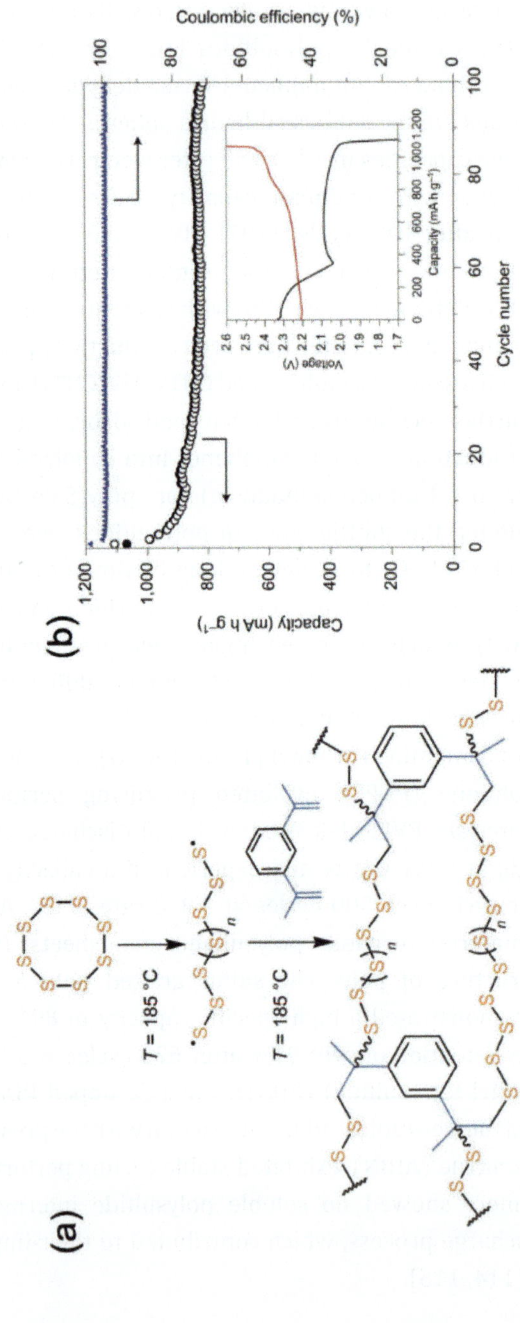
(a)
T = 185 °C
T = 185 °C
Poly(S-r-DIB) copolymer
(b)
Capacity (mA h g−1)
Coulombic efficiency (%)
Cycle number
Voltage (V)
Capacity (mA h g−1)

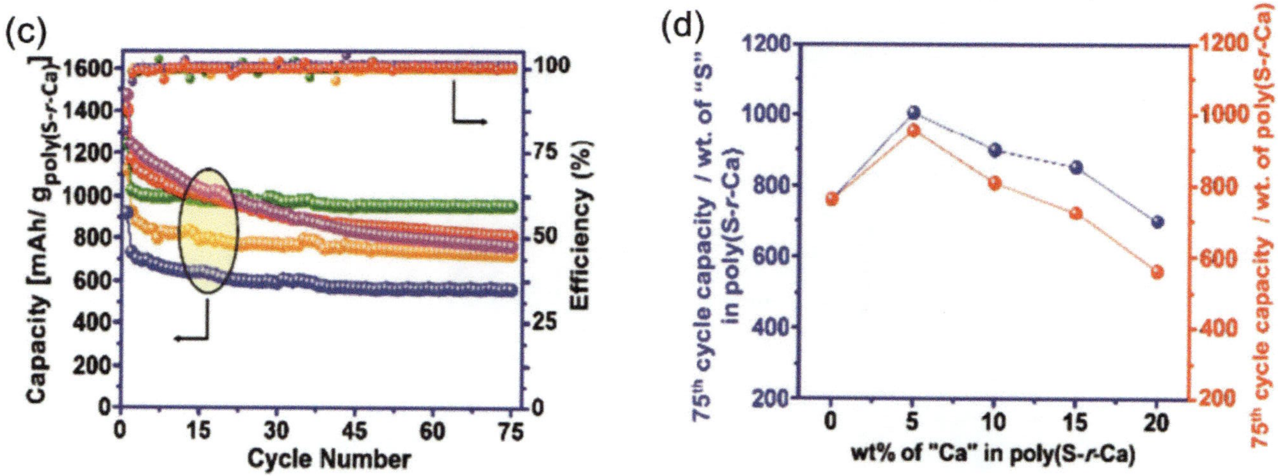
(c)
Capacity [mAh/ g poly(S-r-Ca)]
1600
1400
1200
1000
800
600
400
200
0
Efficiency (%)
100
75
50
25
0
0
15
30
45
60
75
Cycle Number
(d)
75th cycle capacity / wt. of "S" in poly(S-r-Ca)
75th cycle capacity / wt. of poly(S-r-Ca)
1200
1000
800
600
400
200
0
5
10
15
20
wt% of "Ca" in poly(S-r-Ca)

(e)

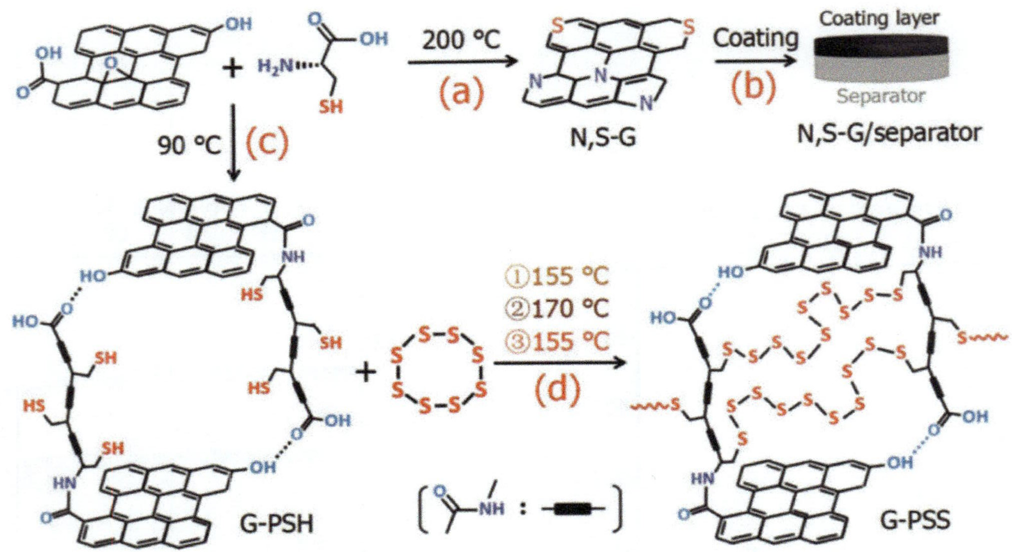

Figure 2.14 (a) Synthetic scheme depicting the copolymerization of S_8 with DIB to form sulfur copolymers with enhanced chemical stability. (b) Comparison of cyclic voltammetry (CV) curves between pure sulfur (solid black line) and poly(S-r-DIB) (10 wt% DIB, open circles) at a scan rate of 20 µV/s. Reproduced with permission from Ref. [99]. Copyright 2013, Nature Publishing Group. (c) Galvanostatic charge–discharge cycling of various sulfur copolymers conducted at a current density of 200 mA/g, calculated based on the mass of the poly(S-r-Ca) copolymer. (d) Plot showing the specific capacity (calculated based on the mass of sulfur and copolymer) measured after the 75th cycle versus that of sulfur copolymer composition. Reproduced with permission from Ref. [116]. Copyright 2013, Elsevier. (e) Schematic illustration for the synthesis of N, S-G, N, S-G/separator, G-PSH, and G-PSS. Reproduced with permission from Ref. [117]. Copyright 2019, Wiley-VCH.

In addition to inverse vulcanized sulfur copolymers, other types of polymers have also shown potential in inhibiting the shuttle effect caused by polysulfide dissolution in LSBs. Poly[(Z)-5-methyl-3-(methylimino)-1,2,4-thiadiazole-2(3H)-carbonitrile] (NS-1) and poly(3,6-diethyl-1,2,3,6-dithiadiazinane) (NS-2) are examples of such polymers. These materials have demonstrated improved battery performance and the ability to suppress the shuttle effect [118]. Another noteworthy polymer is poly(C_6S_6), which is an aromatic organodisulfide polymer. This polymer exhibits a high theoretical capacity of 609 mAh/g, showcasing its potential as a cathode material for LSBs [119].

In recent years, nitrile-based polymers with sulfur, known as SPAN, have received significant attention as cathode materials for LSBs. These polymers offer several advantages over conventional sulfur-based cathodes. They are compatible with carbonate electrolytes, exhibit high conductivity, do not form polysulfides, and have minimal self-discharge [120]. However, despite the growing interest in SPAN, there is still controversy surrounding its exact structure and redox reaction mechanism. Researchers have explored various modifications and proposed electrochemical pathways for SPAN and its composites to gain a better understanding [121–123]. Doping SPAN with heteroatoms such as selenium (Se), tellurium (Te), iodine (I), or cobalt (Co) has been explored as a means to enhance the reaction kinetics and electrochemical performance. Examples include pPAN/SeS_2, $Se_{0.06}$SPAN, $Te_{0.04}S_{0.96}$PAN, I-S@pPAN, and Co_{10}-SPAN-CNT. These doping strategies aim to improve the overall electrochemical behavior and stability of SPAN as a cathode material [123].

To inhibit the formation of long-chain polysulfides, a high-performance material was developed with a nitrogen- and sulfur-doped separator incorporated between the sheets of rGO. The Li/G-PSS cathode displayed a high initial discharge capacity of 1364 mAh/g with a very low capacity decay of 0.037% per cycle [117] (Fig. 2.14f). Studies have also been conducted on the utilization of thiol, thiophene, quinone, and covalent organic frameworks (COFs)-based copolymers, which are summarized in Table 2.2.

Table 2.2 Various sulfur copolymer cathodes and their electrochemical performance in LSBs

Materials	Sulfur Content (wt %)	Initial Capacity (mAh/g) (Current Density)	Capacity Retention (%), Cycles	Ref.
Poly(S-r-DIB)	90	1100 (0.1 C)	74.8, 100	[99]
S-DIB@CNT	~57	898 (1 C)	98, 100	[100]
Poly(S-r-DIB)/ MoS_2	50	1150 (0.2 C)	49.6, 1000	[101]
S-P3HT	17.7	1212 (0.5 C)	65.9, 100	[102]
Poly(S-*co*-EAE)	90	870 (0.1 C)	64, 100	[103]
SVE	90	777.6 (0.5 C)	89.4, 400	[104]
STI	90	707 (1 C)	94, 350	[105]
SAC-350	42.1	810 (2 C)	80, 1000	[107]
20% DEB-2H	80	NA (1 C)	70, 500	[108]
S-*co*-poly-LCNT	58	1040 (0.5 C)	58.7, 200	[109]
Poly(S-*co*-DiPhDY)	90	1050 (0.2 C)	33, 850	[110]
cp(S-TAR)	75	819 (1 C)	79.8, 500	[111]
20PVPS	78	870 (1 C)	~52.9, 500	[112]
OPNS-72	72	891 (1 C)	91, 620	[113]
PDATtS	61.5	700 (200 mA/g)	85, 300	[114]
PDATtSSe	71.6	742 (600 mA/g)	92, 400	[115]
poly(S-r-Ca)	95	630 (1 C)	92.4, 74	[116]
G-PSS	74.4	1364 (0.2 C)	77.2, 700	[117]
NS-2	NA	300 (0.2 C)	50, 100	[118]
SPAN	39.8	~ 600 (0.5 A/g)	96.8, 2000	[122]
S@pPAN	37.64	1729 (200 mA/g)	98.4, 100	[123]
$EP_{10}S_{90}$	83	650 (0.5 C)	60.3, 200	[124]
SPAN/CNT	40.8	1180 (800 mA/g)	NA, 800	[125]
SPAN/GNS	47	1500 (0.1 C)	80, 100	[126]
$pPAN/SeS_2$	63	871 (4 A/g)	72.7, 2000	[127]
$Se_{0.06}SPAN$	47.25	1156 (0.26 C)	79.3, 800	[128]
$Te_{0.04}S_{0.96}PAN$	49.6	1167 (1 A/g)	68.8, 600	[129]

Materials	Sulfur Content (wt %)	Initial Capacity (mAh/g) (Current Density)	Capacity Retention (%), Cycles	Ref.
I-S@pPAN	42.49	1243 (2 C)	85, 1000	[130]
Co_{10}-SPAN-CNT	41.9	1252 (1 C)	81.5, 1500	[131]
S-TTCA	63	1027 (0.2 C)	92, 100	[132]
PPTS	62.9	633 (1 C)	77, 350	[133]
S-CTF−1	62	482.2 (1 C)	85.8, 300	[134]
cPpy-S-CTF	83	702.9 (0.5 C)	86.8, 500	[135]

NA: Not available

Replacements of petro-based polymers with naturally occurring polymers as high-performance sustainable cathodes in LSBs have also been explored. Various cyclic, inherent heteroatom organic units offered several linking sites to improve the loading of sulfur and the cycling stability of cathode materials in LSBs. Shukla et al. [116] reported a facile one-pot synthetic route for the preparation of a sustainable copolymer, poly(S-r-Ca) via ring-opening polymerization reaction of cardanol-aniline (Ca) benzoxazine with elemental sulfur. The resulting copolymer was explored as cathode in the Li-S battery. The resulting cell exhibited a high reversible specific capacity of 1088 mAh g−1 at the second cycle and maintained a capacity of 1006 mAh g−1 after 75 cycles with a very low-capacity loss of 0.01% per cycle (Fig. 2.14 c–e). Another example is clove oil (eugenol, E) benzoxazine-based phosphazene (P) bearing structure, reported by Monisha et al. [136], which demonstrated self-mitigation flame-retardant properties as a cathode material. The $Li/EP_{10}S_{90}$ cell achieved an initial specific capacity of 650 mAh/g at the second cycle with 60% capacity retention after 200 cycles. Bayram et al. [137] explored a hybrid copolymer based on vanillin-benzoxazine and β-ionones with sulfur. The prepared copolymer exhibited a capacity of 400 mAh/g for 100 cycles.

2.6 Conclusion

Overall, a significant development has been made by the exploration of various strategies from physical blending to covalent linking of sulfur with various hosts, small molecules, and even polymers to innovatively design cathode materials for LSBs. A great promise in addressing the challenges associated with the polysulfide shuttling effect, cycle stability, conductivity, and an understanding of the underlying mechanisms to improve the overall electrochemical performance has been made over the years. The ongoing research in this field aims to optimize the material design, adapting greener, safer, and cheaper technologies to achieve commercialization of this next generation of LSBs, the beginning of a new era of high-performance sustainable energy storage devices, beyond LIBs.

References

1. M. Armand and J.M. Tarascon, Building better batteries, *Nature*, **451** (2008) 652–657.
2. J.B. Goodenough and Y. Kim, Challenges for rechargeable Li batteries, *Chem. Mater.*, **22** (2010) 587–603.
3. C.-X. Zu and H. Li, Thermodynamic analysis on energy densities of batteries, *Energy Environ. Sci.*, **4** (2011) 2614–2624.
4. R. Wu, D.P. Wang, X. Rui, B. Liu, K. Zhou, A.W.K. Law, Q. Yan, J. Wei, and Z. Chen, In-situ formation of hollow hybrids composed of cobalt sulfides embedded within porous carbon polyhedra/carbon nanotubes for high-performance lithium-ion batteries, *Adv. Mater.*, **27** (2015) 3038–3044.
5. M.A. Pope and I.A. Aksay, Structural design of cathodes for Li-S batteries, *Adv. Energy Mater.*, **5** (2015) 1500124.
6. Y. Yang, G. Zheng, and Y. Cui, Nanostructured sulfur cathodes, *Chem. Soc. Rev.*, **42** (2013) 3018–3032.
7. A. Manthiram, Y. Fu, S.-H. Chung, C. Zu, and Y.-S. Su, Rechargeable lithium–sulfur batteries, *Chem. Rev.*, **114** (2014) 11751–11787.
8. L. Ma, K.E. Hendrickson, S. Wei, and L.A. Archer, Nanomaterials: Science and applications in the lithium–sulfur battery, *Nano Today*, **10** (2015) 315–338.
9. S. Evers and L. F. Nazar, New approaches for high energy density lithium–sulfur battery cathodes, *Acc. Chem. Res.*, **46** (2013) 1135–1143.
10. Y.-X. Yin, S. Xin, Y.-G. Guo, and L.-J. Wan, Lithium–sulfur batteries: Electrochemistry, materials, and prospects, *Angew. Chem. Int. Ed.*, **52** (2013) 13186–13200.

11. P.G. Bruce, S.A. Freunberger, L.J. Hardwick, and J.-M. Tarascon, Li–O_2 and Li–S batteries with high energy storage, *Nat. Mater*, **11** (2012) 19–29.

12. M.S. Whittingham, Ultimate limits to intercalation reactions for lithium batteries, *Chem. Rev.*, **114** (2014) 11414–11443.

13. J.B. Goodenough, K.-S. Park, The Li-ion rechargeable battery: A perspective, *J. Am. Chem. Soc.*, **135** (2013) 1167–1176.

14. J.R. Dahn, T. Zheng, Y. Liu, and J.S. Xue, Mechanisms for lithium insertion in carbonaceous materials, *Science*, **270** (1995) 590–593.

15. M.S. Whittingham, Electrical energy storage and intercalation chemistry, *Science*, **192** (1976) 1126–1127.

16. P. Adelhelm, P. Hartmann, C.L. Bender, M. Busche, C. Eufinger, and J. Janek, From lithium to sodium: Cell chemistry of room temperature sodium–air and sodium–sulfur batteries, *Beilstein J. Nanotechnol.*, **6** (2015) 1016–1055.

17. S. Chen, X. Huang, H. Liu, B. Sun, W. Yeoh, K. Li, J. Zhang, and G. Wang, 3D hyperbranched hollow carbon nanorod architectures for high-performance lithium-sulfur batteries, *Adv. Energy Mater.*, **4** (2014) 1301761.

18. M. Barghamadi, A.S. Best, A.I. Bhatt, A.F. Hollenkamp, M. Musameh, R.J. Rees, and T. Rüther, Lithium–sulfur batteries—the solution is in the electrolyte, but is the electrolyte a solution? *Energy Environ. Sci.*, **7** (2014) 3902–3920.

19. D.-W. Wang, Q. Zeng, G. Zhou, L. Yin, F. Li, H.-M. Cheng, I.R. Gentle, and G.Q.M. Lu, Carbon–sulfur composites for Li–S batteries: Status and prospects, *J. Mater. Chem. A*, **1** (2013) 9382–9394.

20. H.-S. Ryu, H.-J. Ahn, K.-W. Kim, J.-H. Ahn, J.-Y. Lee, Discharge process of Li/PVdF/S cells at room temperature, *J. Power Sources*, **153** (2006) 360–364.

21. Y. Dong, S. Liu, Z. Wang, Y. Liu, Z. Zhao, and J. Qiu, Sulfur-infiltrated graphene-backboned mesoporous carbon nanosheets with a conductive polymer coating for long-life lithium–sulfur batteries, *Nanoscale*, **7** (2015) 7569–7573.

22. R. Soni, J.B. Robinson, P.R. Shearing, D.J.L. Brett, A.J.E. Rettie, and T.S. Miller, Lithium-sulfur battery diagnostics through distribution of relaxation times analysis, *Energy Storage Mater*, **51** (2022) 97–107.

23. H. Kim, H.-D. Lim, J. Kim, and K. Kang, Graphene for advanced Li/S and Li/air batteries, *J. Mater. Chem. A*, **2** (2014) 33–47.

24. L.F. Nazar, M. Cuisinier, and Q. Pang, Lithium-sulfur batteries, *MRS Bull.*, **39** (2014) 436–442.

25. Y.V. Mikhaylik and J.R. Akridge, Polysulfide shuttle study in the Li/S battery system, *Electrochem Soc.*, **151** (2004) A1969.

26. X. Ji, K.T. Lee, and L.F. Nazar, A highly ordered nanostructured carbon–sulphur cathode for lithium–sulphur batteries, *Nature Mater.*, **8** (2009) 500–506.

27. S.-C. Han, M.-S. Song, H. Lee, H.-S. Kim, H.-J. Ahn, and J.-Y. Lee, Effect of multiwalled carbon nanotubes on electrochemical properties of lithium/sulfur rechargeable batteries, *J. Electrochem. Soc.*, **150** (2003) A889.

28. L. Yuan, H. Yuan, X. Qiu, L. Chen, and W. Zhu, Improvement of cycle property of sulfur-coated multi-walled carbon nanotubes composite cathode for lithium/sulfur batteries, *J. Power Sources*, **189** (2009) 1141–1146.

29. S. Dörfler, M. Hagen, H. Althues, J. Tübke, S. Kaskel, and M.J. Hoffmann, High capacity vertical aligned carbon nanotube/sulfur composite cathodes for lithium–sulfur batteries, *Chem. Commun.*, **48** (2012) 4097–4099.

30. J. Guo, Y. Xu, and C. Wang, Sulfur-impregnated disordered carbon nanotubes cathode for lithium–sulfur batteries, *Nano Lett.*, **11** (2011) 4288–4294.

31. L. Ji, M. Rao, S. Aloni, L. Wang, E.J. Cairns, and Y. Zhang, Porous carbon nanofiber–sulfur composite electrodes for lithium/sulfur cells, *Energy Environ. Sci.*, **4** (2011) 5053–5059.

32. G. Zheng, Y. Yang, J.J. Cha, S.S. Hong, and Y. Cui, Hollow carbon nanofiber-encapsulated sulfur cathodes for high specific capacity rechargeable lithium batteries, *Nano Lett.*, **11** (2011) 4462–4467.

33. G. Zheng, Q. Zhang, J.J. Cha, Y. Yang, W. Li, Z.W. Seh, and Y. Cui, Amphiphilic surface modification of hollow carbon nanofibers for improved cycle life of lithium sulfur batteries, *Nano Lett.*, **13** (2013) 1265–1270.

34. Z.W. Seh, Q. Zhang, W. Li, G. Zheng, H. Yao, and Y. Cui, Stable cycling of lithium sulfide cathodes through strong affinity with a bifunctional binder, *Chem. Sci.*, **4** (2013) 3673–3677.

35. Z. Wang, X. Niu, J. Xiao, C. Wang, J. Liu, F. Gao, First principles prediction of nitrogen-doped carbon nanotubes as a high-performance cathode for Li–S batteries, *RSC Adv.*, **3** (2013) 16775–16780.

36. H.-J. Peng, T.-Z. Hou, Q. Zhang, J.-Q. Huang, X.-B. Cheng, M.-Q. Guo, Z. Yuan, L.-Y. He, and F. Wei, Strongly coupled interfaces between a heterogeneous carbon host and a sulfur-containing guest for highly stable lithium-sulfur batteries: Mechanistic insight into capacity degradation, *Adv. Mater. Interfaces*, **1** (2014) 1400227.

37. R. Fang, S. Zhao, S. Pei, X. Qian, P.-X. Hou, H.-M. Cheng, C. Liu, and F. Li, Toward more reliable lithium–sulfur batteries: An all-graphene cathode structure, *ACS Nano*, **10** (2016) 8676–8682.

38. M.Q. Zhao, Q. Zhang, J.Q. Huang, G.L. Tian, J.Q. Nie, H.J. Peng, and F. Wei, Unstacked double-layer templated graphene for high-rate lithium-sulphur batteries, *Nat. Commun.*, **5** (2014).

39. H. Wang, Y. Yang, Y. Liang, J.T. Robinson, Y. Li, A. Jackson, Y. Cui, and H. Dai, Graphene-wrapped sulfur particles as a rechargeable lithium–sulfur battery cathode material with high capacity and cycling stability, *Nano Lett.*, **11** (2011) 2644–2647.

40. L. Ji, M. Rao, H. Zheng, L. Zhang, Y. Li, W. Duan, J. Guo, E.J. Cairns, and Y. Zhang, Graphene oxide as a sulfur immobilizer in high performance lithium/sulfur cells, *J. Am. Chem. Soc.*, **133** (2011) 18522–18525.

41. L. Zhang, L. Ji, P.-A. Glans, Y. Zhang, J. Zhu, and J. Guo, Electronic structure and chemical bonding of a graphene oxide–sulfur nanocomposite for use in superior performance lithium–sulfur cells, *Phys. Chem. Chem. Phys.*, **14** (2012) 13670–13675.

42. B. Wang, S.M. Alhassan, and S.T. Pantelides, Formation of large polysulfide complexes during the lithium-sulfur battery discharge, *Phys. Rev. Appl.*, **2** (2014) 034004.

43. J. Rong, M. Ge, X. Fang, and C. Zhou, Solution ionic strength engineering as a generic strategy to coat graphene oxide (GO) on various functional particles and its application in high-performance lithium–sulfur (Li–S) batteries, *Nano Lett.*, **14** (2014) 473–479.

44. J. Cao, C. Chen, Q. Zhao, N. Zhang, Q. Lu, X. Wang, Z. Niu, and J. Chen, A flexible nanostructured paper of a reduced graphene oxide–sulfur composite for high-performance lithium–sulfur batteries with unconventional configurations, *Adv. Mater.*, **28** (2016) 9629–9636.

45. H. Chen, C. Wang, W. Dong, W. Lu, Z. Du, and L. Chen, Monodispersed sulfur nanoparticles for lithium–sulfur batteries with theoretical performance, *Nano Lett.*, **15** (2015) 798–802.

46. C. Wang, K. Su, W. Wan, H. Guo, H. Zhou, J. Chen, X. Zhang, and Y. Huang, High sulfur loading composite wrapped by 3D nitrogen-doped graphene as a cathode material for lithium–sulfur batteries, *J. Mater. Chem. A*, **2** (2014) 5018–5023.

47. T.-Z. Hou, H.-J. Peng, J.-Q. Huang, Q. Zhang, and B. Li, The formation of strong-couple interactions between nitrogen-doped graphene and sulfur/lithium (poly)sulfides in lithium-sulfur batteries, *2D Mater.*, **2** (2015) 014011.
48. S.B. Patil, H.J. Kim, H.-K. Lim, S.M. Oh, J. Kim, J. Shin, H. Kim, J.W. Choi, and S.-J. Hwang, Exfoliated 2D lepidocrocite titanium oxide nanosheets for high sulfur content cathodes with highly stable Li–S battery performance, *ACS Energy Lett.*, **3** (2018) 412–419.
49. X. Tao, J. Wan, C. Liu, H. Wang, H. Yao, G. Zheng, Z.W. Seh, Q. Cai, W. Li, G. Zhou, C. Zu, and Y. Cui, Balancing surface adsorption and diffusion of lithium-polysulfides on nonconductive oxides for lithium-sulfur battery design, *Nat. Commun.*, **7** (2016).
50. X. Liang, C. Hart, Q. Pang, A. Garsuch, T. Weiss, and L.F. Nazar, A highly efficient polysulfide mediator for lithium–sulfur batteries, *Nat Commun*, **6** (2015) 5682.
51. H. Lin, S. Zhang, T. Zhang, H. Ye, Q. Yao, G.W. Zheng, and J.Y. Lee, Elucidating the catalytic activity of oxygen deficiency in the polysulfide conversion reactions of lithium–sulfur batteries, *Adv. Energy Mater.*, **8** (2018) 1801868.
52. Y. Guo, Y. Zhang, Y. Zhang, M. Xiang, H. Wu, H. Liu, and S. Dou, Interwoven V_2O_5 nanowire/graphene nanoscroll hybrid assembled as efficient polysulfide-trapping-conversion interlayer for long-life lithium–sulfur batteries, *J. Mater. Chem. A*, **6** (2018) 19358–19370.
53. Z. Hao, J. Chen, X. Lu, L. Kang, C. Tan, R. Xu, L. Yuan, D.J.L. Brett, P.R. Shearing, F.R. Wang, and Y. Huang, Precisely visit the performance modulation of functionalized separator in Li-S batteries via consecutive multiscale analysis, *Energy Storage Mater.*, **49** (2022) 85–92.
54. N. Hu, X. Lv, Y. Dai, L. Fan, D. Xiong, and X. Li, SnO_2/reduced graphene oxide interlayer mitigating the shuttle effect of Li–S batteries, *ACS Appl. Mater. Interfaces*, **10** (2018) 18665–18674.
55. L. Ma, R. Chen, G. Zhu, Y. Hu, Y. Wang, T. Chen, J. Liu, and Z. Jin, Cerium oxide nanocrystal embedded bimodal micromesoporous nitrogen-rich carbon nanospheres as effective sulfur host for lithium–sulfur batteries, *ACS Nano*, **11** (2017) 7274–7283.
56. Y. Zhou, C. Zhou, Q. Li, C. Yan, B. Han, K. Xia, Q. Gao, and J. Wu, Enabling prominent high-rate and cycle performances in one lithium–sulfur battery: Designing permselective gateways for Li+ transportation in holey-CNT/S cathodes, *Adv. Mater.*, **27** (2015) 3774–3781.

57. X.Z. Ma, B. Jin, H.Y. Wang, J.Z. Hou, X.B. Zhong, H.H. Wang, and P.M. Xin, S-TiO_2 composite cathode materials for lithium/sulfur batteries, *J. Electroanal. Chem*, **736** (2015) 127–131.

58. X. Zhao, M. Liu, Y. Chen, B. Hou, N. Zhang, B. Chen, N. Yang, K. Chen, J. Li, and L. An, Fabrication of layered Ti_3C_2 with an accordion-like structure as a potential cathode material for high performance lithium–sulfur batteries, *J. Mater. Chem. A*, **3** (2015) 7870–7876.

59. X. Liang, A. Garsuch, and L.F. Nazar, Sulfur cathodes based on conductive MXene nanosheets for high-performance lithium-sulfur batteries, *Angew. Chem. Int. Ed. Engl.*, **54** (2015) 3907–3911.

60. T. Zhao, P. Zhai, Z. Yang, J. Wang, L. Qu, F. Du, and J. Wang, Self-supporting $Ti_3C_2T_x$ foam/S cathodes with high sulfur loading for high-energy-density lithium-sulfur batteries, *Nanoscale*, **10** (2018) 22954–22962.

61. F. Zhang, Y. Zhou, Y. Zhang, D. Li, and Z. Huang, Facile synthesis of sulfur@titanium carbide Mxene as high-performance cathode for lithium-sulfur batteries, *Nanophotonics*, **9** (2020) 2025–2032.

62. Z. Xiao, Z. Yang, Z. Li, P. Li, and R. Wang, Synchronous gains of areal and volumetric capacities in lithium-sulfur batteries promised by flower-like porous $Ti_3C_2T_x$ matrix, *ACS Nano*, **13** (2019) 3404–3412.

63. W. Bao, L. Liu, C. Wang, S. Choi, D. Wang, and G. Wang, Facile synthesis of crumpled nitrogen-doped mxene nanosheets as a new sulfur host for lithium-sulfur batteries, *Adv. Energy Mater.*, **8** (2018) 1870060.

64. Y. Song, Z. Sun, Z. Fan, W. Cai, Y. Shao, G. Sheng, M. Wang, L. Song, Z. Liu, Q. Zhang, and J. Sun, Rational design of porous nitrogen-doped Ti_3C_2 MXene as a multifunctional electrocatalyst for Li–S chemistry, *Nano Energy*, **70** (2020) 104555.

65. D. Zhang, S. Wang, R. Hu, J. Gu, Y. Cui, B. Li, W. Chen, C. Liu, J. Shang, and S. Yang, Catalytic conversion of polysulfides on single atom zinc implanted MXene toward high-rate lithium–sulfur batteries, *Adv. Funct. Mater.*, **30** (2020) 2002471.

66. W. Bao, X. Xie, J. Xu, X. Guo, J. Song, W. Wu, D. Su, and G. Wang, Confined sulfur in 3D MXene/reduced graphene oxide hybrid nanosheets for lithium-sulfur battery, *Chemistry*, **23** (2017) 12613–12619.

67. J. Song, X. Guo, J. Zhang, Y. Chen, C. Zhang, L. Luo, F. Wang, and G. Wang, Rational design of free-standing 3D porous MXene/rGO hybrid aerogels as polysulfide reservoirs for high-energy lithium–sulfur batteries, *J. Mater. Chem. A*, **7** (2019) 6507–6513.

68. J. Wang, Z. Zhang, X. Yan, S. Zhang, Z. Wu, Z. Zhuang, and W.Q. Han, Rational design of porous N-Ti_3C_2 MXene@CNT microspheres for high cycling stability in Li-S battery, *Nanomicro. Lett.*, **12** (2019) 4.

69. C. Xiong, G.Y. Zhu, H.R. Jiang, Q. Chen, and T.S. Zhao, Achieving multiplexed functionality in a hierarchical MXene-based sulfur host for high-rate, high-loading lithium-sulfur batteries, *Energy Storage Mater.*, **33** (2020) 147–157.

70. D.-Y. Wang, Y. Si, J. Li, and Y. Fu, Tuning the electrochemical behavior of organodisulfides in rechargeable lithium batteries using N-containing heterocycles, *J. Mater. Chem. A*, **7** (2019) 7423–7429.

71. Y. Zhang, Z. Mu, C. Yang, Z. Xu, S. Zhang, X. Zhang, Y. Li, J. Lai, Z. Sun, Y. Yang, Y. Chao, C. Li, X. Ge, W. Yang, and S. Guo, Rational design of MXene/1T-2H MoS2-C nanohybrids for high-performance lithium-sulfur batteries, *Adv. Funct. Mater.*, **28** (2018) 1707578.

72. S.Y. Qiu, C. Wang, Z.X. Jiang, L.S. Zhang, L.L. Gu, K.X. Wang, J. Gao, X.D. Zhu, and G. Wu, Rational design of MXene@TiO_2 nanoarray enabling dual lithium polysulfide chemisorption towards high-performance lithium-sulfur batteries, *Nanoscale*, **12** (2020) 16678–16684.

73. A. Manthiram, S.-H. Chung, and C. Zu, Lithium–sulfur batteries: Progress and prospects, *Adv. Mater.*, **27** (2015) 1980–2006.

74. Y. Huang, L. Lin, C. Zhang, L. Liu, Y. Li, Z. Qiao, J. Lin, Q. Wei, L. Wang, Q. Xie, and D.-L. Peng, Recent advances and strategies toward polysulfides shuttle inhibition for high-performance Li–S batteries, *Adv. Sci.*, **9** (2022) 2106004.

75. Z. Shadike, S. Tan, Q.-C. Wang, R. Lin, E. Hu, D. Qu, and X.-Q. Yang, Review on organosulfur materials for rechargeable lithium batteries, *Mater. Horiz.*, **8** (2021) 471–500.

76. Q. Zhang, Q. Ma, R. Wang, Z. Liu, Y. Zhai, Y. Pang, Y. Tang, Q. Wang, K. Wu, H. Wu, Y. Zhang, L. Zhang, C. Zhang, L. Fu, S. Eliseeva, V. Kondratiev, and Y. Wu, Recent progress in advanced organosulfur cathode materials for rechargeable lithium batteries, *Mater. Today*, **65** (2023) 100–121.

77. B. Boateng, Y. Han, C. Zhen, G. Zeng, N. Chen, D. Chen, C. Feng, J. Han, J. Xiong, X. Duan, and W. He, Organosulfur compounds enable uniform lithium plating and long-term battery cycling stability, *Nano Lett.*, **20** (2020) 2594–2601.

78. P. Sang, Q. Chen, D.-Y. Wang, W. Guo, and Y. Fu, Organosulfur materials for rechargeable batteries: Structure, mechanism, and application, *Chem. Rev.*, **123** (2023) 1262–1326.

79. P. Sang, D.-Y. Wang, W. Guo, Y. Fu, Organosulfide cathodes, in: A. Manthiram and Y. Fu (Eds.) *Advances in Rechargeable Lithium–Sulfur Batteries*, Springer International Publishing, Cham, 2022, pp. 247–294.

80. S.J. Visco, C.C. Mailhe, L.C. De Jonghe, and M.B. Armand, A novel class of organosulfur electrodes for energy storage, *J. Electrochem. Soc.*, **136** (1989) 661.

81. T. Inamasu, D. Yoshitoku, Y. Sumi-otorii, H. Tani, and N. Ono, Electrochemical behaviors of naphtho [1,8-cd][1, 2] dithiol, dibenzo [c, e][1, 2] dithiin, and naphtho [1, 8-cd: 4, 5-c′ d′] bis [1, 2] dithiol, *J. Electrochem. Soc.*, **150** (2002) A128.

82. A. Bhargav, S.V. Patil, and Y. Fu, A phenyl disulfide@CNT composite cathode for rechargeable lithium batteries, *Sustainable Energy Fuels*, **1** (2017) 1007–1012.

83. F. Li, Y. Si, B. Liu, Z. Li, and Y. Fu, Lithium benzenedithiolate catholytes for rechargeable lithium batteries, *Adv. Funct. Mater.*, **29** (2019) 1902223.

84. M. Tang, S. Zhu, Z. Liu, C. Jiang, Y. Wu, H. Li, B. Wang, E. Wang, J. Ma, and C. Wang, Tailoring π-conjugated systems: From π-π stacking to high-rate-performance organic cathodes, *Chem*, **4** (2018) 2600–2614.

85. P. Hu, X. He, M.-F. Ng, J. Ye, C. Zhao, S. Wang, K. Tan, A. Chaturvedi, H. Jiang, C. Kloc, W. Hu, and Y. Long, Trisulfide-bond acenes for organic batteries, *Angew. Chem. Int. Ed.*, **58** (2019) 13513–13521.

86. J. Wang and W.-Q. Han, A review of heteroatom doped materials for advanced lithium–sulfur batteries, *Adv. Funct. Mater.*, **32** (2022) 2107166.

87. A. Bhargav, Y. Ma, K. Shashikala, Y. Cui, Y. Losovyj, and Y. Fu, The unique chemistry of thiuram polysulfides enables energy dense lithium batteries, *J. Mater. Chem. A*, **5** (2017) 25005–25013.

88. M. Wu, Y. Cui, A. Bhargav, Y. Losovyj, A. Siegel, M. Agarwal, Y. Ma, and Y. Fu, Organotrisulfide: A high-capacity cathode material for rechargeable lithium batteries, *Angew. Chem. Int. Ed.*, **55** (2016) 10027–10031.

89. A. Bhargav and A. Manthiram, Xanthogen polysulfides as a new class of electrode material for rechargeable batteries, *Adv. Energy Mater.*, **10** (2020) 2001658.

90. D.-Y. Wang, Y. Si, W. Guo, and Y. Fu, Electrosynthesis of 1,4-bis(diphenylphosphanyl) tetrasulfide via sulfur radical addition as cathode material for rechargeable lithium battery, *Nat. Commun.*, **12** (2021) 3220.

91. Z. Shadike, H.-S. Lee, C. Tian, K. Sun, L. Song, E. Hu, I. Waluyo, A. Hunt, S. Ghose, Y. Hu, J. Zhou, J. Wang, P. Northrup, S.-M. Bak, and X.-Q. Yang, Synthesis and characterization of a molecularly designed high-performance organodisulfide as cathode material for lithium batteries, *Adv. Energy Mater.*, **9** (2019) 1900705.

92. W. Guo, Z.D. Wawrzyniakowski, M.M. Cerda, A. Bhargav, M.D. Pluth, Y. Ma, and Y. Fu, Bis(aryl) tetrasulfides as cathode materials for rechargeable lithium batteries, *Chem.: A Eur. J.*, **23** (2017) 16941–16947.

93. J. Zhou, S. Qian, B. Hao, J. Liu, X. Zhou, C. Yan, and T. Qian, Small molecules, great powers: Chemistry of small organo-chalcogenide molecules in rechargeable Li-sulfur batteries, *Adv. Funct. Mater.*, **33** (2023) 2213966.

94. W. Guo, A. Bhargav, J.D. Ackerson, Y. Cui, Y. Ma, and Y. Fu, Mixture is better: Enhanced electrochemical performance of phenyl selenosulfide in rechargeable lithium batteries, *Chem. Commun.*, **54** (2018) 8873–8876.

95. Y. Cui, J.D. Ackerson, Y. Ma, A. Bhargav, J.A. Karty, W. Guo, L. Zhu, and Y. Fu, Phenyl selenosulfides as cathode materials for rechargeable lithium batteries, *Adv. Funct. Mater.*, **28** (2018) 1801791.

96. Q. Chen, Y. Si, W. Guo, and Y. Fu, A phenyl S–Te bond with unique redox activity in dilute electrolyte of a lithium battery, *Chem. Commun.*, **58** (2022) 10993–10996.

97. S.-H. Chung, C.-H. Chang, and A. Manthiram, Progress on the critical parameters for lithium–sulfur batteries to be practically viable, *Adv. Funct. Mater.*, **28** (2018) 1801188.

98. M. Liu, S.J. Visco, and L.C. De Jonghe, Novel solid redox polymerization electrodes: All-solid-state, thin-film, rechargeable lithium batteries, *J. Electrochem. Soc.*, **138** (1991) 1891.

99. W.J. Chung, J.J. Griebel, E.T. Kim, H. Yoon, A.G. Simmonds, H.J. Ji, P.T. Dirlam, R.S. Glass, J.J. Wie, N.A. Nguyen, B.W. Guralnick, J. Park, Á. Somogyi, P. Theato, M.E. Mackay, Y.-E. Sung, K. Char, and J. Pyun, The use of elemental sulfur as an alternative feedstock for polymeric materials, *Nat. Chem.*, **5** (2013) 518–524.

100. G. Hu, Z. Sun, C. Shi, R. Fang, J. Chen, P. Hou, C. Liu, H.-M. Cheng, and F. Li, A sulfur-rich copolymer@CNT Hybrid cathode with dual-confinement of polysulfides for high-performance lithium–sulfur batteries, *Adv. Mater.*, **29** (2017) 1603835.

101. P.T. Dirlam, J. Park, A.G. Simmonds, K. Domanik, C.B. Arrington, J.L. Schaefer, V.P. Oleshko, T.S. Kleine, K. Char, R.S. Glass, C.L. Soles, C. Kim, N. Pinna, Y.-E. Sung, and J. Pyun, Elemental sulfur and molybdenum disulfide composites for Li–S batteries with long cycle life and high-rate capability, *ACS Appl. Mater. Interf.*, **8** (2016) 13437–13448.

102. B. Oschmann, J. Park, C. Kim, K. Char, Y.-E. Sung, and R. Zentel, Copolymerization of polythiophene and sulfur to improve the electrochemical performance in lithium–sulfur batteries, *Chem. Mater.*, **27** (2015) 7011–7017.

103. A. Hoefling, D.T. Nguyen, Y.J. Lee, S.-W. Song, and P. Theato, A sulfur–eugenol allyl ether copolymer: A material synthesized via inverse vulcanization from renewable resources and its application in Li–S batteries, *Mater. Chem. Front.*, **1** (2017) 1818–1822.

104. T. Zhang, F. Hu, W. Shao, S. Liu, H. Peng, Z. Song, C. Song, N. Li, and X. Jian, Sulfur-rich polymers based cathode with epoxy/ally dual-sulfur-fixing mechanism for high stability lithium–sulfur battery, *ACS Nano*, **15** (2021) 15027–15038.

105. X. Li, L. Yuan, D. Liu, Z. Li, J. Chen, K. Yuan, J. Xiang, and Y. Huang, High sulfur-containing organosulfur polymer composite cathode embedded by monoclinic S for lithium sulfur batteries, *Energy Storage Mater.*, **26** (2020) 570–576.

106. Q. Pang, A. Shyamsunder, B. Narayanan, C.Y. Kwok, L.A. Curtiss, and L.F. Nazar, Tuning the electrolyte network structure to invoke quasi-solid state sulfur conversion and suppress lithium dendrite formation in Li–S batteries, *Nat. Energy*, **3** (2018) 783–791.

107. W. Chang, J. Qu, Y. Sui, Q.-Y. Ji, T.-T. Zhang, X.-Z. Zhai, Y.-Q. Jing, and Z.-Z. Yu, Constructing mesoporous hollow polysulfane spheres bonded with short-chain sulfurs (Sx, x≤3) as high-performance sulfur cathodes in both ether and ester electrolytes, *Energy Storage Mater.*, **27** (2020) 426–434.

108. Z. Sun, M. Xiao, S. Wang, D. Han, S. Song, G. Chen, and Y. Meng, Sulfur-rich polymeric materials with semi-interpenetrating network structure as a novel lithium–sulfur cathode, *J. Mater. Chem. A*, **2** (2014) 9280–9286.

109. V.K. Tiwari, H. Song, Y. Oh, and Y. Jeong, Synthesis of sulfur-co-polymer/porous long carbon nanotubes composite cathode by chemical and physical binding for high performance lithium-sulfur batteries, *Energy*, **195** (2020) 117034.

110. P.T. Dirlam, A.G. Simmonds, T.S. Kleine, N.A. Nguyen, L.E. Anderson, A.O. Klever, A. Florian, P.J. Costanzo, P. Theato, M.E. Mackay, R.S. Glass, K. Char, and J. Pyun, Inverse vulcanization of elemental sulfur with 1,4-diphenylbutadiyne for cathode materials in Li–S batteries, *RSC Adv.*, **5** (2015) 24718–24722.

111. S. Zeng, L. Li, J. Yu, N. Wang, and S. Chen, Highly crosslinked organosulfur copolymer nanosheets with abundant mesopores as cathode materials for efficient lithium-sulfur batteries, *Electrochim. Acta*, **263** (2018) 53–59.

112. B. Long, J. Ma, T. Song, L. Liu, X. Wang, S. Song, and Y. Tong, Bifunctional polyvinylpyrrolidone generates sulfur-rich copolymer and acts as "residence" of polysulfide for advanced lithium-sulfur battery, *Chem. Eng. J.*, **414** (2021) 128799.

113. H. Hu, B. Zhao, H. Cheng, S. Dai, N. Kane, Y. Yu, and M. Liu, A robust 2D organic polysulfane nanosheet with grafted polycyclic sulfur for highly reversible and durable lithium-organosulfur batteries, *Nano Energy*, **57** (2019) 635–643.

114. J. Zhou, X. Zhou, Y. Sun, X. Shen, T. Qian, and C. Yan, Insight into the reaction mechanism of sulfur chains adjustable polymer cathode for high-loading lithium-organosulfur batteries, *J. Energy Chem.*, **56** (2021) 238–244.

115. J. Zhou, T. Qian, N. Xu, M. Wang, X. Ni, X. Liu, X. Shen, and C. Yan, Selenium-doped cathodes for lithium–organosulfur batteries with greatly improved volumetric capacity and coulombic efficiency, *Adv. Mater.*, **29** (2017) 1701294.

116. S. Shukla, A. Ghosh, P.K. Roy, S. Mitra, and B. Lochab, Cardanol benzoxazines: A sustainable linker for elemental sulphur based copolymers via inverse vulcanisation, *Polymer*, **99** (2016) 349–357.

117. H. Li, Y. Zhou, M. Zhao, B. Jin, Z. Wen, H. Xie, S. Dou, and Q. Jiang, Suppressed shuttle via inhibiting the formation of long-chain lithium polysulfides and functional separator for greatly improved lithium–organosulfur batteries performance, *Adv. Energy Mater.*, **10** (2020) 1902695.

118. M. Baloch, H. Ben Youcef, C. Li, O. Garcia-Calvo, L.M. Rodriguez, D. Shanmukaraj, T. Rojo, and M. Armand, New redox polymers that exhibit reversible cleavage of sulfur bonds as cathode materials, *ChemSusChem*, **9** (2016) 3206–3212.

119. M.B. Preefer, B. Oschmann, C.J. Hawker, R. Seshadri, and F. Wudl, High sulfur content material with stable cycling in lithium-sulfur batteries, *Angew. Chem. Int. Ed.*, **56** (2017) 15118–15122.

120. J. Wang, J. Yang, J. Xie, and N. Xu, A novel conductive polymer–sulfur composite cathode material for rechargeable lithium batteries, *Adv. Mater.*, **14** (2002) 963–965.

121. J. Fanous, M. Wegner, J. Grimminger, Ä. Andresen, and M.R. Buchmeiser, Structure-related electrochemistry of sulfur-poly(acrylonitrile) composite cathode materials for rechargeable lithium batteries, *Chem. Mater.*, **23** (2011) 5024–5028.

122. W. Wang, Z. Cao, G.A. Elia, Y. Wu, W. Wahyudi, E. Abou-Hamad, A.-H. Emwas, L. Cavallo, L.-J. Li, and J. Ming, Recognizing the mechanism of sulfurized polyacrylonitrile cathode materials for Li–S batteries and beyond in Al–S batteries, *ACS Energy Lett.*, **3** (2018) 2899–2907.

123. Z.-Q. Jin, Y.-G. Liu, W.-K. Wang, A.-B. Wang, B.-W. Hu, M. Shen, T. Gao, P.-C. Zhao, and Y.-S. Yang, A new insight into the lithium storage mechanism of sulfurized polyacrylonitrile with no soluble intermediates, *Energy Storage Mater.*, **14** (2018) 272–278.

124. M. Monisha, P. Permude, A. Ghosh, A. Kumar, S. Zafar, S. Mitra, and B. Lochab, Halogen-free flame-retardant sulfur copolymers with stable Li–S battery performance, *Energy Stor. Mater.*, **29** (2020) 350–360.

125. X. Wang, Y. Qian, L. Wang, H. Yang, H. Li, Y. Zhao, and T. Liu, Sulfurized polyacrylonitrile cathodes with high compatibility in both ether and carbonate electrolytes for ultrastable lithium–sulfur batteries, *Adv. Funct. Mater.*, **29** (2019) 1902929.

126. L. Yin, J. Wang, F. Lin, J. Yang, and Y. Nuli, Polyacrylonitrile/graphene composite as a precursor to a sulfur-based cathode material for high-rate rechargeable Li–S batteries, *Energy Environ. Sci.*, **5** (2012) 6966–6972.

127. Z. Li, J. Zhang, Y. Lu, and X.W. Lou, A pyrolyzed polyacrylonitrile/selenium disulfide composite cathode with remarkable lithium and sodium storage performances, *Sci. Adv.*, **4** (2018) eaat1687.

128. X. Chen, L. Peng, L. Wang, J. Yang, Z. Hao, J. Xiang, K. Yuan, Y. Huang, B. Shan, L. Yuan, and J. Xie, Ether-compatible sulfurized polyacrylonitrile cathode with excellent performance enabled by fast kinetics via selenium doping, *Nat. Commun.*, **10** (2019) 1021.

129. S. Li, Z. Han, W. Hu, L. Peng, J. Yang, L. Wang, Y. Zhang, B. Shan, and J. Xie, Manipulating kinetics of sulfurized polyacrylonitrile with tellurium as eutectic accelerator to prevent polysulfide dissolution in lithium-sulfur battery under dissolution-deposition mechanism, *Nano Energy*, **60** (2019) 153–161.

130. S. Ma, Z. Zhang, Y. Wang, Z. Yu, C. Cui, M. He, H. Huo, G. Yin, and P. Zuo, Iodine-doped sulfurized polyacrylonitrile with enhanced electrochemical performance for lithium sulfur batteries in carbonate electrolyte, *Chem. Eng. J.*, **418** (2021) 129410.

131. A. Abdul Razzaq, G. Chen, X. Zhao, X. Yuan, J. Hu, Z. Li, Y. Chen, J. Xu, R. Shah, J. Zhong, Y. Peng, and Z. Deng, Cobalt coordination with pyridines in sulfurized polyacrylonitrile cathodes to form conductive pathways and catalytic M-N_4S sites for accelerated Li-S kinetics, *J. Energy Chem.*, **61** (2021) 170–178.

132. H. Kim, J. Lee, H. Ahn, O. Kim, and M.J. Park, Synthesis of three-dimensionally interconnected sulfur-rich polymers for cathode materials of high-rate lithium–sulfur batteries, *Nat. Commun.*, **6** (2015) 7278.

133. A. Bhargav, M.E. Bell, Y. Cui, and Y. Fu, Polyphenylene tetrasulfide as an inherently flexible cathode material for rechargeable lithium batteries, *ACS Appl. Energy Mater.*, **1** (2018) 5859–5864.

134. S.N. Talapaneni, T.H. Hwang, S.H. Je, O. Buyukcakir, J.W. Choi, and A. Coskun, Elemental-sulfur-mediated facile synthesis of a covalent triazine framework for high-performance lithium–sulfur batteries, *Angew. Chem. Int. Edition*, **55** (2016) 3106–3111.

135. J. Kim, A. Elabd, S.-Y. Chung, A. Coskun, and J.W. Choi, Covalent triazine frameworks incorporating charged polypyrrole channels for high-performance lithium–sulfur batteries, *Chem. Mater.*, **32** (2020) 4185–4193.

136. M. Monisha, P. Permude, A. Ghosh, A. Kumar, S. Zafar, S. Mitra, and B. Lochab, Halogen-free flame-retardant sulfur copolymers with stable Li–S battery performance, *Energy Storage Mater.*, **29** (2020) 350–360.

137. O. Bayram, B. Kiskan, E. Demir, R. Demir-Cakan, and Y. Yagci, Advanced thermosets from sulfur and renewable benzoxazine and ionones via inverse vulcanization, *ACS Sustain. Chem. Eng.*, **8** (2020) 9145–9155.

Chapter 3

Dielectric Relaxation and Transport Dynamics of Solid-State Polymer Electrolytes

Stavros X. Drakopoulos
Princeton Materials Institute, Princeton University, Princeton, NJ 08544, USA
sd5541@princeton.edu

3.1 Introduction

It is well-understood by now that one of the greatest challenges of the 21st century is energy consumption and its relation to climate change [1]. Climate change has resulted from the excessive burning of fossil fuels over the last 100 years, which has increased the CO_2 concentration in the atmosphere and thus raised the average temperature of the planet. Indeed, according to the Climate Action of the United Nations, over 75% of global greenhouse gas emissions and almost 90% of all CO_2 emissions occur due to coal, oil, and gas consumption [2]. The use of fossil fuels led to an industrial explosion and the rapid production of energy (electricity or heat) and goods,

Batteries: The Future of Energy Storage
Edited by Subashani Maniam

ISBN 978-981-5129-38-0 (Hardcover), 978-1-003-51288-2 (eBook)
www.jennystanford.com

which establishes a direct link between climate change and energy. Indeed, modern society depends on the production, transfer, and storage of electrical energy, from everyday life to the most advanced technological applications. Thus, there is a high demand for the efficient use of energy and the invention of renewable and cleaner ways of producing, storing, and transferring it, without having a negative impact on the average standards of living and quality of life.

Sustainable methods of green energy production involve the utilization of solar power, hydropower, wind power, nuclear power, and tidal power, with the major challenge being the efficient store and recovery of the energy produced [3]. The efficient recovery of energy is particularly important for the viability of renewable energy production. For instance, the production of solar and wind energies relies heavily on natural conditions, thus producing non-continuous power. Therefore, high-performing energy-storage systems are required to effectively store the generated power and supply it to the electricity grid in a stable and reliable manner [4]. This can be achieved by the efficient conversion of one form of energy to another, such as thermal or electrochemical, by employing thermoelectric systems or battery devices, respectively. Materials science is at the forefront of innovation related to electrical energy storage and has led to the continuous development of energy materials. For example, materials science is employed in the three major types of electrical storage devices: dielectric capacitors, supercapacitors, and batteries. These three types are briefly discussed below and compared in Fig. 3.1 [5].

A dielectric capacitor, in principle, is a very simple device with a dielectric material, i.e., an insulator placed between two conductors (electrodes). A capacitor stores energy because of the polarization effects produced by the reorientation of dipoles within the dielectric materials upon the application of an external electric field and the concentration of charge carriers at the dielectric–electrode interface. When the external electric field is removed, the equilibrium state of the dipoles changes, leading to the release of charge carriers from the dielectric–electrode interface and, thus, the recovery of energy. Since the inception of the dielectric capacitor 250 years ago, the design and development of dielectric materials have been the subjects of continuous improvement and innovation [6]. More information can be found in Section 3.2.1.1 and in the literature [7–11].

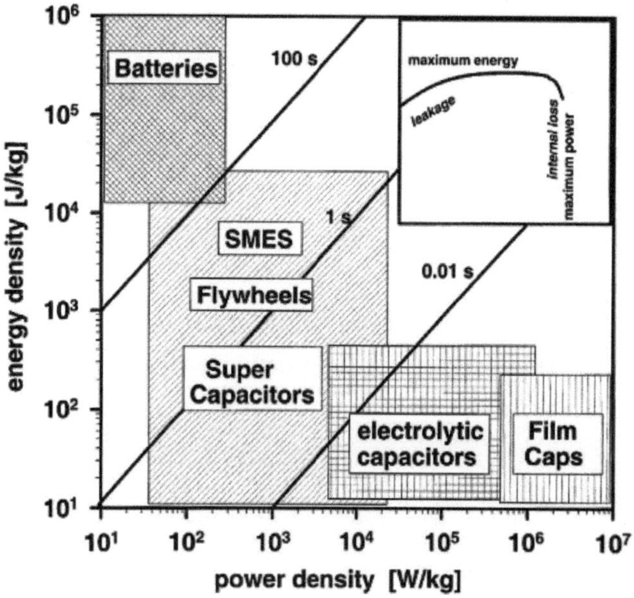

Figure 3.1 Ragone plot to compare batteries, supercapacitors, and dielectric (film) capacitors. Reprinted with permission from Ref. [5]. Copyright (2000) Elsevier.

Consisting of two electrodes, an electrolyte, and a separator, battery devices are used to store chemical energy and convert it into electrical energy. Through an external circuit, electrons flow from one electrode to the other because of the chemical reactions inside the battery cell that cause the transport of ions through the ionically conducting electrolyte solution. Different types of electrodes and electrolytes possess different electrochemical properties and thus are capable of a range of energy and power performance, efficiency, and operational voltage range. There are two major types of electrochemical cells: primary and secondary batteries. Primary batteries are designed to be used only once and are gradually discharged over the course of their lifetime. Secondary batteries can be recharged several times, in some cases yielding cycle lives in the order of 1000 cycles.

According to Fig. 3.1, supercapacitors (or electrochemical capacitors) are between dielectric capacitors and batteries in terms of energy and power density. Their position in the Ragone plot gives them a unique ability to possess both high power (compared to batteries) and high energy (compared to dielectric capacitors) as well as be associated with long cycle life. Like electrochemical battery cells, supercapacitors consist of two electrodes, an ion-conducting electrolyte, and a separator. Nanoporous materials are used as active electrodes, leading to gigantic capacitances as opposed to dielectric capacitors. Supercapacitors utilize the mechanism of an electric double layer to store energy. More information on supercapacitors can be found elsewhere [5, 12, 13].

3.1.1 Dominant Battery Technology: Liquid-Electrolyte Lithium-Ion Battery

Since its commercialization from Sony Co. back in 1991, the Li-ion battery has revolutionized portable devices and is still one of the most active areas of research [14]. Its applicability in devices from laptops and cell phones to now electric vehicles makes it the dominant technology in secondary (rechargeable) batteries [15]. The market size of Li-ion batteries alone is expected to reach US$92 billion globally by 2025, thus making it one of the prominent and profitable inventions of the last 30 years [16]. Excluding plug-in hybrid vehicles, by 2030 more than 14 million electric cars are expected to be on the road, more than seven times that of 2020 [17, 18]. Taking into account all types of four-wheel electric vehicles, the total fleet is expected to far exceed 200 million by 2030 [18, 19].

The conventional Li-ion battery cell consists of a liquid Li-conducting electrolyte, a separator, a metal oxide as the cathode electrode, and graphite as the anode material [20, 21]. An ideal electrolyte is a perfect ionic conductor and an electronic insulator. The operation of Li-ion batteries relies on the intercalation and de-intercalation phenomena of the lithium ions within the porous electrodes, as depicted in Fig. 3.2. The traditional choice for the anode material in Li-ion batteries is graphite, which possesses a low specific capacity of 372 mAh/g. Despite its low specific capacity, graphite is the mainstream choice due to its volumetric stability over intercalation and de-intercalation of lithium ions. An intense volumetric expansion accelerates the deterioration of the electrodes due to cracking and an eventual fracture of the electrode particles. The mechanical failure of the electrode leads to rapid capacity fading, thus minimizing the cycle life of the battery device. The design of new electrodes for Li-ion batteries with improved specific capacity and attractive volume expansion has been the subject of intense research over the past decade [22–27].

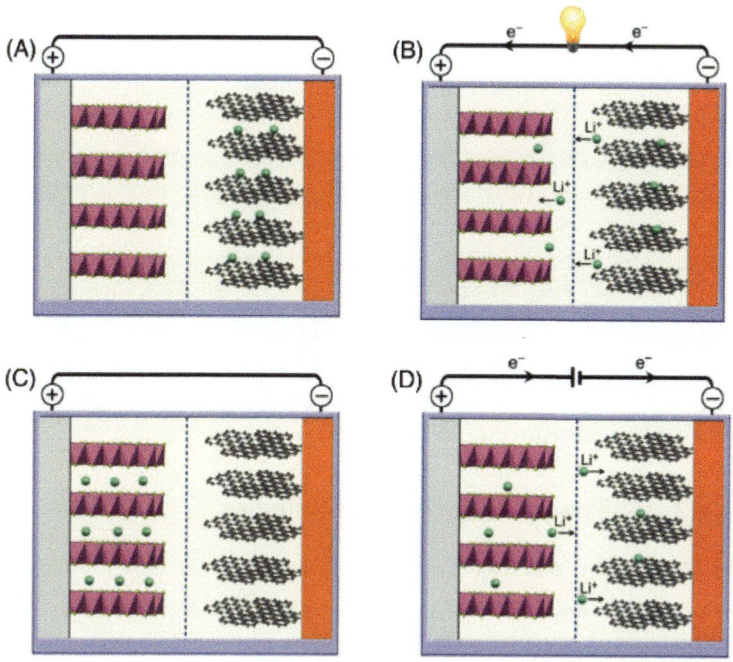

Figure 3.2 Schematic depiction of an Li-ion battery in different operational conditions: (A) fully charged state, (B) discharging process, (C) fully discharged state, and (D) charging process. The Li ions are depicted as green spheres; on the right (positive electrode), the cathode is shown, while on the left (negative electrode), the graphite based anode is shown. Reprinted with permission from Ref. [15]. Copyright (2019) The Authors.

Despite their commercial success, liquid-electrolyte Li-ion batteries also possess some disadvantages and hazards. The temperature window at which a liquid electrolyte is operational is limited by its phase stability; at lower temperatures, the frozen electrolytes are resistive, while at higher temperatures, the electrolyte boils or decomposes [28]. Safety is a major drawback of liquid electrolytes, which results into operational failure, fires or explosions, and the release of hazardous materials in the environment, causing concern or damage for public health reasons [29]. Leakage is a running problem of liquid electrolytes, which produces a range of risks, from biomedical issues to explosions [30]. In high-density rechargeable batteries, often lithium metal is used as the anode since it is the most electropositive metal. However, Li-metal batteries suffer from a major drawback: short-circuiting due to the growth of dendrites that protrude from the lithium surface [31]. One of the most prominent alternative technologies is solid-state batteries (SSBs), which employ nonflammable, nontoxic, and mechanically durable solid-state electrolytes (ceramic or polymeric in nature). Their mechanical durability (higher mechanical modulus) compared to traditional liquid-electrolyte batteries is responsible for the endurance of SSBs against the formation and evolution of dendrites. Dendrite-induced failure is caused when dendrites grow through the separator toward the cathode, short-circuiting the cell [32]. Additionally, employing solid electrolytes simplifies the cell design and thus its cost since it eliminates the requirement of a vessel to contain any liquid parts [33].

3.1.2 SSBs as Alternative Technologies

Solid-state electrolytes can be divided into two general classes: inorganic ceramics and organic polymers. Depending on the direct application, each possesses advantages or disadvantages against the other mostly in terms of processability and cost. Polymer-based materials are processed at significantly lower temperatures, thus decreasing the manufacturing costs and can easily take complex shapes toward flexible battery designs. On the other hand, ceramics possess higher values of elastic modulus and exhibit higher resistance to corrosive environments and higher temperatures [33]. Inorganic ceramic electrolytes have the potential to operate on a broader temperature range from –50 to 200 °C or higher, and their ionic transport is characterized by low activation energy, indicating small dependence from temperature, thus ensuring stability [34]. There are three general categories of ceramic electrolytes: oxides like perovskites [35, 36], sulfides [37, 38], and phosphates [39, 40]. Finally, hybrid solid-state electrolytes consisting of organic–inorganic composite materials have been developed and tested over the last decade [41, 42].

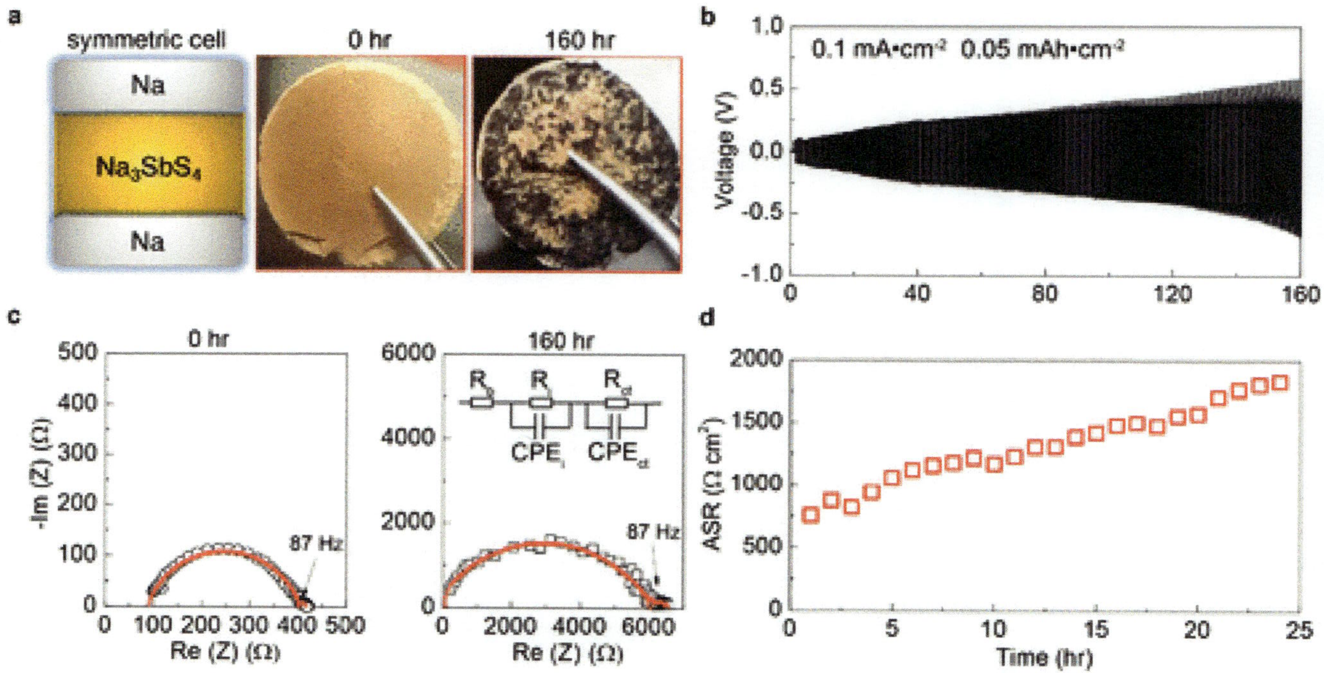

Figure 3.3 Na/Na_3SbS_4 interface instability in a symmetrical sodium cell. (a) Comparison before and after 160 h of operation. (b) Galvanostatic voltage profile of the cell cycled at a current density of 0.1 mA/cm^2 for 0.5 h. (c) Nyquist plots at 60 °C, including equivalent circuits fittings, and (d) evolution of total area specific resistance (ASR) during the first 25 h. Reprinted with permission from Ref. [43]. Copyright (2019) American Chemical Society.

There are still challenges that hinder the performance and commercialization of SSBs (Fig. 3.3). The electrode–electrolyte interface plays a vital role in ionic transport, which can be hindered when ionically resistive by-products of electrolyte decomposition are progressively grown. Another challenge faced when developing SSBs is the deficient physical contact between individual components to assure charge transport [43]. Poor wetting of surfaces eliminates areas of the device from participating in ionic diffusion. Moreover, during cycling, mechanical stresses build up at the points of contact between solid particles, increasing the risk of cracking, delamination, and eventual failure [44, 45]. Finally, despite the superior mechanical modulus of solid electrolytes, the formation of dendrites and thus dendrite-induced failure cannot be completely avoided in SSBs [46, 47].

3.2 Polyelectrolytes

As mentioned previously, SSBs are an alternative to liquid-electrolyte batteries. They employ nonflammable, nontoxic, and mechanically durable solid-state electrolytes and can be divided into ceramic, polymeric, and hybrid types. Although ceramic electrolytes are operational at very broad temperatures, they are not stable in the presence of moisture or CO_2. This leads to weaker chemical endurance and a need to be processed at elevated temperatures, which increases production costs. Contrary to ceramic solid electrolytes, polymers are more chemically stable and softer (less viscous). Thus, they are less sensitive to volume changes that increase the interfacial resistance and friction. To date, several polymers have been found to be ionically conductive when mixed with ionic salts, particularly those of lithium. Some of these most prominent polymers include poly(ethylene oxide) (PEO), poly(vinylidene fluoride) (PVDF), poly(vinylide fluoride-hexafluoropropylene) (PVDF-HFP), poly(methyl methacrylate) (PMMA), and poly(acrylonitrile) (PAN) [48].

Nevertheless, for SSBs to be fully commercialized, they require an operational temperature of 70–80 °C due to their low ionic conductivity [49]. At elevated temperatures, the cations have more thermal energy to flow between the anode and the cathode, and in return the mobility and ionic conductivity increase. The solid-state electrolyte in polymer SSBs consists of a semi-crystalline polymer, a lithium salt, and carbon black, all residing in the amorphous parts of the polymer. For the successful commercialization of polymer SSBs, it is crucial to reduce the operational temperature, and to achieve this, scientists need to discern the physical nature of resistance contributions that currently reduce the mobility and thus the ionic conductivity at room temperature.

Obviously, ions are charged particles; thus, their movement produces an electrical response that can be detected and measured by various experimental techniques [50]. Over the last decade, dynamic measurements have been performed on a great range of polyelectrolytes. To investigate features related to polymers and ionic transport, the technique of broadband dielectric spectroscopy (BDS) is employed in broad frequency and temperature ranges. As it will be explained later, by means of BDS, the connection of ionic transport with the dynamic glass-to-rubber transition (segmental relaxation) has been experimentally proven. In Section 3.2.1, general information on the technique is provided.

3.2.1 Polymer Physics: BDS

The term "dielectric material" refers to materials that can be polarized under the application of an external electric field. The aforementioned polarization corresponds to a separation of positive and negative charges in a molecular or atomic level and induces an internal electric field that is opposite to the external. The observed polarization due to the applied electric field occurs because of the reorientation of dipoles. Such dipoles are most often of molecular nature, and they relate to the chemical structure of the dielectric material under investigation. The reorientation of dipoles produces the dielectric dispersion and absorption phenomena, which can be examined in the framework of dielectric relaxation that obeys the linear response theory for small electric field strengths.

In an ideal dielectric material, no macroscopic migration of charges is observed, which is expected from a conductor or semiconductor above the activation temperature. Hence, all the insulators and high energy gap semiconductors ($E_g > 3$ eV) are classified as "dielectric materials" or simply "dielectrics" [51]. In reality, charges often exist either due to ionic/metallic impurities within a dielectric matrix, or as a result of doping in a composite system. Indeed, the electrical properties of dielectric materials can benefit from the presence of fillers of various shapes and electrical characteristics, and heterogeneous materials tailored for different applications can be produced. For example, the field of nanodielectrics for capacitive energy storage in capacitor applications is based on the development of heterogeneous materials with improved electrical/dielectric properties [52, 53].

To examine the dielectric relaxation response and charge carrier dynamics of dielectrics, the go-to technique is BDS, which is one of the most powerful techniques in molecular spectroscopy and materials physics. BDS can be employed in broad temperature and frequency ranges, similar to electrochemical impedance spectroscopy (EIS), a technique that most battery scientists are accustomed [51, 54, 55]. To perform the dielectric measurements in solid-state samples, a circular parallel-plate capacitor configuration is usually employed, as seen in Fig. 3.4 [56]. The dielectric material is placed between two symmetrical metallic current collectors. The distance between the current collectors is expected to be constant, thus resulting in a homogeneous electric field of parallel field lines.

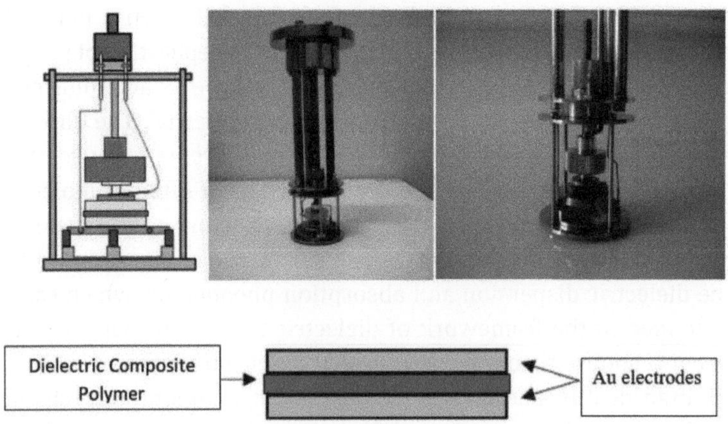

Figure 3.4 Schematic representation of a dielectric cell often used when performing BDS experiments. Reprinted with permission from Ref. [56]. Copyright (2019) MDPI.

To represent and understand frequency-dependent data obtained from BDS, scientists use dielectric formalisms, most prominently the complex dielectric permittivity function ε^*:

$$\varepsilon^*(\omega) = \varepsilon'(\omega) - i\varepsilon''(\omega) \tag{3.1}$$

where ω is the angular frequency equal to $2\pi f$ with f being the frequency of the externally applied electric field, and ε' and ε'' are the real and imaginary parts of the dielectric permittivity. Of equal importance are also the complex electric modulus M^*, complex impedance Z^*, and ac conductivity σ^*, shown below as a function of the real and imaginary parts of dielectric permittivity:

$$\begin{aligned} M^*(\omega) &= \frac{1}{\varepsilon^*(\omega)} = \frac{\varepsilon'(\omega)}{[\varepsilon'(\omega)]^2 + [\varepsilon''(\omega)]^2} + i\frac{\varepsilon''(\omega)}{[\varepsilon'(\omega)]^2 + [\varepsilon''(\omega)]^2} \\ &= M'(\omega) + iM''(\omega) \end{aligned} \tag{3.2}$$

$$\begin{aligned} Z^*(\omega) &= \frac{L}{\varepsilon_0 \omega A \varepsilon^*(\omega)} \\ &= \left(\frac{L}{\varepsilon_0 \omega A}\right)\frac{\varepsilon''(\omega)}{[\varepsilon'(\omega)]^2 + [\varepsilon''(\omega)]^2} - i\left(\frac{L}{\varepsilon_0 \omega A}\right)\frac{\varepsilon'(\omega)}{[\varepsilon'(\omega)]^2 + [\varepsilon''(\omega)]^2} \\ &= Z'(\omega) - iZ''(\omega) \end{aligned} \tag{3.3}$$

$$\begin{aligned} \sigma^*(\omega) &= i\varepsilon_0 \omega \varepsilon^*(\omega) = \varepsilon_0 \omega \varepsilon''(\omega) - i\varepsilon_0 \omega \varepsilon'(\omega) \\ &= \sigma'(\omega) - i\sigma''(\omega) \end{aligned} \tag{3.4}$$

$$\tan\delta = \frac{\varepsilon''}{\varepsilon'} = \frac{M''}{M'} = \frac{Z'}{Z''} = \frac{\sigma'}{\sigma''} \tag{3.5}$$

where ε_0 is the dielectric constant of vacuum (8.854×10^{-12} F/m), A is the electrode cross-sectional area, and L is the distance between the electrodes and corresponds to the thickness of the dielectric film. The loss tangent (tan δ) is also defined according to Eq. (3.5) as a function of dielectric formalisms. Hence, the use of various dielectric formalisms can be complementary and emphasize different aspects of the same processes [57–59].

While analyzing the dielectric spectrum, scientists most often try to discern the various dipolar contributions and charge transport based on their frequency and temperature dependencies. To achieve that, they often employ the Havriliak–Negami (HN) function model isothermally, presented here in its generalized complex permittivity form for N relaxation processes and leakage conductivity contribution:

$$\varepsilon_{HN}^{*}(\omega)=\varepsilon_{\infty}+\sum_{i=1}^{N}\left[\frac{\Delta\varepsilon}{\left[1+\left(i\omega\langle\tau\rangle\right)^{m}\right]^{n}}\right]+\frac{\sigma_{dc}}{i\varepsilon_{0}\omega^{s}} \tag{3.6}$$

$$\tau=\frac{1}{\omega_{\max}}\left[\frac{\sin\left(\frac{m\pi}{2+2n}\right)}{\sin\left(\frac{mn\pi}{2+2n}\right)}\right]^{\frac{1}{m}} \tag{3.7}$$

For each dielectric process, $\langle\tau\rangle$ corresponds to the average value of the HN relaxation time and $\Delta\varepsilon$ is the dielectric relaxation strength where $\Delta\varepsilon = \varepsilon_s - \varepsilon_\infty$ with ε_s and ε_∞ being the static and infinite frequency permittivity values, respectively ($\varepsilon_s = \lim_{\omega\to 0}\varepsilon'$ and $\varepsilon_\infty = \lim_{\omega\to\infty}\varepsilon'$). As m and n are the shape parameters and correspond to the symmetrical and asymmetrical distributions of relaxation times, respectively, with $0 < m \leq 1$ and $0 < mn \leq 1$; for $m = n = 1$, Eq. (3.6) reduces to the Debye model [60–62] for a single relaxation time.

In Eq. (3.6), σ_{dc} is considered the pure leakage conductivity that does not induce polarization. The parameter s depends on the examined material and the underlying conduction mechanism and takes values between 0.5 and 1.0 with unity being considered pure Ohmic [63]. In disordered materials, the current density is an additive value composed of Ohmic conductivity, diffusion, and a displacement current density being equal to $\partial D/\partial t$, where D is the electric displacement [64]. Under low electric fields, at steady state ($\partial D/\partial t = 0$) and constant charge carrier density in the bulk material, Ohmic conductivity appears to be the only contribution. The deviation from the pure Ohmic behavior can be expressed when the semi-empirical parameter s is below unity where the diffusion current and time-varying fields ($\partial D/\partial t \neq 0$) contribute to the system's conductivity.

To express the dielectric data with the various formalisms, a simulated case of $N = 2$ dielectric process and a dc conductivity contribution following the HN function model are shown in Fig. 3.5. By employing the real part of the dielectric permittivity formalism,

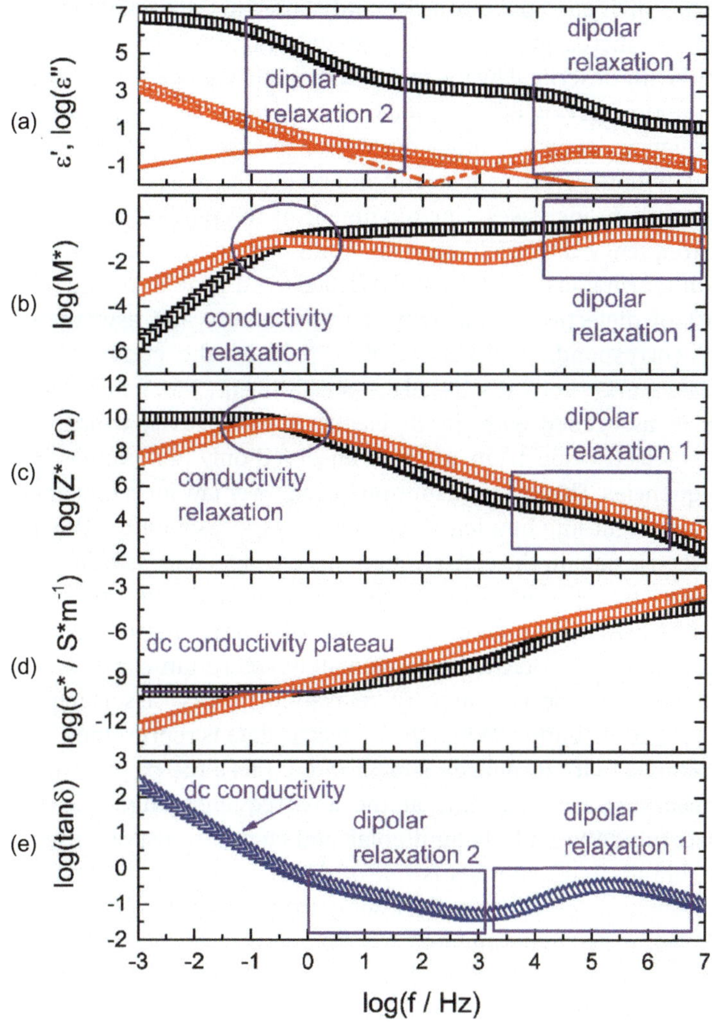

Figure 3.5 The simulated dielectric response as a function of frequency is presented via (a) dielectric permittivity, (b) electric modulus, (c) impedance, (d) ac conductivity, and (e) tanδ. In Figs. (a–d), the black open squares correspond to the real part and the red open squares correspond to the imaginary part. Two dipolar relaxations and the dc conductivity contribution were simulated based on Eq. (3.6) discussed previously. The fitting parameters are ε_∞ = 1. For the dipolar relaxation 1 (dashed line), they are $\Delta\varepsilon_1$ = 2, m_1 = n_1 = 0.75, and $f_{max,1}$ = 10^5 Hz, while for dipolar 2 (dash-dotted line), they are $\Delta\varepsilon_2$ = 4, m_2 = 0.5, n_2 = 1, and $f_{max,2}$ = 10^0 Hz. The fitting parameters for the dc conductivity contribution (dotted line) were σ_{dc} = 10^{-10} S/m and s = 1. Original graphs prepared for the present chapter.

each individual dipolar process contributed to the ε' values in a step-like way, increasing them from high to low frequencies. This effect is physically described by the fact that frequency is equal to the inverse of the time period of the applied electric field. Therefore, when the frequency decreases, more dipoles of the specific dipolar process align with the electric field, and as a result, the real part of dielectric permittivity increases. The alignment of the dipoles as a function of frequency and their corresponding retardation are associated with thermal losses that are described via a peak in the imaginary part of dielectric permittivity. The broadness and asymmetry of the corresponding peaks are described by the empirical shape parameters m and n discussed previously in Eq. (3.6). When electrical losses associated with the dc electrical conductivity contribute to the dielectric spectrum, a rise is observed only in ε'' values at low frequencies. This σ_{dc} contribution has a power law form that appears as a straight line in a log ε'' versus log f representation. When the dielectric data in the electric modulus or impedance is transformed using Eqs. (3.2) and (3.3), respectively, an additional peak in M'' and Z'' arises apart from the dipolar processes. Its frequency loss peak position is located approximately where $\tan\delta = 1$, and it is identified as the conductivity relaxation process associated with the σ_{dc} contribution. When the dielectric data is represented via the ac conductivity formalism shown in Eq. (3.4), the σ_{dc} contribution appears as a plateau line at the low-frequency edge (Fig. 3.5d). Discerning the underlying dipolar and charge transport processes in complicated materials is not often an easy task, particularly in polyelectrolytes, which are yet not entirely understood.

Dielectric experiments are often performed at a broad temperature range, which allows the temperature dependence of relaxation times, commonly referred to as relaxation dynamics. Usually, the relaxation times follow the Arrhenius law that allows the determination of the activation energy, an intrinsic property of dipolar reorientation. The Arrhenius law expressed for the relaxation time can be seen here:

$$\langle\tau_{HN}\rangle = \tau_0 \exp\left(\frac{E_A}{k_B T}\right) \tag{3.8}$$

where $\langle\tau_{HN}\rangle$ is the average value of the HN function model discussed previously. As τ_0 is the relaxation time of the process at infinite temperature, E_A is the activation energy, k_B is the Boltzmann constant, and T is the absolute temperature. Most dielectric processes follow the Arrhenius law although the segmental relaxation (dynamic glass-to-rubber transition process) attributed to the T_g famously deviates from the Arrhenius behavior [65]. The origin of the segmental relaxation lies on the cooperative rearrangement of each amorphous segment affected by its neighboring segments [66–68]. The segmental relaxation is usually best described by the Vogel–Fulcher–Tammann equation:

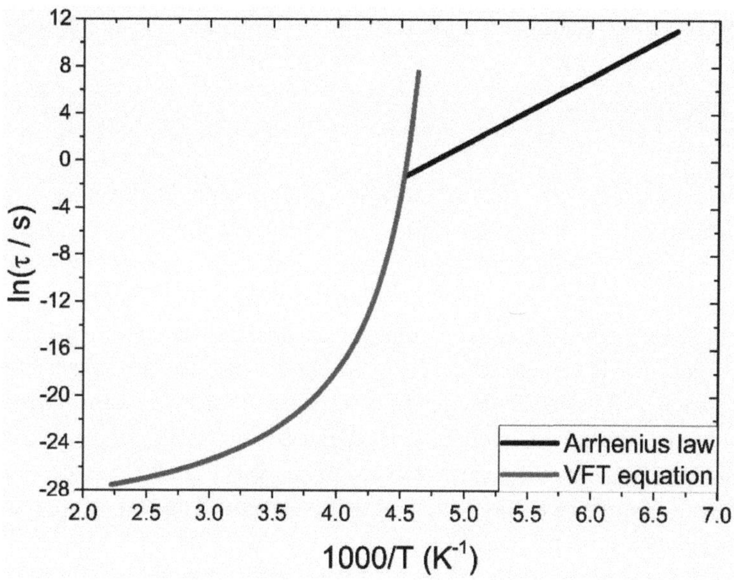

Figure 3.6 Simulated temperature dependencies of the relaxation time following the Arrhenius law and the VFT equation. The Arrhenius law follows Eq. (3.8), and the VFT equation follows Eq. (3.9). The fitting parameters for the Arrhenius temperature dependence are $\tau_0 = 10^{-12}$ s and $E_A = 0.5$ eV. For the VFT temperature dependence, the fitting parameters are $\tau_0 = 10^{-13}$ s, $D = 3$, and $T_V = 200$ K. An original graph was prepared for the present chapter.

$$\langle \tau_{HN} \rangle = \tau_0 \exp\left(\frac{DT_V}{T - T_V} \right) \tag{3.9}$$

where D and T_V are a dimensionless parameter that can be related with the steepness (fragility) of the relaxation process and the Vogel temperature, which corresponds to the ideal glass-to-rubber transition temperature, respectively. The Vogel temperature is around 40 K below the calorimetric glass-to-rubber transition [54]. The segmental relaxation's deviation from the Arrhenius behavior at temperatures over the T_g is a long-standing scientific problem. The underlying mechanism is associated with the change in unoccupied volume accompanying thermal expansion [69–71]. The free volume changes as the temperature decreases toward the T_g, resulting in a dramatic slowing down of the relaxation dynamics. This effect is represented as an increase in the segmental relaxation time, which is more intense than the Arrhenius law predicts [72, 73]. A comparative example between the Arrhenius and VFT equations is shown in Fig. 3.6.

3.2.2 Dielectric Relaxations of Various Polyelectrolytes

Particularly in the case of polyelectrolytes, due to the very high values of ionic conductivity, the dielectric spectrum is dominated by high losses mainly at frequencies associated with charge transport. Thus, compared to typical low-loss dielectric materials (as seen in Fig. 3.5), the dielectric spectrum of a polyelectrolyte is more difficult to discern. However, since ionic transport is a thermally assisted process (for more details see Section 3.2.4), at low temperatures, the contribution of mobile ions at the dielectric response is very limited. Thus, the dielectric relaxation analysis of polyelectrolytes at low temperatures is not different from the typical dielectric polymer materials. Like polymer composites, the presence of an ionic salt affects the temperature dependence and intensity of the host polymer's dipolar relaxations and can even decrease its crystallinity. For example, as opposed to plain PEO, the crystallinity and crystal size of the host polymer are affected by the presence of the LiTFSI salt, which acts as a plasticizer [74]. Reducing the crystallinity of the host polymer in the presence of plasticizers is preferrable as it increases the ionic mobility and thus the ionic conductivity [75]. As a result, the dielectric response of the host polymer in the presence of the ionic salt is expected to change, even at low temperatures where the ion mobility is reduced or hindered. Secondary relaxations associated with local dynamics of the polymer host are detectable. Figure 3.7 provides an example of the dielectric response of a polyelectrolyte for HBC-TEG$_6$, both neat and in the presence of an ionic salt ($LiCF_3SO_3$), including its chemical structure.

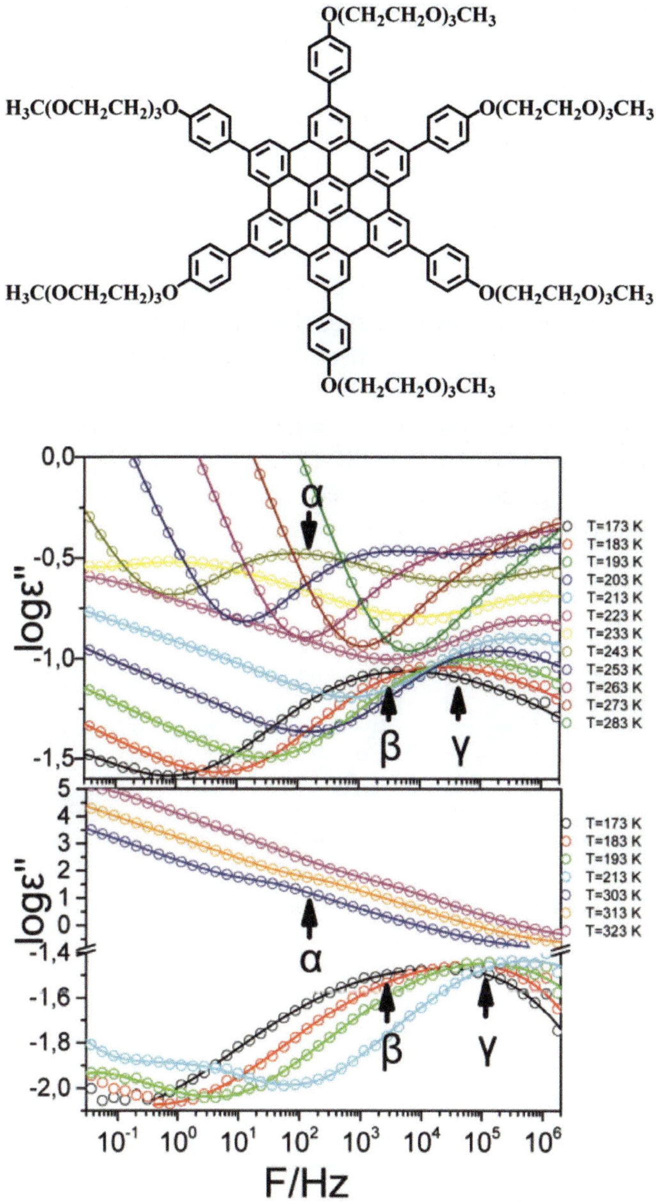

Figure 3.7 (Top) The chemical structure of HBC-TEG$_6$, and the imaginary part of dielectric permittivity as a function of frequency for (middle) neat HBC-TEG$_6$ and (bottom) HBC-TEG$_6$:$LiCF_3SO_3$ with salt concentration [EO]:[Li^+] = 8:1. Reprinted with permission from Ref. [76]. Copyright (2017) American Chemical Society.

According to Fig. 3.7 and the work of Pipertzis et al., two secondary relaxations, namely β- and γ-processes, are observed for HBC-TEG$_6$ due to local fluctuations of ethylene glycol at the glassy state [76]. At higher temperatures, the emergence of the α-process is observed and is attributed to the dynamic glass-to-rubber transition process, commonly known also as the segmental relaxation. This dielectric process is directly linked to the T_g of the polymer and its relaxation times follow a VFT temperature dependence, as briefly discussed in Section 3.2.1 and Eq. (3.9) previously. With the addition of an ionic salt, it is evident that the relaxation characteristics of each dielectric process are affected. Especially above the T_g, the dielectric response changes as the Li^+ ions become mobile [76]. The temperature dependence of the relaxation times of the observed dielectric processes for the HBC-TEG$_6$-based samples is shown in Fig. 3.8. The presence of ionic salt upon the polymer's relaxation dynamics can be observed in the relaxation map.

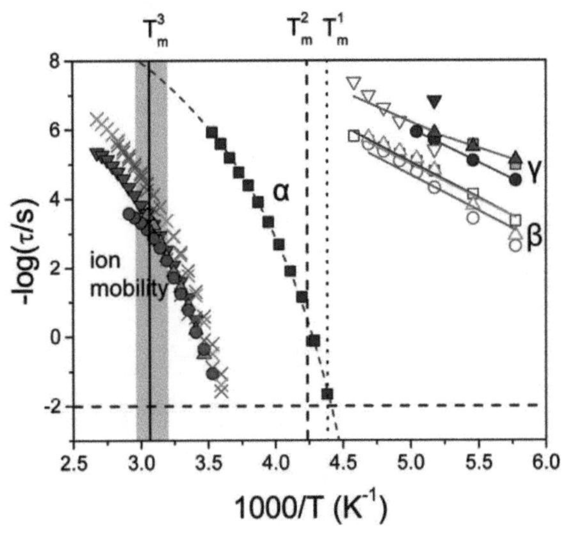

Figure 3.8 Relaxation map of the HBC-TEG$_6$-based polyelectrolytes with the vertical T_m lines being melting temperatures obtained from differential scanning calorimetry. With squares, circles, up-triangles and down-triangles are the neat and salt concentrations [EO]:[Li^+] = 12:1 or 8:1 or 3:1, respectively. The α- and γ-processes are represented with filled symbols, while the β-process with open symbols. Reprinted with permission from Ref. [76]. Copyright (2017) American Chemical Society.

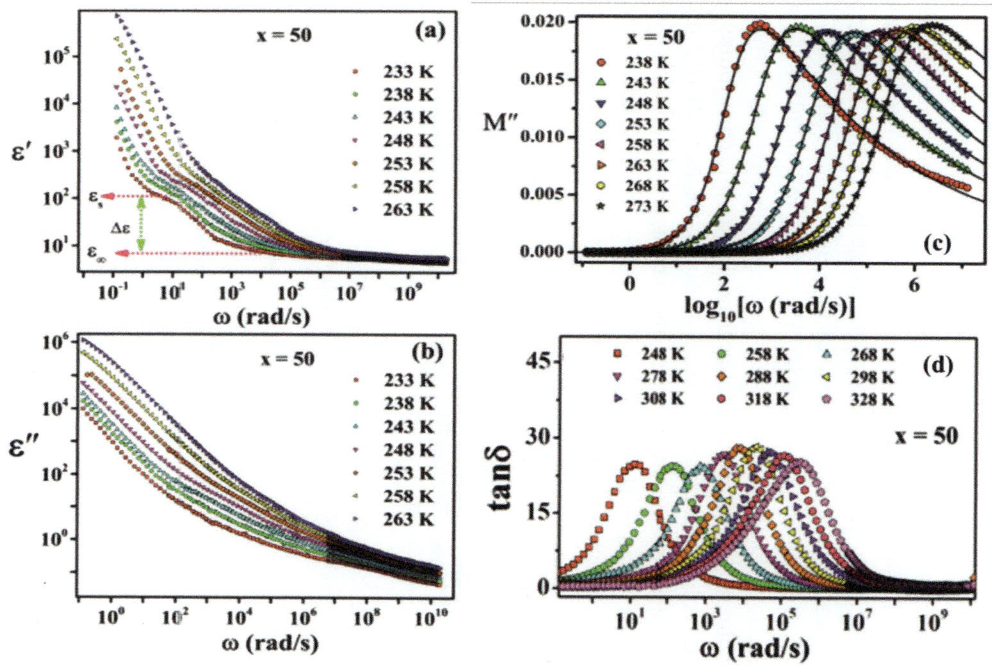

Figure 3.9 The dielectric response of PMMA-LiTFSI-50%BMPTFSI polymer electrolytes with salt concentration [MMA]:[Li^+] = 4:1, represented via (a) the real part and (b) the imaginary part of dielectric permittivity, (c) the imaginary part of electric modulus, and (d) loss tangent, all varying the temperature. Reprinted with permission from Ref. [77]. Copyright (2020) AIP Publishing.

As shown in Section 3.2.1, the use of various dielectric formalisms can be beneficial toward the understanding of dielectric materials and specifically polyelectrolytes. Figure 3.9 shows the representation of the dielectric response of PMMA-LiTFSI-50% BMPTFSI via dielectric permittivity, electric modulus, and loss tangent [77].

Finally, the low-frequency dielectric response is dominated by the so-called "electrode polarization" effects or EP, as shown in Fig. 3.10 [78]. It is evident that the EP is orders of magnitude more intense that the host polymer's dielectric relaxation response. The EP effect is the result of the formation of space-charge layers that produce a rapid voltage drop and thus very strong electrical polarization at the low-frequency edge. The physical mechanism behind EP is the concentration of immobile charge carriers at the interface between the sample and the electrode (current collector). It contributes to both the real and imaginary parts of dielectric permittivity and reduces the ac conductivity with decreasing frequency [79–81]. When EP is particularly strong (which is the case in conductive polymers like polyelectrolytes), it dominates the dielectric spectra, thus masking the dipolar relaxation process and hindering the dielectric analysis. For that purpose, EP is most often considered a parasitic effect.

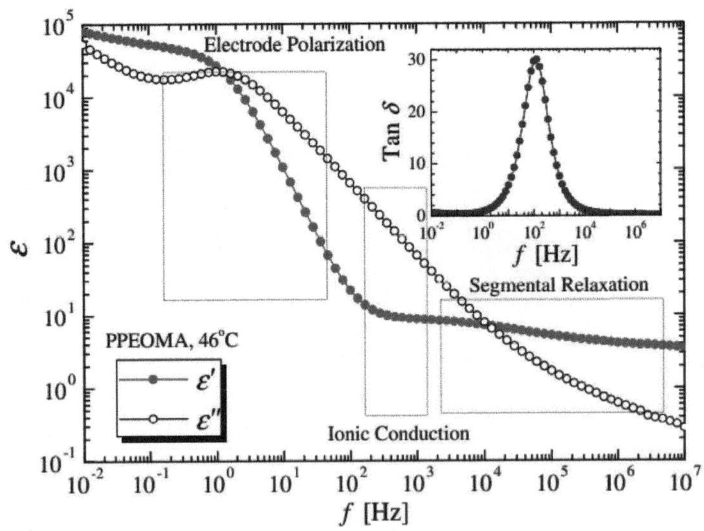

Figure 3.10 Real and imaginary parts of the dielectric permittivity of PPEOMA as a function of frequency at 46 °C with the inset showing the loss tangent. The electrode polarization, ionic conduction, and segmental relaxation are highlighted in the graph. Reprinted with permission from Ref. [78]. Copyright (2012) American Physical Society.

3.2.3 Charge Transport Dynamics

The dielectric process briefly discussed earlier, called EP, is the central process for the analysis of ion mobility and mobile ion concentration from the dielectric spectra. Polyelectrolytes are most often a composite material of an ionically conductive polymer and an ionic salt that dissolves partially or completely in the polymer host. In the following discussion of EP, the mobile species are assumed to be cations (like Li^+) and the anionic charges are perceived as fixed in space. In their ground-breaking paper, Klein et al. [82] considered ionic generation and recombination over space and time to be at an equilibrium state, thus be dependent only on temperature. Figure 3.11 shows a schematic describing the charge density distribution under the application of a dc electric field.

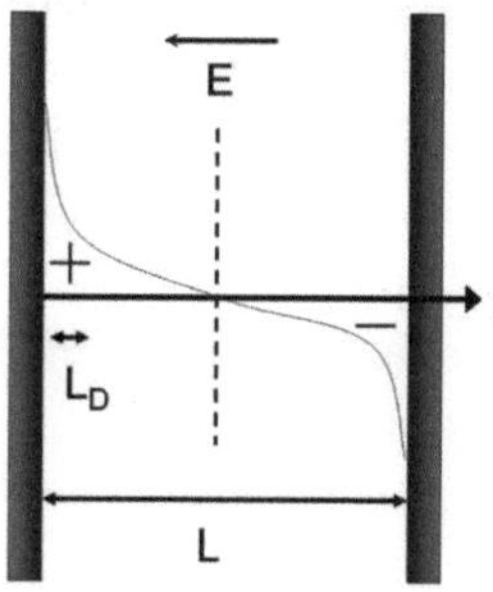

Figure 3.11 Parallel-plate capacitor of thickness L under the application of an externally applied dc electric field. L_D is the Debye length, which will be discussed later. Reprinted with permission from Ref. [82]. Copyright (2006) AIP Publishing.

The dielectric constant is defined as the ability of a dielectric to be polarized and, thus, is important in cation–anion dissociation and charge transport. In a polyelectrolyte, the dielectric constant is equal to the static dielectric permittivity ε_s (where $\varepsilon_s = \lim_{\omega \to 0} \varepsilon'$) prior to the occurrence of EP [83]. Under the application of an external electric field, a charge moves inside the dielectric material, driven by the Coulomb force. The electric field produced by the charge itself is sensed in the surrounding matter at a distance equal to the Debye length L_D or screening length and follows Eq. (3.10):

$$L_D = \frac{1}{q_e}\sqrt{\frac{\varepsilon_s \varepsilon_0 k_B T}{p_0}} \tag{3.10}$$

where q_e is equal to the ionic charge and p_0 is the equilibrium number density of mobile ions. For the description of EP, L_D is regarded as the length scale of the electrostatic double layer formed at the interface between the dielectric material and the electrode (current collector). The number density of mobile ions follows Eq. (3.11):

$$p_0 = \frac{\sigma_{ion}}{\mu q_e} \quad (3.11)$$

where μ is the mobility of the free ions and follows Eq. (3.12):

$$\mu = \frac{q_e L L_D}{2\tau_{EP} k_B T} \quad (3.12)$$

where τ_{EP} is the time constant of the EP process and can be estimated as the inverse angular frequency obtained from the experimental data, i.e., $\tau_{EP} = (w_{max})^{-1} = (2pf_{max})^{-1}$. However, this relation applies only when EP can accurately be described by the Debye or Cole–Cole function models. The frequency-dependent description of EP with the Debye model has proven to be challenging experimentally despite its theoretical foundation [84–86]. By estimating the τ_{EP} from the experimental data and using Eq. (3.13) and obtaining the ionic conductivity from σ' and σ'' or Z' and Z'', every other physical parameter can be calculated.

$$\tau_{EP} = \frac{L\varepsilon_s \varepsilon_0}{2L_D \sigma_{ion}} \quad (3.13)$$

All the above calculations are performed isothermally; however, one of the great advantages of BDS is that it can cover a broad temperature range too. Like the discussed temperature dependence of the relaxation time, the ionic conductivity σ_{ion} can be modeled as a function of temperature. The ionic conductivity can follow either the Arrhenius law or the VFT equation as seen below:

$$\sigma_{ion} = \sigma_0 \exp\left(-\frac{E_A}{k_B T}\right) \quad (3.14)$$

$$\sigma_{ion} = \sigma_0 \exp\left(-\frac{DT_V}{T - T_V}\right) \quad (3.15)$$

where σ_0 is the ionic conductivity at infinite temperature. Both equations indicate that ionic conductivity is a thermally assisted process. It has been reported experimentally that above the glass-to-rubber transition process of polyelectrolytes, the ionic conductivity follows a similar temperature dependence described by the VFT equation. This has led to the association of ion hopping to the segmental motion of the polyelectrolyte host material that assists ion diffusion [87–89]. The coupling between ionic conductivity and segmental relaxation can be expressed via the proportionality between the mobility of the free ions μ and the inverse value of the segmental relaxation time. The observed proportionality indicates that the rate of ion transport in polymers is defined by the rate of segmental reorientation [90].

The coupling of ionic conductivity to segmental relaxation indicates that the ionic transport is limited from the T_g of the host polymer. As a result, the research efforts have been mainly focused on developing polyelectrolytes characterized by a very low T_g. A low T_g translates to fast segmental dynamics at room temperature, thus high ionic conductivity [91]. Another concept to reach the same result (σ_{ion} ~10^{-3} S/cm at room temperature) is by decoupling the ionic conductivity from segmental relaxation and thus produce superionic polyelectrolytes. About 20 years ago, Angell et al. [92] designed polyelectrolytes heavily doped with ionic salts. In these systems, the polymer is required only for its mechanical properties; thus, its concentration was kept minimum. However, the resulting mechanical properties were inferior to high-polymer content polyelectrolytes [91]. This approach was applied to several polymers, including polypropylene glycol [92], poly(acrylonitrile) [93–95], poly(1-vinyl pyrrolidone), and poly(*N,N*-dimethyl acrylamide) [95]. An alternative idea is employing rigid polymer structures to increase the free volume and thus promote ionic transport [96, 97]. The aforementioned strategies have increased significantly the ionic conductivities even at room temperature, reaching 10^{-4} to 10^{-3} S/cm.

In semiconducting materials, including ionically conductive polymers, the dominating charge transport mechanism is hopping conductivity. Hopping is defined as the sudden displacement of a charge carrier from one position to another close by. One of the many hopping models, the symmetrical hopping model (also known as the random free-energy barrier model) developed by Dyre, accurately describes the ac conduction in various semiconducting materials [98, 99]. The symmetrical hopping model (shown schematically in

Fig. 3.12) allows only one charge carrier per energy minimum and ignores the interaction between charges. One prominent success of the symmetrical hopping model is that it serves as the foundation behind the Barton, Nakajima, and Namikawa (BNN) relation:

$$\sigma_{dc} = p\varepsilon_0\Delta\varepsilon\omega_{max} \tag{3.16}$$

where p is a temperature-independent numerical constant close to unity, and $\Delta\varepsilon$ and ω_{max} are the dielectric relaxation strength and angular frequency loss peak position of the dielectric process associated with charge transport [98]. Here, ω_{max} is associated with the lowest effective jump frequency of the hopping charge transport.

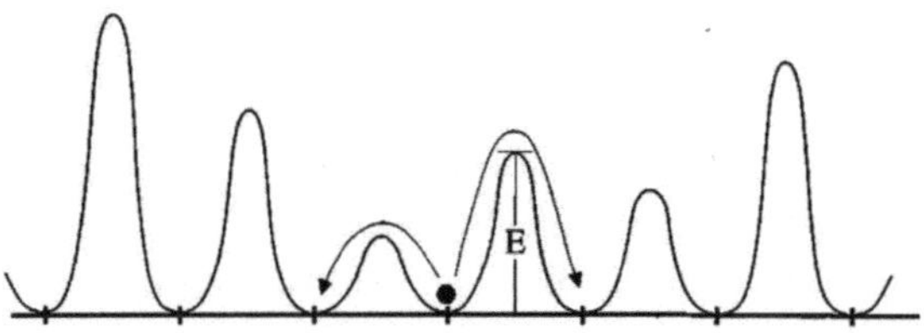

Figure 3.12 Simple representation of the symmetric hopping model in one dimension showing that a charge carrier can go either forward or backward across an energy barrier *E* with an equal jump rate (probability). Reprinted with permission from Ref. [57]. Copyright (2000) American Physical Society.

In addition to other semiconducting materials, the symmetrical hopping model and the BNN relation have found application also in the dielectric analysis of polyelectrolytes. The BNN relation connects ion conductivity with dielectric loss caused by ion migration (hopping frequency or electrical relaxation frequency). The lowest effective jump frequency of the hopping charge transport can be obtained directly from the symmetrical hopping model by fitting $\sigma'(\omega)$ as a function of frequency [99], as shown below:

$$\sigma'(\omega) = \frac{\sigma_{dc}\omega\tau\left[\arctan(\omega\tau)\right]}{\frac{1}{4}\left\{\ln\left[1+(\omega\tau)^2\right]\right\}^2 + \left[\arctan(\omega\tau)\right]^2} \tag{3.17}$$

where τ is the characteristic time of the symmetrical hopping model and equals the inverse lowest effective jump frequency. By obtaining the characteristic time, a proportionality between τ and ionic conductivity can be obtained, as shown by Zardalidis et al. [100]. This can be appreciated in Fig. 3.13.

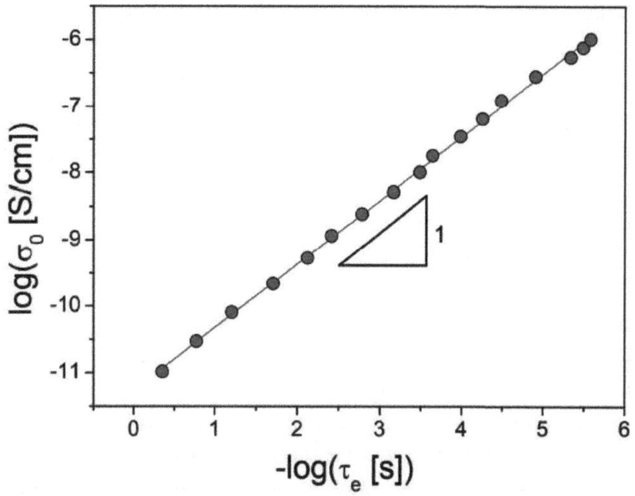

Figure 3.13 Proportionality between ionic conductivity and the inverse characteristic time of the symmetrical hopping model for PEO:LiTf. Reprinted with permission from Ref. [100]. Copyright (2013) American Chemical Society.

It has been discussed via ab initio molecular dynamics simulations that either ion jumps happen collectively between several ions or individual ion hopping affects its neighboring ions, indicating a cooperative nature in either case [50, 101]. The atomic-scale mechanisms that promote and contribute to ionic transport in crystalline solids are better understood due to the periodicity of the crystalline structure [102]. In crystalline materials, the available vacancies for an ion to migrate into are strongly temperature dependent due to thermally activated defect formation [50]. In amorphous materials such as glass, the concentration of free vacancies is a more complicated matter as it depends on thermal events such as T_g and the presence of surrounding energy barriers that produce a distribution of site energies [103, 104]. In glass, the vacancy concentration is expected to be greater than in crystals and vacancy–vacancy interactions are not excluded, reinforcing the complicated nature of charge transport in amorphous ionic conductors. The aforementioned reasons also affect the temperature dependence of ionic conductivity and contribute to the deviations from the Arrhenius law, which are often observed for amorphous solid-state electrolytes [105].

3.3 Future Prospects and Challenges

It is sensible to consider that the field of solid-state polymer electrolytes is still at its infancy. Therefore, it is certain that the future holds many advancements toward both deeper physical understanding, chemical innovations, and ultimately successful, completely liquid-free polymer-based batteries for commercial use. To achieve a commercially available, polymer-based, and completely solid-state battery, there are several factors to be addressed, both from an electrochemical performance point of view and sustainability.

Sustainability is of great importance to battery research, and as such, eco-friendly polymers are examined in various studies. Cellulose, the most naturally abundant polysaccharide, is considered a great candidate as it is easy to be chemically modified [106]. Indeed, scientific efforts have shown that modified cellulose-based materials can be used in solid-state polyelectrolytes and exhibit outstanding cycle life performance [107, 108]. Along the lines of safety, efforts are also concentrated on replacing flammable polymers such as PEO, the benchmark in polyelectrolytes, without compromising any electrochemical properties [109]. Solid-state polymer-based batteries still suffer from low operational voltages that require innovations in polymer chemistry [110] and require high temperatures to exhibit high enough values of ionic conductivity [111].

References

1. Krauss L. M. *Physics of Climate Change*. S.l.: Apollo; 2021. https://search.ebscohost.com/login.aspx?direct=true&scope=site&db=nlebk&db=nlabk&AN=2604229.
2. World Meteorological Organization. *Provisional State of the Global Climate 2022*. World Meteorological Organization; 2022. WMO. https://library.wmo.int/doc_num.php?explnum_id=11359.
3. Liu C, Li F, Ma L-P, and Cheng H-M. Advanced materials for energy storage. *Advanced Materials*, 2010; **22**(8): E28–E62, doi:10.1002/adma.200903328.
4. Divya KC and Østergaard J. Battery energy storage technology for power systems: An overview. *Electric Power Systems Research*, 2009; **79**(4): 511–520, doi:10.1016/j.epsr.2008.09.017.

5. Christen T and Carlen MW. Theory of Ragone plots. *Journal of Power Sources*, 2000; **91**(2): 210–216, doi:10.1016/S0378-7753(00)00474-2.

6. Ho J, Jow TR, and Boggs S. Historical introduction to capacitor technology. *IEEE Electrical Insulation Magazine*, 2010; **26**(1): 20–25, doi:10.1109/MEI.2010.5383924.

7. Wallace RM and Wilk GD. High-κ dielectric materials for microelectronics. *Critical Reviews in Solid State and Materials Sciences*, 2003; **28**(4): 231–285, doi:10.1080/714037708.

8. Dang Z-M, Yuan J-K, Zha J-W, Zhou T, Li S-T, and Hu G-H. Fundamentals, processes and applications of high-permittivity polymer–matrix composites. *Progress in Materials Science*, 2012; **57**(4): 660–723, doi:10.1016/j.pmatsci.2011.08.001.

9. Hao X. A review on the dielectric materials for high energy-storage application. *Journal of Advanced Dielectrics*, 2013; **03**(01):1330001, doi:10.1142/S2010135X13300016.

10. Dang Z-M, Yuan J-K, Yao S-H, and Liao R-J. Flexible nanodielectric materials with high permittivity for power energy storage. *Advanced Materials*, 2013; **25**(44): 6334–6365, doi:10.1002/adma.201301752.

11. Li Q, Yao F-Z, Liu Y, Zhang G, Wang H, and Wang Q. High-temperature dielectric materials for electrical energy storage. *Annual Review of Materials Research*, 2018; **48**(1): 219–243, doi:10.1146/annurev-matsci-070317-124435.

12. González A, Goikolea E, Barrena JA, and Mysyk R. Review on supercapacitors: Technologies and materials. *Renewable and Sustainable Energy Reviews*, 2016; **58**: 1189–1206, doi:10.1016/j.rser.2015.12.249.

13. Poonam, Sharma K, Arora A, and Tripathi SK. Review of supercapacitors: Materials and devices. *Journal of Energy Storage*, 2019; **21**: 801–825, doi:10.1016/j.est.2019.01.010.

14. Nagaura T. Lithium ion rechargeable battery. *Progress in Batteries & Solar Cells*, 1990; **9**: 209.

15. Liang Y, Zhao C, Yuan H, Chen Y, Zhang W, Huang J, Yu D, Liu Y, Titirici M, Chueh Y, et al. A review of rechargeable batteries for portable electronic devices. *InfoMat*, 2019; **1**(1): 6–32, doi:10.1002/inf2.12000.

16. Miao Y, Liu L, Zhang Y, Tan Q, and Li J. An overview of global power lithium-ion batteries and associated critical metal recycling. *Journal of Hazardous Materials*, 2022; **425**: 127900, doi:10.1016/j.jhazmat.2021.127900.

17. Lander L, Tagnon C, Nguyen-Tien V, Kendrick E, Elliott RJR, Abbott AP, Edge JS, and Offer GJ. Breaking it down: A techno-economic assessment of the impact of battery pack design on disassembly costs. *Applied Energy*, 2023; **331**: 120437, doi:10.1016/j.apenergy.2022.120437.

18. Global EV Outlook. 2021. Paris: International Energy Agency; 2021. Report No.: IEA (2021). https://www.iea.org/reports/global-ev-outlook-2021.

19. Miao Y, Liu L, Zhang Y, Tan Q, and Li J. An overview of global power lithium-ion batteries and associated critical metal recycling. *Journal of Hazardous Materials*, 2022; **425**: 127900, doi:10.1016/j.jhazmat.2021.127900.

20. Wakihara M. Recent developments in lithium ion batteries. *Materials Science and Engineering R Reports*, 2001; **33**(4): 109–134, doi:10.1016/S0927-796X(01)00030-4.

21. Drakopoulos SX, Gholamipour-Shirazi A, MacDonald P, Parini RC, Reynolds CD, Burnett DL, Pye B, O'Regan KB, Wang G, Whitehead TM, et al. Formulation and manufacturing optimization of lithium-ion graphite-based electrodes via machine learning. *Cell Reports Physical Science*, 2021; **2**(12): 100683, doi:10.1016/j.xcrp.2021.100683.

22. Drakopoulos SX, Cowell T, and Kendrick E. Graphite-SiO_x electrodes with a biopolymeric binder for Li-ion batteries: Predicting the cycle life performance from physical properties. *ACS Applied Energy Materials*, 2023; **6**(12): 6543–6553, doi:10.1021/acsaem.3c00488.

23. Lakhdar Y, Geary H, Houck M, Gastol D, Groombridge AS, Slater PR, and Kendrick E. Optimization of electrode and cell design for ultrafast-charging lithium-ion batteries based on molybdenum niobium oxide anodes. *ACS Applied Energy Materials*, 2022; **5**(9): 11229–11240, doi:10.1021/acsaem.2c01814.

24. Park S-W, Ha JH, Cho BW, and Choi H-J. Designing of high capacity Si nanosheets anode electrodes for lithium batteries. *Surface and Coatings Technology*, 2021; **421**: 127358, doi:10.1016/j.surfcoat.2021.127358.

25. Wang T, Shi S, Kong F, Yang G, Qian B, and Yin F. The role of stable interface in nano-sized $FeNbO_4$ as anode electrode for lithium-ion batteries. *Electrochimica Acta*, 2016; **203**: 206–212, doi:10.1016/j.electacta.2016.04.045.

26. Sharma SK, Kim M-S, Kim DY, and Yu J-S. Al nanorod thin films as anode electrode for Li ion rechargeable batteries. *Electrochimica Acta*, 2013; **87**: 872–879, doi:10.1016/j.electacta.2012.09.028.

27. Gao L, Wu G, Ma J, Jiang T, Chang B, Huang Y, and Han S. SnO_2 Quantum Dots@Graphene Framework as a high-performance flexible anode electrode for lithium-ion batteries. *ACS Applied Materials & Interfaces*, 2020; **12**(11): 12982–12989, doi:10.1021/acsami.9b22679.

28. Kundu S, Kraytsberg A, and Ein-Eli Y. Recent development in the field of ceramics solid-state electrolytes: I—Oxide ceramic solid-state electrolytes. *Journal of Solid State Electrochemistry*, 2022; **26**(9): 1809–1838, doi:10.1007/s10008-022-05206-x.

29. Bandhauer TM, Garimella S, and Fuller TF. A critical review of thermal issues in lithium-ion batteries. *Journal of the Electrochemical Society*, 2011; **158**(3): R1, doi:10.1149/1.3515880.

30. Park YJ, Kim MK, Kim HS, and Lee BM. Risk assessment of lithium-ion battery explosion: Chemical leakages. *Journal of Toxicology and Environmental Health Part B*, 2018; **21**(6–8): 370–381, doi:10.1080/10937404.2019.1601815.

31. Harry KJ, Hallinan DT, Parkinson DY, MacDowell AA, and Balsara NP. Detection of subsurface structures underneath dendrites formed on cycled lithium metal electrodes. *Nature Materials*, 2014; **13**(1): 69–73, doi:10.1038/nmat3793.

32. Mu W, Liu X, Wen Z, and Liu L. Numerical simulation of the factors affecting the growth of lithium dendrites. *Journal of Energy Storage*, 2019; **26**: 100921, doi:10.1016/j.est.2019.100921.

33. Fergus JW. Ceramic and polymeric solid electrolytes for lithium-ion batteries. *Journal of Power Sources*, 2010; **195**(15): 4554–4569, doi:10.1016/j.jpowsour.2010.01.076.

34. Famprikis T, Canepa P, Dawson JA, Islam MS, and Masquelier C. Fundamentals of inorganic solid-state electrolytes for batteries. *Nature Materials*, 2019; **18**(12): 1278–1291, doi:10.1038/s41563-019-0431-3.

35. Mei A, Wang X, Feng Y, Zhao S, Li G, Geng H, Lin Y, and Nan C. Enhanced ionic transport in lithium lanthanum titanium oxide solid state electrolyte by introducing silica. *Solid State Ionics*, 2008; **179**(39): 2255–2259, doi:10.1016/j.ssi.2008.08.013.

36. Bohnke O. The fast lithium-ion conducting oxides $Li_{3x}La_{2/3-x}TiO_3$ from fundamentals to application. *Solid State Ionics*, 2008; **179**(1–6): 9–15, doi:10.1016/j.ssi.2007.12.022.

37. Trevey J, Jang JS, Jung YS, Stoldt CR, and Lee S-H. Glass–ceramic Li_2S–P_2S_5 electrolytes prepared by a single step ball billing process and their

application for all-solid-state lithium–ion batteries. *Electrochemistry Communications*, 2009; **11**(9): 1830–1833, doi:10.1016/j.elecom.2009.07.028.

38. Maeder T. Review of Bi_2O_3 based glasses for electronics and related applications. *International Materials Reviews*, 2013; **58**(1): 3–40, doi:10.1179/1743280412Y.0000000010.

39. Kumar B, Thomas D, and Kumar J. Space-charge-mediated superionic transport in lithium ion conducting glass–ceramics. *Journal of the Electrochemical Society*, 2009; **156**(7): A506, doi:10.1149/1.3122903.

40. Kosova NV, Devyatkina ET, Stepanov AP, and Buzlukov AL. Lithium conductivity and lithium diffusion in NASICON-type $Li_{1+x}Ti_{2-x}Al_x(PO_4)_3$ (x = 0; 0.3) prepared by mechanical activation. *Ionics*, 2008; **14**(4): 303–311, doi:10.1007/s11581-007-0197-5.

41. Cheng M, Jiang Y, Yao W, Yuan Y, Deivanayagam R, Foroozan T, Huang Z, Song B, Rojaee R, Shokuhfar T, et al. Elevated-temperature 3D printing of hybrid solid-state electrolyte for Li-ion batteries. *Advanced Materials*, 2018; **30**(39): 1800615, doi:10.1002/adma.201800615.

42. Liang J, Luo J, Sun Q, Yang X, Li R, and Sun X. Recent progress on solid-state hybrid electrolytes for solid-state lithium batteries. *Energy Storage Materials*, 2019; **21**: 308–334, doi:10.1016/j.ensm.2019.06.021.

43. Hu P, Zhang Y, Chi X, Kumar Rao K, Hao F, Dong H, Guo F, Ren Y, Grabow LC, and Yao Y. Stabilizing the interface between sodium metal anode and sulfide-based solid-state electrolyte with an electron-blocking interlayer. *ACS Applied Materials & Interfaces*, 2019; **11**(10): 9672–9678, doi:10.1021/acsami.8b19984.

44. Zhang W, Schröder D, Arlt T, Manke I, Koerver R, Pinedo R, Weber DA, Sann J, Zeier WG, and Janek J. (Electro)chemical expansion during cycling: Monitoring the pressure changes in operating solid-state lithium batteries. *Journal of Materials Chemistry A*, 2017; **5**(20): 9929–9936, doi:10.1039/C7TA02730C.

45. Bucci G, Talamini B, Renuka Balakrishna A, Chiang Y-M, and Carter WC. Mechanical instability of electrode-electrolyte interfaces in solid-state batteries. *Physical Review Materials*, 2018; **2**(10): 105407, doi:10.1103/PhysRevMaterials.2.105407.

46. Dollé M, Sannier L, Beaudoin B, Trentin M, and Tarascon J-M. Live scanning electron microscope observations of dendritic growth in lithium/polymer cells. *Electrochemical and Solid-State Letters*, 2002; **5**(12): A286, doi:10.1149/1.1519970.

47. Porz L, Swamy T, Sheldon BW, Rettenwander D, Frömling T, Thaman HL, Berendts S, Uecker R, Carter WC, and Chiang Y. Mechanism of lithium metal penetration through inorganic solid electrolytes. *Advanced Energy Materials*, 2017; **7**(20): 1701003, doi:10.1002/aenm.201701003.

48. Fan L, Wei S, Li S, Li Q, and Lu Y. Recent progress of the solid-state electrolytes for high-energy metal-based batteries. *Advanced Energy Materials*, 2018; **8**(11): 1702657, doi:10.1002/aenm.201702657.

49. Vadhva P, Hu J, Johnson MJ, Stocker R, Braglia M, Brett DJL, and Rettie AJE. Electrochemical impedance spectroscopy for all-solid-state batteries: Theory, methods and future outlook. *ChemElectroChem*, 2021; **8**(11): 1930–1947, doi:10.1002/celc.202100108.

50. Dyre JC, Maass P, Roling B, and Sidebottom DL. Fundamental questions relating to ion conduction in disordered solids. *Reports on Progress in Physics*, 2009; **72**(4): 046501, doi:10.1088/0034-4885/72/4/046501.

51. Psarras GC. Fundamentals of dielectric theories. In: *Dielectric Polymer Materials for High-Density Energy Storage*. Elsevier; 2018. pp. 11–57. https://linkinghub.elsevier.com/retrieve/pii/B9780128132159000026, doi:10.1016/B978-0-12-813215-9.00002-6.

52. Schadler LS and Nelson JK. Polymer nanodielectrics: Short history and future perspective. *Journal of Applied Physics*, 2020; **128**(12): 120902, doi:10.1063/5.0019865.

53. Lau KY, Vaughan AS, and Chen G. Nanodielectrics: Opportunities and challenges. *IEEE Electrical Insulation Magazine*, 2015; **31**(4): 45–54, doi:10.1109/MEI.2015.7126073.

54. Kremer F and Schönhals A (Eds). *Broadband Dielectric Spectroscopy*. Berlin, Heidelberg: Springer Berlin Heidelberg; 2003. http://link.springer.com/10.1007/978-3-642-56120-7, doi:10.1007/978-3-642-56120-7.

55. Psarras GC. Conductivity and dielectric characterization of polymer nanocomposites. In: *Physical Properties and Applications of Polymer Nanocomposites*. Elsevier; 2010. pp. 31–69. https://linkinghub.elsevier.com/retrieve/pii/B9781845696726500029, doi:10.1533/9780857090249.1.31.

56. Asimakopoulos I, Psarras G, and Zoumpoulakis L. Nanocomposites of barium titanate nanoparticles embedded in thermosetting polymer matrices (novolac resin/unsaturated polyesters/epoxy resin): A comparative study. *ChemEngineering*, 2019; **3**(1): 12, doi:10.3390/chemengineering3010012.

57. Dyre JC and Schrøder TB. Universality of ac conduction in disordered solids. *Reviews of Modern Physics*, 2000; **72**(3): 873–892, doi:10.1103/RevModPhys.72.873.

58. Tsangaris GM, Psarras GC, and Kouloumbi N. Electric modulus and interfacial polarization in composite polymeric systems. *Journal of Materials Science*, 1998; **33**(8): 2027–2037, doi:10.1023/A:1004398514901.

59. Ioannou EF, Mountrichas G, Pispas S, Kamitsos EI, and Floudas G. Lithium ion induced nanophase ordering and ion mobility in ionic block copolymers. *Macromolecules*, 2008; **41**(16): 6183–6190, doi:10.1021/ma8008542.

60. Cole KS and Cole RH. Dispersion and absorption in dielectrics I. Alternating current characteristics. *The Journal of Chemical Physics*, 1941; **9**(4): 341–351, doi:10.1063/1.1750906.

61. Davidson DW and Cole RH. Dielectric relaxation in glycerin. *The Journal of Chemical Physics*, 1950; **18**(10): 1417–1417, doi:10.1063/1.1747496.

62. Havriliak S and Negami S. A complex plane representation of dielectric and mechanical relaxation processes in some polymers. *Polymer*, 1967; **8**: 161–210, doi:10.1016/0032-3861(67)90021-3.

63. Hartmann L, Fukao K, and Kremer F. Molecular dynamics in thin polymer films. In: Kremer F and Schönhals, A (Eds), *Broadband Dielectric Spectroscopy*. Heidelberg, Berlin: Springer Berlin Heidelberg; 2003. pp. 433–473.

64. Böttger H and Bryskin VV. *Hopping Conduction in Solids*. Berlin: Akademie-Verlag; 1985.

65. Angell CA. Why C1 = 16–17 in the WLF equation is physical—and the fragility of polymers. *Polymer*, 1997; **38**(26): 6261–6266.

66. Adam G and Gibbs JH. On the temperature dependence of cooperative relaxation properties in glass-forming liquids. *The Journal of Chemical Physics*, 1965; **43**(1): 139–146, doi:10.1063/1.1696442.

67. Berthier L, Biroli G, Bouchaud J-P, Cipelletti L, Masri DE, L'Hôte D, Ladieu F, and Pierno M. Direct experimental evidence of a growing length scale accompanying the glass transition. *Science*, 2005; **310**(5755): 1797–1800, doi:10.1126/science.1120714.

68. Bauer Th, Lunkenheimer P, and Loidl A. Cooperativity and the freezing of molecular motion at the glass transition. *Physical Review Letters*, 2013; **111**(22): 225702, doi:10.1103/PhysRevLett.111.225702.

69. Paluch M, Casalini R, and Roland CM. Relative contributions of thermal energy and free volume to the temperature dependence of structural relaxation in fragile glass-forming liquids. *Physical Review B*, 2002; **66**(9): 092202, doi:10.1103/PhysRevB.66.092202.

70. Paluch M, Roland CM, Casalini R, Meier G, and Patkowski A. The relative contributions of temperature and volume to structural relaxation of van der Waals molecular liquids. *The Journal of Chemical Physics*, 2003; **118**(10): 4578–4582, doi:10.1063/1.1545449.

71. Roland CM and Casalini R. Temperature dependence of local segmental motion in polystyrene and its variation with molecular weight. *The Journal of Chemical Physics*, 2003; **119**(3): 1838–1842, doi:10.1063/1.1581850.

72. Paluch M. Effect of temperature, pressure and volume on long time relaxation dynamics in fragile glass-forming liquid. *The Journal of Chemical Physics*, 2001; **115**(21): 10029–10035, doi:10.1063/1.1415442.

73. Ngai KL, Casalini R, Capaccioli S, Paluch M, and Roland CM. Do theories of the glass transition, in which the structural relaxation time does not define the dispersion of the structural relaxation, need revision? *The Journal of Physical Chemistry B*, 2005; **109**(37): 17356–17360, doi:10.1021/jp053439s.

74. Atchia S, Gorecki W, Armand M, and Deroo D. Ionic conduction in PEO-perfluorosulphonimide divalent salt complexes. *Electrochimica Acta*, 1992; **37**(9): 1743–1745, doi:10.1016/0013-4686(92)80151-B.

75. Klongkan S and Pumchusak J. Effects of nano alumina and plasticizers on morphology, ionic conductivity, thermal and mechanical properties of PEO-$LiCF_3SO_3$ solid polymer electrolyte. *Electrochimica Acta*, 2015; **161**: 171–176, doi:10.1016/j.electacta.2015.02.074.

76. Pipertzis A, Zardalidis G, Wunderlich K, Klapper M, Müllen K, and Floudas G. Ionic conduction in poly(ethylene glycol)-functionalized hexa-*peri*-hexabenzocoronene amphiphiles. *Macromolecules*, 2017; **50**(5): 1981–1990, doi:10.1021/acs.macromol.7b00224.

77. Pal P and Ghosh A. Broadband dielectric spectroscopy of BMPTFSI ionic liquid doped solid-state polymer electrolytes: Coupled ion transport and dielectric relaxation mechanism. *Journal of Applied Physics*, 2020; **128**(8): 084104, doi:10.1063/5.0022027.

78. Wang Y, Agapov AL, Fan F, Hong K, Yu X, Mays J, and Sokolov AP. Decoupling of ionic transport from segmental relaxation in polymer electrolytes. *Physical Review Letters*, 2012; **108**(8): 088303, doi:10.1103/PhysRevLett.108.088303.

79. Serghei A, Tress M, Sangoro JR, and Kremer F. Electrode polarization and charge transport at solid interfaces. *Physical Review B*, 2009; **80**(18): 184301, doi:10.1103/PhysRevB.80.184301.

80. Ishai PB, Talary MS, Caduff A, Levy E, and Feldman Y. Electrode polarization in dielectric measurements: A review. *Measurement Science and Technology*, 2013; **24**(10): 102001, doi:10.1088/0957-0233/24/10/102001.

81. Emmert S, Wolf M, Gulich R, Krohns S, Kastner S, Lunkenheimer P, and Loidl A. Electrode polarization effects in broadband dielectric spectroscopy. *The European Physical Journal B*, 2011; **83**(2): 157–165, doi:10.1140/epjb/e2011-20439-8.

82. Klein RJ, Zhang S, Dou S, Jones BH, Colby RH, and Runt J. Modeling electrode polarization in dielectric spectroscopy: Ion mobility and mobile ion concentration of single-ion polymer electrolytes. *The Journal of Chemical Physics*, 2006; **124**(14): 144903, doi:10.1063/1.2186638.

83. Coelho R. On the static permittivity of dipolar and conductive media: An educational approach. *Journal of Non-Crystalline Solids*, 1991; **131–133**: 1136–1139, doi:10.1016/0022-3093(91)90740-W.

84. Karan N, Pradhan D, Thomas R, Natesan B, and Katiyar R. Solid polymer electrolytes based on polyethylene oxide and lithium trifluoro- methane sulfonate (PEO–$LiCF_3SO_3$): Ionic conductivity and dielectric relaxation. *Solid State Ionics*, 2008; **179**(19–20): 689–696, doi:10.1016/j.ssi.2008.04.034.

85. Fragiadakis D, Dou S, Colby RH, and Runt J. Molecular mobility, ion mobility, and mobile ion concentration in poly(ethylene oxide)-based polyurethane ionomers. *Macromolecules*, 2008; **41**(15): 5723–5728, doi:10.1021/ma800263b.

86. Fragiadakis D, Dou S, Colby RH, and Runt J. Molecular mobility and Li+ conduction in polyester copolymer ionomers based on poly(ethylene oxide). *The Journal of Chemical Physics*, 2009; **130**(6): 064907, doi:10.1063/1.3063659.

87. Mongcopa KIS, Tyagi M, Mailoa JP, Samsonidze G, Kozinsky B, Mullin SA, Gribble DA, Watanabe H, and Balsara NP. Relationship between segmental dynamics measured by quasi-elastic neutron scattering and conductivity in polymer electrolytes. *ACS Macro Letters*, 2018; **7**(4): 504–508, doi:10.1021/acsmacrolett.8b00159.

88. Utpalla P, Sharma SK, Sudarshan K, Deshpande SK, Sahu M, and Pujari PK. Investigating the correlation of segmental dynamics, free volume characteristics, and ionic conductivity in poly(ethylene oxide)-based electrolyte: A broadband dielectric and positron annihilation

spectroscopy study. *The Journal of Physical Chemistry C*, 2020; **124**(8): 4489–4501, doi:10.1021/acs.jpcc.9b11722.

89. Sinha K and Maranas JK. Segmental dynamics and ion association in PEO-based single ion conductors. *Macromolecules*, 2011; **44**(13): 5381–5391, doi:10.1021/ma2005074.
90. Wang Y, Fan F, Agapov AL, Saito T, Yang J, Yu X, Hong K, Mays J, and Sokolov AP. Examination of the fundamental relation between ionic transport and segmental relaxation in polymer electrolytes. *Polymer*, 2014; **55**(16): 4067–4076, doi:10.1016/j.polymer.2014.06.085.
91. Wang Y and Sokolov AP. Design of superionic polymer electrolytes. *Current Opinion in Chemical Engineering*, 2015; **7**: 113–119, doi:10.1016/j.coche.2014.09.002.
92. Angell CA, Liu C, and Sanchez E. Rubbery solid electrolytes with dominant cationic transport and high ambient conductivity. *Nature*, 1993; **362**(6416): 137–139, doi:10.1038/362137a0.
93. Ferry A, Edman L, Forsyth M, MacFarlane DR, and Sun J. Connectivity, ionic interactions, and migration in a fast-ion-conducting polymer-in-salt electrolyte based on poly(acrylonitrile) and $LiCF_3SO_3$. *Journal of Applied Physics*, 1999; **86**(4): 2346–2348, doi:10.1063/1.371053.
94. Ferry A, Edman L, Forsyth M, MacFarlane DR, and Sun J. NMR and Raman studies of a novel fast-ion-conducting polymer-in-salt electrolyte based on $LiCF_3SO_3$ and PAN. *Electrochimica Acta*, 2000; **45**(8–9): 1237–1242, doi:10.1016/S0013-4686(99)00386-2.
95. Forsyth M, Jiazeng S, and MacFarlane DR. Novel high salt content polymer electrolytes based on high Tg polymers. *Electrochimica Acta*, 2000; **45**(8–9): 1249–1254, doi:10.1016/S0013-4686(99)00328-X.
96. Yamamoto T, Inami M, and Kanbara T. Preparation and properties of polymer solid electrolytes using poly(vinyl alcohol) and thermally resistive poly[arylene(1,3-imidazolidine-2,4,5-trione-1,3-diyl)] as matrix polymers. *Chemistry of Materials*, 1994; **6**(1): 44–50, doi:10.1021/cm00037a011.
97. Wei X and Shriver DF. Highly conductive polymer electrolytes containing rigid polymers. *Chemistry of Materials*, 1998; **10**(9): 2307–2308, doi:10.1021/cm980170z.
98. Dyre JC. The random free-energy barrier model for ac conduction in disordered solids. *Journal of Applied Physics*, 1988; **64**(5): 2456–2468, doi:10.1063/1.341681.
99. Psarras GC. Hopping conductivity in polymer matrix–metal particles composites. *Composites Part A: Applied Science and Manufacturing*, 2006; **37**(10): 1545–1553, doi:10.1016/j.compositesa.2005.11.004.

100. Zardalidis G, Ioannou E, Pispas S, and Floudas G. Relating structure, viscoelasticity, and local mobility to conductivity in PEO/LiTf electrolytes. *Macromolecules*, 2013; **46**(7): 2705–2714, doi:10.1021/ma400266w.

101. Smith JG and Siegel DJ. Low-temperature paddlewheel effect in glassy solid electrolytes. *Nature Communications*, 2020; **11**(1): 1483, doi:10.1038/s41467-020-15245-5.

102. Wang Y, Richards WD, Ong SP, Miara LJ, Kim JC, Mo Y, and Ceder G. Design principles for solid-state lithium superionic conductors. *Nature Materials*, 2015; **14**(10): 1026–1031, doi:10.1038/nmat4369.

103. Dyre JC. Is there a "native" band gap in ion conducting glasses? *Journal of Non-Crystalline Solids*, 2003; **324**(1–2): 192–195, doi:10.1016/S0022-3093(03)00237-0.

104. Schäfer M and Weitzel K-M. Site energy distribution of ions in the potential energy landscape of amorphous solids. *Materials Today Physics*, 2018; **5**: 12–19, doi:10.1016/j.mtphys.2018.05.002.

105. Ohtomo T, Mizuno F, Hayashi A, Tadanaga K, and Tatsumisago M. Mechanochemical synthesis of lithium ion conducting glasses and glass–ceramics in the system Li_2S–P–S. *Solid State Ionics*, 2005; **176**(31–34): 2349–2353, doi:10.1016/j.ssi.2005.03.025.

106. Moon RJ, Martini A, Nairn J, Simonsen J, and Youngblood J. Cellulose nanomaterials review: Structure, properties and nanocomposites. *Chemical Society Reviews*, 2011; **40**(7): 3941, doi:10.1039/c0cs00108b.

107. Wang S, Zhang L, Zeng Q, Liu X, Lai W-Y, and Zhang L. Cellulose microcrystals with brush-like architectures as flexible all-solid-state polymer electrolyte for lithium-ion battery. *ACS Sustainable Chemistry & Engineering*, 2020; **8**(8): 3200–3207, doi:10.1021/acssuschemeng.9b06658.

108. Flouda P, Bukharina D, Pierce KJ, Stryutsky AV, Shevchenko VV, and Tsukruk VV. Flexible sustained ionogels with ionic hyperbranched polymers for enhanced ion-conduction and energy storage. *ACS Applied Materials & Interfaces*, 2022; **14**(23): 27028–27039, doi:10.1021/acsami.2c04502.

109. Liu X, Zhang C, Gao S, Cai S, Wang Q, Liu J, and Liu Z. A novel polyphosphonate flame-retardant additive towards safety-reinforced all-solid-state polymer electrolyte. *Materials Chemistry and Physics*, 2020; **239**: 122014, doi:10.1016/j.matchemphys.2019.122014.

110. Zhao C, Zhao Q, Liu X, Zheng J, Stalin S, Zhang Q, and Archer LA. Rechargeable lithium metal batteries with an in-built solid-state polymer electrolyte and a high voltage/loading Ni-rich layered cathode. *Advanced Materials*, 2020; **32**(12): 1905629, doi:10.1002/adma.201905629.

111. Zhao X, Wang C, Liu H, Liang Y, and Fan L. A review of polymer-based solid-state electrolytes for lithium-metal batteries: Structure, kinetic, interface stability, and application. *Batteries & Supercaps*, 2023; **6**(4): e202200502, doi:10.1002/batt.202200502.

Chapter 4

Organic-Based Batteries for the Future of Energy Storage

Ryota Akai,[a] Norimitsu Tohnai,[a] and Kouki Oka[a,b]
[a]*Graduate School of Engineering, Osaka University U1E-1101,*
2-1 Yamadaoka, Suita, Osaka 565-0871, Japan
[b]*Institute of Multidisciplinary Research for Advanced Materials,*
Tohoku University, 2-1-1 Katahira, Aoba-ku, Sendai, Miyagi 980-8577, Japan
oka@tohoku.ac.jp

4.1 Introduction

Collectively, 195 countries have pledged to realize the goal of reducing 81% of greenhouse gas emissions by 2050, which was signed at the COP21/CMP11 in Paris in 2015 [1]. The shift from reliance on conventional fossil fuels to the utilization of renewable energy sources for power supply is happening steadily, paralleled by the expansion of electric mobility initiatives [1]. Therefore, rechargeable batteries, which can store electrical energy, play an important role in this transition, and their demand is expected to increase exponentially [2–4]. Among rechargeable batteries, lead–acid and lithium-ion batteries accounted for 95% of the U.S. rechargeable battery market share in 2019 [5]. Lead–acid batteries

Batteries: The Future of Energy Storage
Edited by Subashani Maniam

ISBN 978-981-5129-38-0 (Hardcover), 978-1-003-51288-2 (eBook)
www.jennystanford.com

are inexpensive, but their power and energy densities are low, and the lead used in the electrode-active material is classified as highly toxic by many environmental regulations [6]. However, lithium-ion batteries are widely used and can be found in various devices, including mobile phones and electric vehicles, owing primarily to their high energy density [7, 8]. The Nobel Prize in Chemistry for 2019 was awarded to J. B. Goodenough [9], M. S. Whittingham [10], and A. Yoshino [11] for their work on the development of lithium-ion batteries. Furthermore, with the spread of renewable energy, lithium-ion batteries, which can store electrical energy in a relatively eco-friendly manner (unlike lead–acid batteries), are generally considered to contribute toward realizing a fossil-fuel-free society [7, 12]. The demand for lithium-ion batteries is increasing rapidly, and further large consumption of transition metal oxides such as cobalt and nickel used in cathodes is expected to continue [2]. However, transition metal materials have limited resources and a relatively low recyclability [2, 13]. In addition, transition metal materials (e.g., Co, Mn, Ni) are expensive to produce and are responsible for 30–50% of the global warming originating from the battery-manufacturing process [14–16].

To depart from conventional transition-metal-based electrode-active materials and to realize a sustainable society, researchers have now focused on all-organic rechargeable batteries that use organic redox-active materials as electrode-active materials [1, 2, 17]. Organic redox-active materials, which are used as the electrode-active materials, are composed of earth-abundant building blocks (C, H, N, O, and S) and are a promising alternative to conventional transition-metal–based electrodes [1, 2, 17]. Furthermore, almost all-organic redox-active materials can be synthesized under mild conditions of less than 200 °C [18, 19]. The molecular structure of organic redox-active materials can be modified by organic synthesis, and a rational design of these materials may lead to batteries with high energy density [20]. The fundamental research on organic redox-active materials has led to the establishment of a group of materials that exhibit equivalent capacity and cycle life to conventional transition-metal-based electrode-active materials [20]. All-organic rechargeable batteries that use these organic redox-active materials represent a promising next-generation battery concept that overcomes the disadvantages of conventional metal-based batteries (i.e., limited resources, relatively low recyclability, and high toxicity).

In this chapter, Section 4.2 introduces organic redox-active materials used as electrode-active materials, Section 4.3 describes all-organic rechargeable batteries and their recyclability, and Section 4.4 is the conclusion.

4.2 Organic Redox-Active Materials as Electrode-Active Materials

4.2.1 Organic Redox-Active Molecules

Organic redox-active materials facilitate charge storage by undergoing electron-releasing (oxidation) or electron-accepting (reduction) reactions [2]. For example, in the former, a molecule exhibits a reversible reaction by transitioning from a neutral state to a cationic state via oxidation and then transitions back to the neutral state via reduction. However, in the latter, a molecule exhibits a reversible reaction by transitioning from a neutral state to a reduced state via reduction and then back to the neutral state via oxidation. Some organic molecules have bipolar redox moieties that exhibit both oxidation and reduction from the neutral state [21].

The redox properties of organic redox-active materials depend on the redox moieties in the molecule. In the following paragraphs, representative organic redox-active molecules and organic redox-active materials composed of those molecules will be introduced according to the type of redox molecule.

Organic redox-active molecules include redox moieties roughly classified into oxygen-, nitrogen-, sulfur-, and radical-containing moieties. As shown in Fig. 4.1, each molecule has a unique redox reaction (redox mechanism and potential) for each redox moiety, which is accompanied by electron transfer and charge compensation.

Since dichloroisocyanuric acid was used as a cathode-active material for lithium-ion batteries in 1969 [22], quinone derivatives with oxygen-containing moieties have been reported and exhibit a wide range of redox potentials ranging from below +1.0 V (versus Li/Li^+) to above +4.0 V (versus Li/Li^+), depending on the type of molecular structures and their electronic states (Fig. 4.1, blue dots).

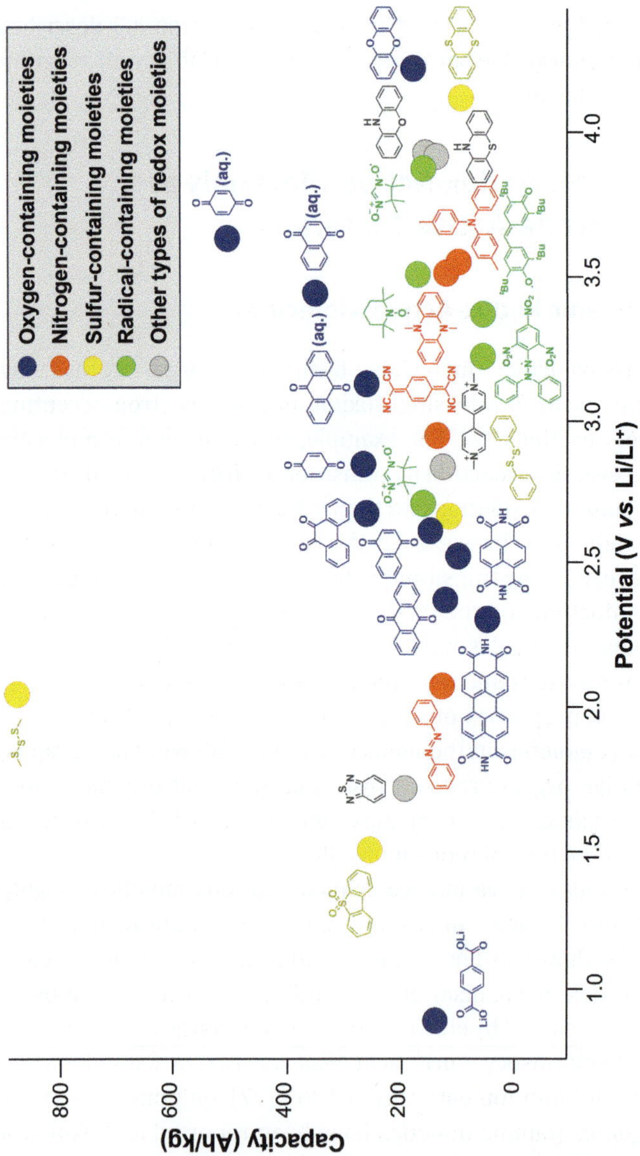

Figure 4.1 Organic redox-active molecules projected on redox potential and capacity axes.

Nitrogen-containing moieties (except those containing radicals) exhibit a wide range of redox potentials ranging from about +2.0 V (versus Li/Li$^+$) to about +4.0 V (versus Li/Li$^+$) (Fig. 4.1, orange dots). Since disulfide-based molecules were first used as cathode-active materials in 1988 [23], sulfur-containing moieties have been aggressively investigated and were found to exhibit a wide range of redox potentials ranging from about +1.5 V (versus Li/Li$^+$) to about +4.5 V (versus Li/Li$^+$), which has been linked to the structural diversity of sulfur atoms that form two or six covalent bonds (Fig. 4.1, yellow dots). Since the initial utilization of sulfur-containing moieties as cathode-active materials in lithium-ion batteries in 2002 [24], nitroxyl radical polymers have paved the way for the synthesis of a diverse array of stable radical molecules and their respective polymers. These materials demonstrated a broad spectrum of redox potentials, ranging approximately from +2.5 V (versus Li/Li$^+$) to +3.5 V (versus Li/Li$^+$) (Fig. 4.1, green dots). The following sections describe each type of redox moiety with examples of representative organic redox-active molecules.

4.2.1.1 Oxygen-containing moieties

Table 4.1 A summary of organic redox-active molecules with oxygen-containing moieties

Entry	Redox-Active Organic Molecule	Redox Potential (V versus Li/Li$^+$)	Redox Capacity (mAh/g)	Electrolyte[a]	Ref.
1	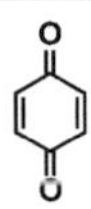	+2.8	248	0.5 M LiTFSI in DMA	[25]
	1,4-Benzoquinone	+3.7	496	1 M *p*-TsOH	[26]
2	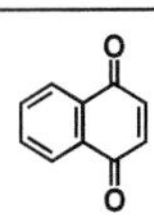	+2.6	169	0.5 M LiTFSI in DMA	[25]
	1,4-Naphthoquinone	+3.4	339	1 M *p*-TsOH	[26]
3	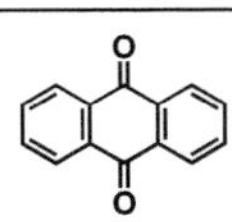	+2.3	129	0.5 M LiTFSI in DMA	[25]
	9,10-Anthraquinone	+3.1	257	1 M *p*-TsOH	[26]

(Continued)

Table 4.1 (*Continued*)

Entry	Redox-Active Organic Molecule	Redox Potential (V versus Li/Li$^+$)	Redox Capacity (mAh/g)	Electrolyte[a]	Ref.
4	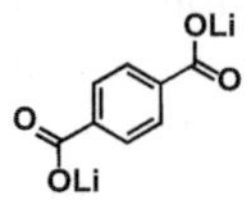 Di-lithium telephthalate	(+) 0.7–1.0	151	1 M $LiPF_6$ in EC/DMC	[27]
5	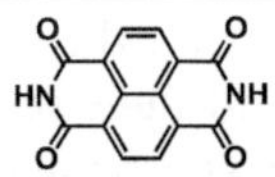 Naphthalene-1,4,5,8-tetracarboxylic diimide	+2.5	101	0.1 M $TBAPF_6$ in CH_2Cl_2	[28]
6	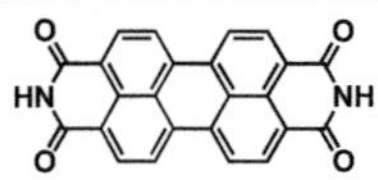 Perylene-3,4,9,10-tetracarboxylic diimide	+2.3	69	1 M $NaPF_6$ in EC/DEC/PC	[29]
7	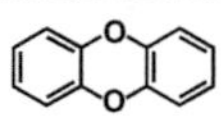 Dibenzo-1,4-dioxin	+4.2	146	5 M $LiClO_4$ in EC/DMC	[30]

[a] LiTFSI: Lithium bis(trifluoromethanesulfonyl)imide, *p*-TsOH: *p*-toluenesulfonic acid, $TBAPF_6$: tetrabutylammonium hexafluorophosphate, DMA: *N,N*-dimethylacetamide, EC: ethylene carbonate, DMC: dimethyl carbonate, DEC: diethyl carbonate, PC: propylene carbonate.

The most common organic redox-active materials are oxygen-containing moieties, which are represented by 1,4-benzoquinone and 1,4-hydroquinone. They are usually characterized by the chemical bistability of their neutral and reduced species. Representative redox-active groups are carbonyl and ether groups, and various carbonyl organic redox-active materials, including quinone derivatives [31–33] (Table 4.1, entries 1–3), carboxylate salts [27, 34, 35] (Table 4.1, entry 4), and imides [28, 36, 37] (Table 4.1, entries 5 and 6).

The carbonyl moiety accepts one electron and undergoes an enolization reaction, causing the excess electron to delocalize on the conjugated structure to form a radical monoanion, which is a reversible redox reaction characterized by electron acceptance [39] (Fig. 4.2). As shown in Fig. 4.3, quinones generally have a one-step two-electron redox reaction in acidic aqueous electrolytes due to a fast proton-coupled electron transfer and a small activation energy needed to reorganize the aromatic ring via an outer-sphere electron transfer reaction [39] (Table 4.1, entries 1–3). Organic redox-active molecules with carbonyl moieties have a wide range of redox potentials ranging from under +2.0 V (versus Li/Li$^+$) to above +4.0 V (versus Li/Li$^+$) and can be used in both cathodes and anodes.

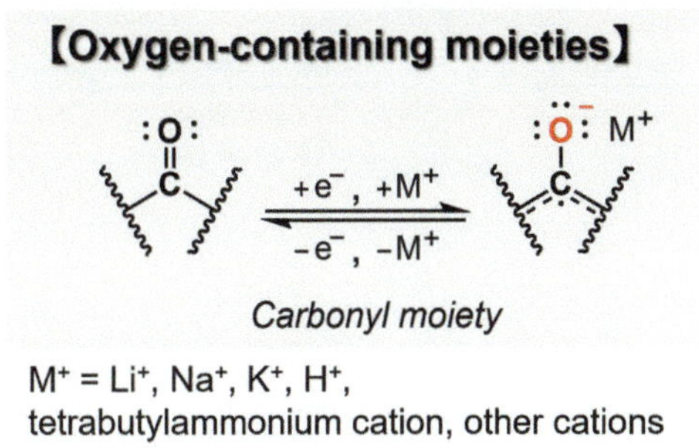

Figure 4.2 Redox reaction of carbonyl moiety.

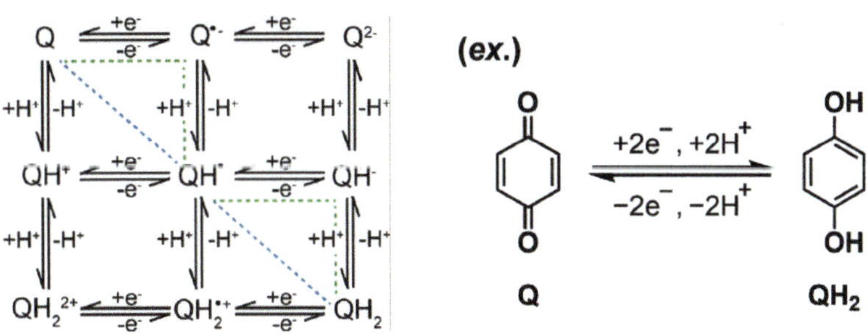

Figure 4.3 Quinone (Q) nine-membered scheme of squares, showing the proton (H$^+$)-dependent stepwise (green dotted line) and concerted (blue dotted line) proton-coupled electron transfer routes. Reprinted with permission from [38]. Copyright 2019 American Chemical Society.

As shown in Fig. 4.4, carboxylate compounds with C=O bonds in the molecular structure can undergo superlithiation, a process characterized by the insertion of lithium ions into the unsaturated C–C double bond sites at negative potentials following the enolization of the carbonyl groups [1, 40, 41]. Di-lithium terephthalate discharges at a negative potential of approximately +0.8 V (versus Li/Li$^+$) in organic electrolytes ($LiPF_6$ in EC/DMC) [27, 34, 35] (Table 4.1, entry 4, and Fig. 4.4). In addition, the maximum theoretical capacity drastically increases to 1809 mAh/g due to superlithiation.

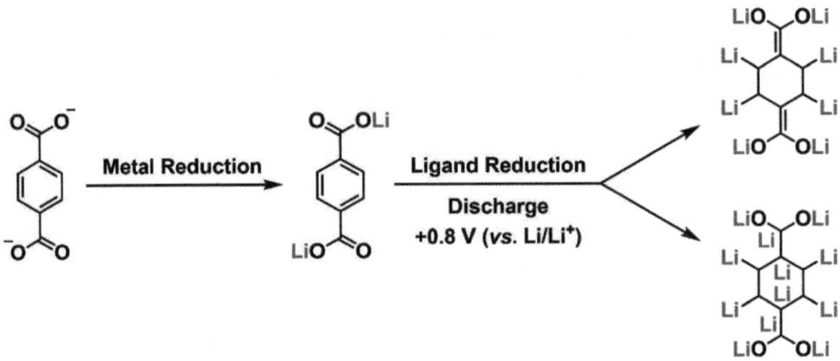

Figure 4.4 Superlithiation mechanism of an anhydrous transition-metal-based coordination compound upon discharging.

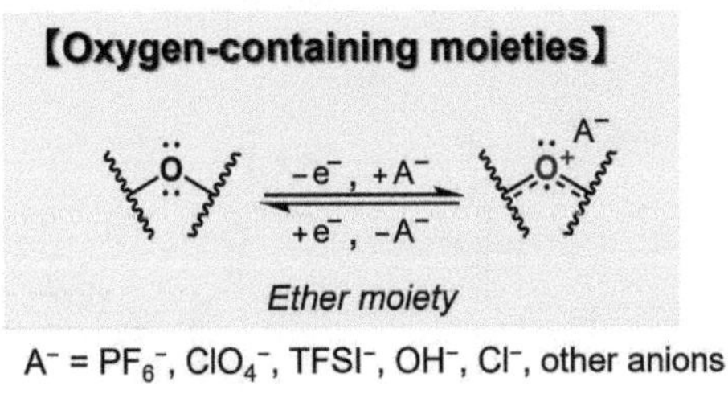

Figure 4.5 Redox reaction of ether moiety.

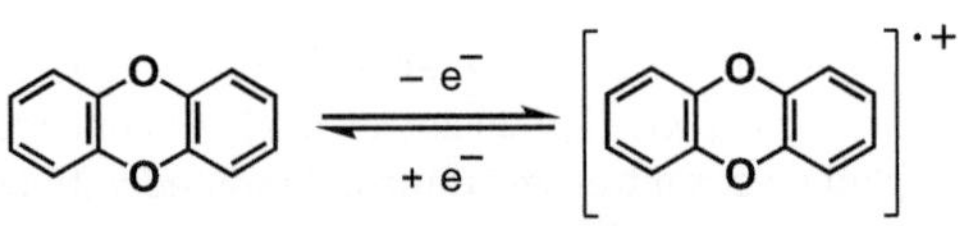

Figure 4.6 Redox reaction of dibenzo-1,4-dioxin.

Ether moiety (C–O–C) reversibly forms a radical cation, causing redox reactions with electron release at relatively positive potentials (Fig. 4.5). For example, dibenzo-1,4-dioxin has been reported to have a redox capability at a very positive redox potential of +4.2 V (versus Li/Li^+) in organic electrolytes ($LiClO_4$ in EC/DMC) [30] (Table 4.1, entry 7 and Fig. 4.6).

4.2.1.2 Nitrogen-containing moieties

Nitrogen-containing moieties have redox moieties such as tertiary amines (Table 4.2, entries 8 and 9), azos (Table 4.2, entry 10), imines, and nitriles [2] (Table 4.2, entry 11). Nitrogen atoms can form three covalent bonds, allowing for a greater variety of redox moieties than oxygen-containing moieties.

Table 4.2 Organic redox-active molecules with nitrogen-containing moieties

Entry	Redox-Active Organic Molecule	Redox Potential (V versus Li/Li^+)	Redox Capacity (mAh/g)	Electrolyte[a]	Ref.
8	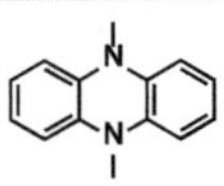 5,10-Dihydro-5,10-dimethylphenazine	+3.6	127	0.1 M $LiClO_4$ in AN	[42]
9	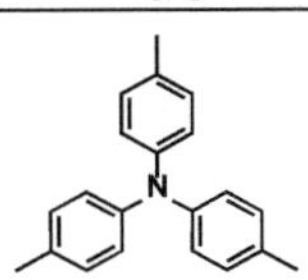 Tri-*N*-tolylaminc	+3.5	93	0.2 M TBAP in THF	[43]
10	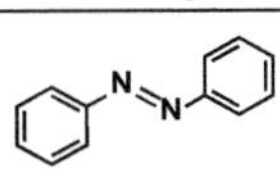 Azobenzene	+2.1	147	0.5 M LiTFSI in DMF	[44]
11	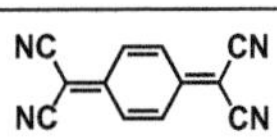 7,7,8,8-Tetracyano-quinodimethane	+3.0	131	0.1 M $LiClO_4$ in AN	[43]

[a] TBAP: Tetrabutylammonium perchlorate, LiTFSI: lithium bis(trifluoromethane-sulfonyl)imide, AN: acetonitrile, THF: tetrahydrofuran, DMF: *N*,*N*-dimethylformamide.

Among redox-active molecules with nitrogen-containing moieties, 5,10-dihydro-5,10-dimethylphenazine [42, 46, 47] (Table 4.2, entry 8) and triphenylamine derivatives [48, 49] (Table 4.2, entry 9), which have tertiary amine moieties, undergo redox reactions with electron release at positive potentials of above +3.5 V (versus Li/Li^+) (Fig. 4.7). As shown in Fig. 4.8, triphenylamine exhibits an irreversible redox reaction because it easily combines with other triphenylmethylamines to form polymers immediately after one-electron oxidation [45]. Tri-*N*-tolylamine with methyl groups introduced into the *para*-position of phenyl groups, which are the polymerization sites, has been reported to have a very positive redox potential of +3.5 V (versus Li/Li^+) in an organic electrolyte (TBAP in THF) [43] (Table 4.2, entry 9).

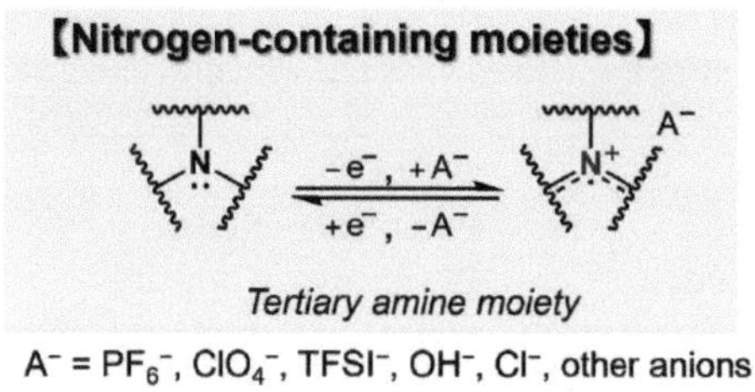

Figure 4.7 Redox reaction of tertiary amine moiety.

Redox moieties of azo (Table 4.2, entry 10), imines, and nitriles (Table 4.2, entry 11) exhibit electron-accepting redox reactions based on multiple bonds between C and N [2] (Fig. 4.9). Azo compounds have been reported to have fast redox responses and high capacities, which originate from the maximum two-electron-accepting redox reaction in an organic electrolyte (LiTFSI in DMF) [50] (Table 4.2, entry 10). Imine compounds are often found in biological charge-transfer systems such as in pteridine and pyridinium derivatives. These systems are characterized by one-electron redox reactions [2, 51] (Fig. 4.9). Among redox molecules with nitrile moiety, the 7,7,8,8-tetracyanoquinodimethane (TCNQ) derivative has been the most investigated (Table 4.2, entry 11). TCNQ undergoes redox reactions by accepting a maximum of two electrons, and owing to the strong electron-accepting characteristics originating from four electron-withdrawing nitrile groups, it exhibits the most positive redox potential of +3.0 V (versus Li/Li^+) in an organic electrolyte ($LiClO_4$ in AN) among organic redox-active materials that contain electron-withdrawing groups [30, 52–54] (Table 4.2, entry 11).

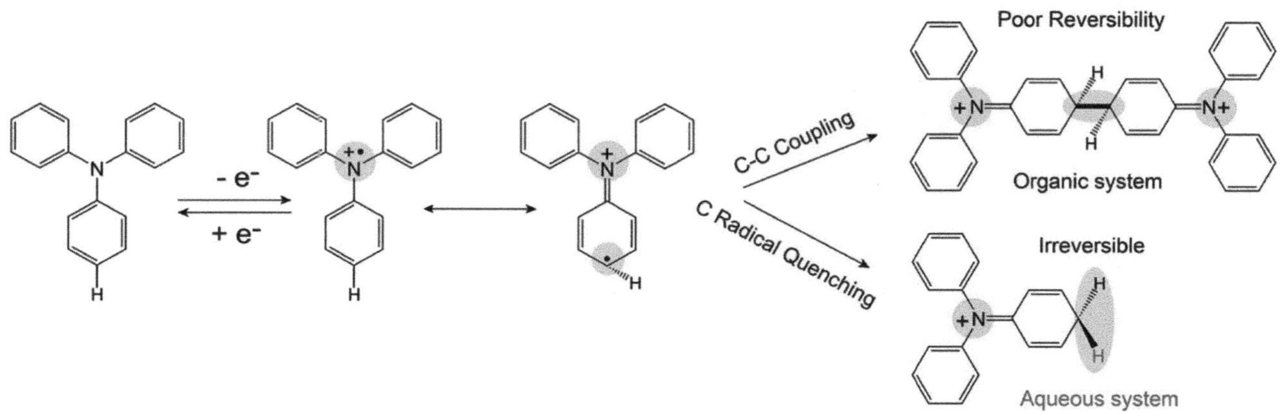

Figure 4.8 Mechanistic origin of triphenylamine redox instability in organic and aqueous media, and radical quenching processes. Reprinted with permission from [45]. Copyright 2021 American Chemical Society.

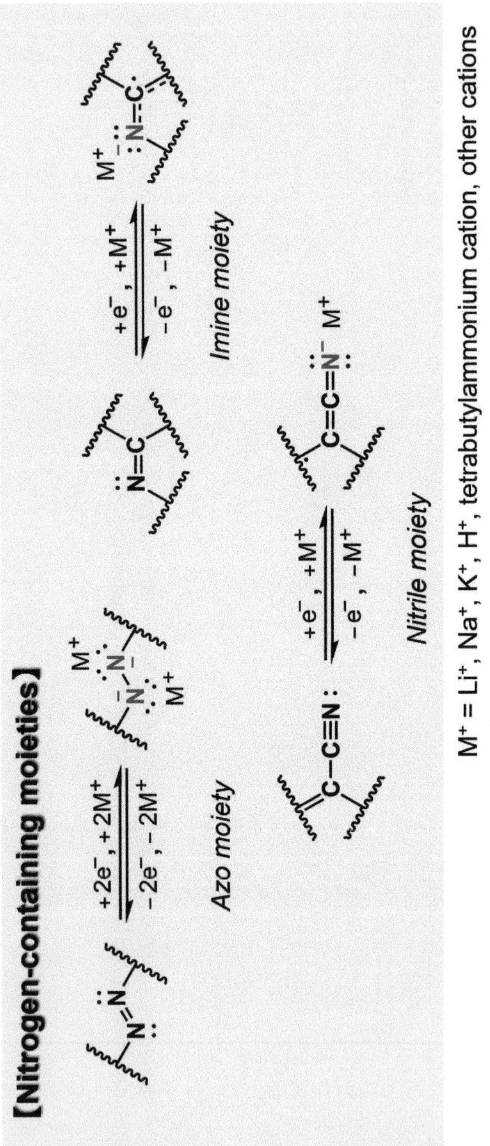

Figure 4.9 Redox reaction of azo, imine, and nitrile moieties.

4.2.1.3 Sulfur-containing moieties

Sulfur-containing moieties exhibit redox reactions similar to those in oxygen-containing moieties because sulfur is a homolog of oxygen. The thioether moiety (R_1–S–R_2) consists of two single bonds between S and C and undergoes electron-release redox reactions with a similar mechanism to ether moiety (Fig. 4.10). Among sulfur-containing moieties, thianthrene exhibits the maximum two-electron-release redox reaction, in which the redox potential in the first step is extremely positive at +4.1 V (versus Li/Li^+) in an organic electrolyte ($TBABF_4$ in AN) [55, 59–61] (Table 4.3, entry 12). The polysulfide accepts the electrons to cleave the S–S bond, which results in stabilization by the counter ion [62] (Fig. 4.10). For example, dimethyl trisulfide exhibits an extremely high theoretical capacity of 849 mAh/g in an organic electrolyte ($LiNO_3$ and LiTFSI in DOL/DME), stemming from a four-electron-accepting redox reaction [57] (Table 4.3, entry 14). Polysulfide has been actively investigated for use as a cathode-active material in Li-sulfur batteries because of its high specific capacity [62–65]. However, the binding and dissociation of polysulfide during redox reactions decrease the redox rate and cause a larger peak separation [66]. Furthermore, a dissolution of redox-active molecules formed by the dissociation of bonds is another problem that prohibits its use as an electrode-active material [66].

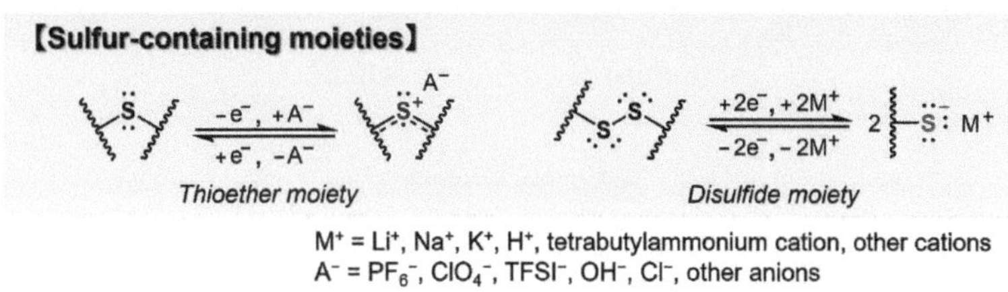

Figure 4.10 Redox reaction of thioether and disulfide moieties.

Dibenzothiophenesulfone with an electron-withdrawing sulfone moiety exhibits a two-electron-accepting redox reaction [58] (Fig. 4.11). In addition, it exhibits one-step two-electron redox reactions in an organic electrolyte ($TBAClO_4$ in AN) at a very negative potential of +1.5 V (versus Li/Li^+) [58] (Table 4.3, entry 15) [67].

Table 4.3 Organic redox-active molecules with sulfur-containing moieties

Entry	Redox-Active Organic Molecule	Redox Potential (V versus Li/Li$^+$)	Redox Capacity (mAh/g)	Electrolyte[a]	Ref.
12	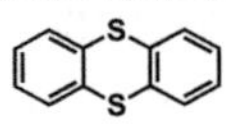 Thianthrene	+4.1	124	0.1 M $TBABF_4$ in AN	[55]
13	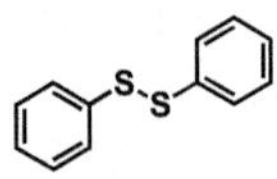 Diphenyl disulfide	+2.7	123	0.2 M $TBAPF_6$ in AN	[56]
14	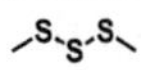 Dimethyl trisulfide	+2.2	849	0.1 M $LiNO_3$ and 1 M LiTFSI in DOL/DME	[57]
15	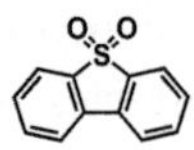 Dibenzothiophene sulfone	+1.5	248	0.1 M $TBAClO_4$ in AN	[58]

[a] $TBABF_4$: Tetrabutylammonium tetrafluoroborate, $TBAPF_6$: tetrabutylammonium hexafluorophosphate, LiTFSI: lithium bis(trifluoromethanesulfonyl)imide, $TBAClO_4$: tetrabutylammonium perchlorate, AN: acetonitrile, DOL: 1,3-dioxolane, DME: 1,2-dimethoxyethane.

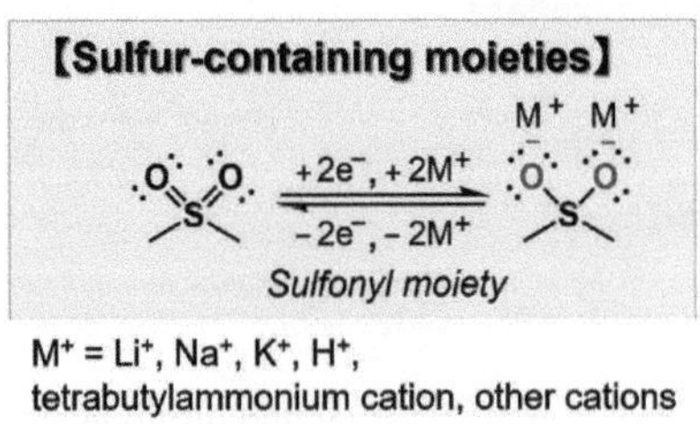

Figure 4.11 Redox reaction of sulfonyl moiety.

4.2.1.4 Radical-containing moieties

Radicals are defined as organic molecules with one or more unpaired electrons and an open-shell electronic configuration [4]. The nitroxide radical, 2,2,6,6-tetramethylpiperidinyloxyl (TEMPO) [4, 55, 71–73] (Table 4.4, entry 16), the nitrogen radical, 1,1-diphenyl-2-picrylhydrazyl [68, 74, 75] (Table 4.4, entry 17), and the oxygen-centered radical, 2,4,6-tri-*t*-butylphenoxy radical [76] (Table 4.4, entry 18) have been reported as stable redox-active moieties. TEMPO is one of the bipolar redox molecules that form an unstable reduced species by accepting one electron, and a relatively stable oxidized species by releasing one electron [4] (Fig. 4.12). TEMPO is often employed as a redox material in electron-release reactions because the oxidation reaction of TEMPO occurs at a very positive redox potential of approximately +3.5 V (versus Li/Li^+) in an organic electrolyte (LiTFSI in EC/DEC) [55] (Table 4.4, entry 16). Most of TEMPO-based redox-active materials are reported as polymers (organic radical polymers), which are made to overcome the high solubility of TEMPO and its derivatives and to create high-power electrode-active materials capable of fast charge transport [4, 55, 71–73]. In addition, as shown in Fig. 4.13, the polymer is reported to facilitate a fast charge transport in its aggregation layer owing to a fast charge transport between the adjacent radical molecules [72].

The galvinoxyl radical has been reported as a redox-active radical-containing moiety that undergoes stable electron-accepting redox reactions [69] (Table 4.4, entry 19). The molecule undergoes a one-electron-accepting redox reaction at +3.4 V (versus Li/Li^+) in an organic electrolyte ($TBAClO_4$ in DMF) and can be used in high-power batteries owing to its fast reaction rate [4, 69].

Table 4.4 Organic redox-active molecules with radical-containing moieties

Entry	Redox-Active Organic Molecule	Redox Potential (V versus Li/Li^+)	Redox Capacity (mAh/g)	Electrolyte[a]	Ref.
16	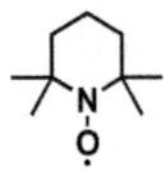 2,2,6,6-Tetramethyl-piperidinyloxyl	+3.5	172	1 M LiTFSI in EC/DEC	[55]
17	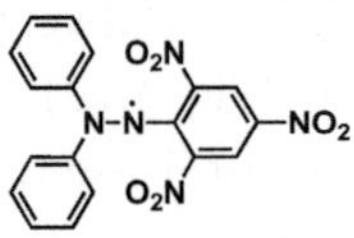 1,1-Diphenyl-2-picryl-hydrazyl radical	+3.3	68	0.33 M KCl in 0.055 M KH_2PO_4 and 0.055 M Na_2HPO_4 (v/v = 1/4) with EtOH 40% (v/v)	[68]
18	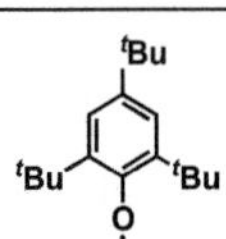 2,4,6-Tri-*t*-butylphenoxy radical	+2.3	181	0.1 M TEAP in DMSO	[43]
19	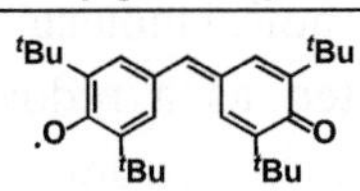 Galvinoxyl radical	+3.4	64	0.1 M $TBAClO_4$ in DMF	[69]
20[b]	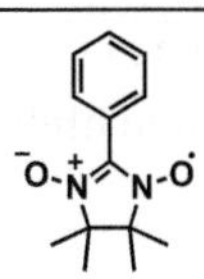 Nitronylnitroxyl radical	+2.7 (*n*) +3.9 (*p*)	103	0.2 M $TBAClO_4$ in AN with 0.01 M TBAOH	[70]

[a] LiTFSI: Lithium bis(trifluoromethanesulfonyl)imide, TEAP: tetraethylammonium perchlorate, $TBAClO_4$: tetrabutylammonium perchlorate, EC: ethylene carbonate, DEC: diethyl carbonate, EtOH: ethanol, DMSO: dimethyl sulfoxide, DMF: *N,N*-dimethylformamide, AN: acetonitrile.

[b] The electrochemical properties were evaluated for a polyvinyl polymer.

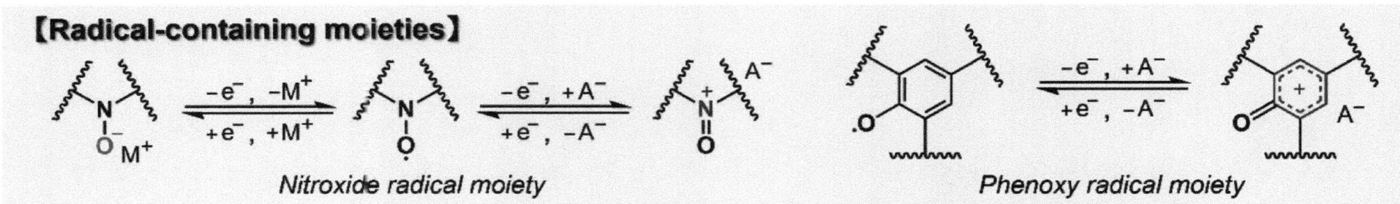

M^+ = Li^+, Na^+, K^+, H^+, tetrabutylammonium cation, other cations
A^- = PF_6^-, ClO_4^-, $TFSI^-$, OH^-, Cl^-, other anions

Figure 4.12 Redox reaction of nitroxide and phenoxy radical moieties.

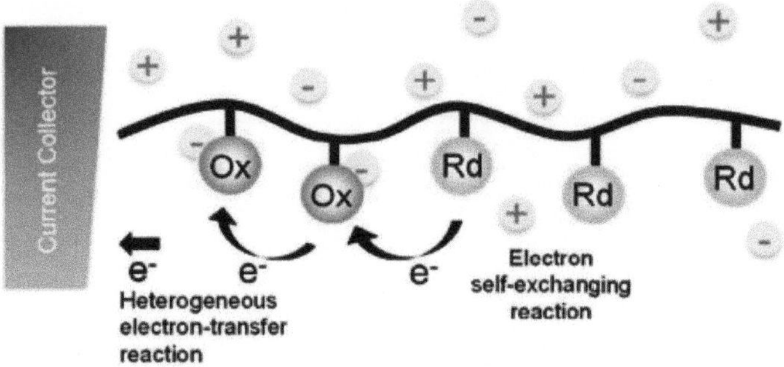

Figure 4.13 Scheme depicting rapid and reversible redox reactions among densely populated redox sites along a polymer backbone. Ox: oxidized molecules. Rd: Reduced molecules. Reprinted with permission from Ref. [71]. Copyright 2022 Royal Society of Chemistry.

In addition, nitronylnitroxide has been reported as a redox-active radical-containing moiety that undergoes both electron-releasing and electron-accepting redox reactions [70] (Table 4.4, entry 20). In one study, as shown in Fig. 4.14, a polymer containing this molecule was reported to undergo one-electron redox reactions at +2.7 V (versus Li/Li^+) and +3.9 V (versus Li/Li^+) in an organic electrolyte ($TBAClO_4$ in AN) [70].

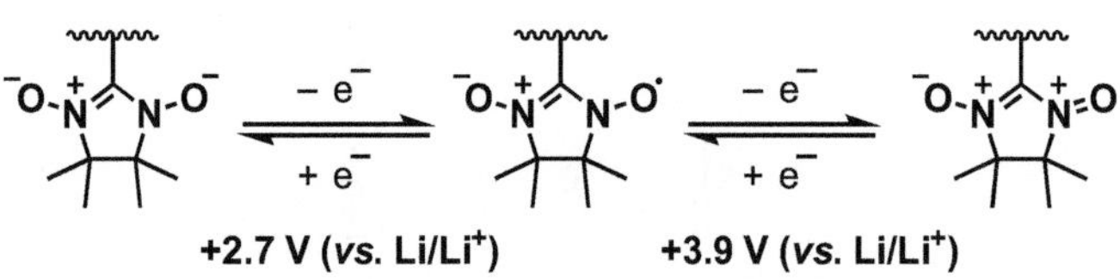

Figure 4.14 Redox reaction of poly(nitronylnitroxylstyrene).

4.2.1.5 Other types of redox moieties

Viologen undergoes reversible one-electron redox reactions, oscillating between dication and cation states at a potential of +3.0 V (versus Li/Li^+) when suspended in an organic electrolyte ($TBAPF_6$ in AN) [77, 82–84] (Table 4.5, entry 21 and Fig. 4.15). As shown in Fig. 4.16, the dication salts of viologen are nearly colorless, contrasting starkly with its deep-blue cation radical salts. These states are thermodynamically stable [85], facilitating a high cyclability in the redox reactions [83]. This, in turn, renders the dication and cation radicals of viologen electrochemically bistable, which allows them to be used in practical applications such as in antiglare rear-view mirrors and the auto-dimming smart window of the Boeing 787 aircraft [83, 84].

Ether-containing organic polymers such as phenoxazine derivatives have been reported to exhibit a redox capability at very positive redox potentials of +3.5 V (versus Li/Li^+) in an organic electrolyte ($TBAPF_6$ in AN), owing primarily to their strong electronegativity [78, 86, 87] (Table 4.5, entry 22).

Table 4.5 Organic redox-active molecules with other types of redox moieties

Entry	Redox-Active Organic Molecule	Redox Potential (V versus Li/Li^+)	Redox Capacity (mAh/g)	Electrolyte[a]	Ref.
21	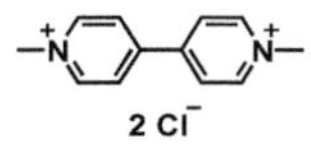 Methyl viologen dichloride	+3.0	104	0.1 M $TBAPF_6$ in AN	[77]
22	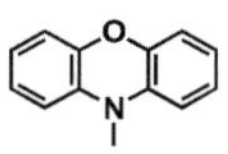 *N*-Methyl-phenoxazine	+3.5	136	0.1 M $TBAPF_6$ in AN	[78]
23	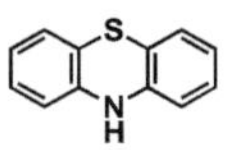 Phenothiazine	+3.9	134	0.1 M $TBAClO_4$ in DMF	[79]
24	 2,1,3-Benzo-thiadiazole	+1.8	136	0.1 M $TBAPF_6$ in AN	[80]

[a] $TBAPF_6$: Tetrabutylammonium hexafluorophosphate, $TBAClO_4$: tetrabutylammonium perchlorate, AN: acetonitrile, DMF: *N*,*N*-dimethylformamide.

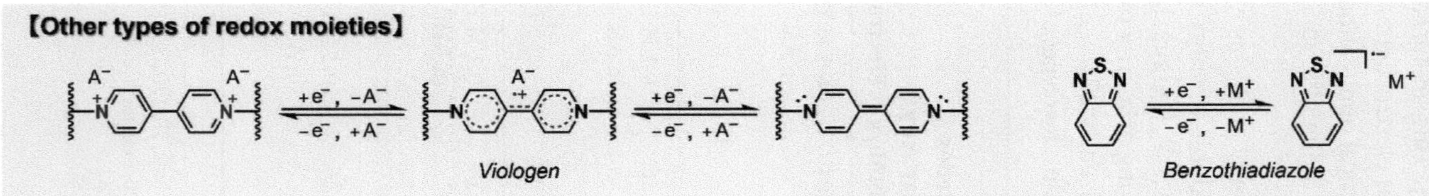

M^+ = Li^+, Na^+, K^+, H^+, tetrabutylammonium cation, other cations
A^- = PF_6^-, ClO_4^-, TFSI⁻, OH^-, Cl^-, other anions

Figure 4.15 Redox reaction of other types of redox moieties.

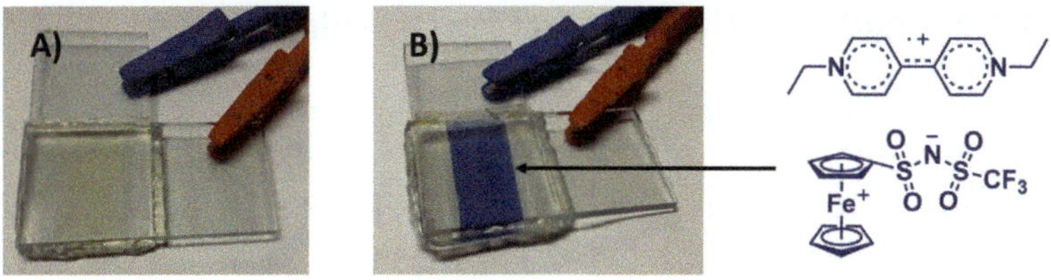

Figure 4.16 Images of [ferrocenylsulfonyl(trifluoromethylsulfonyl)imide anion]/[ethyl viologen] (50 mM/50 mM) in (A) a bleached state (left; 0 V versus Fc/Fc^+) and (B) a colored state (right; +2 V versus Fc/Fc^+). Reprinted with permission from Ref. [81]. Copyright 2017 American Chemical Society.

Phenothiazine is a molecule rich in electrons, a characteristic attributed to its sulfur and nitrogen atoms. It undergoes a one-electron-releasing redox reaction at +3.9 V (versus Li/Li^+) in an organic electrolyte (TBAP in AN) [79] (Table 4.5, entry 23). As shown in Fig. 4.17, for example, poly(3-vinyl-*N*-methylphenothiazine) has been reported to have a high cyclability in redox reactions because adjacent *N*-methylphenothiazine molecules stabilize their charges by π–π interactions [88, 89].

In addition, 2,1,3-benzothiadiazole is a strong electron acceptor with an electron deficiency and, therefore, undergoes a one-electron-accepting redox reaction at very negative potentials of +1.8 V (versus Li/Li^+) in an organic electrolyte ($TBAPF_6$ in AN), exhibiting a redox capacity of 136 mAh/g [90] (Table 4.5, entry 24 and Fig. 4.15).

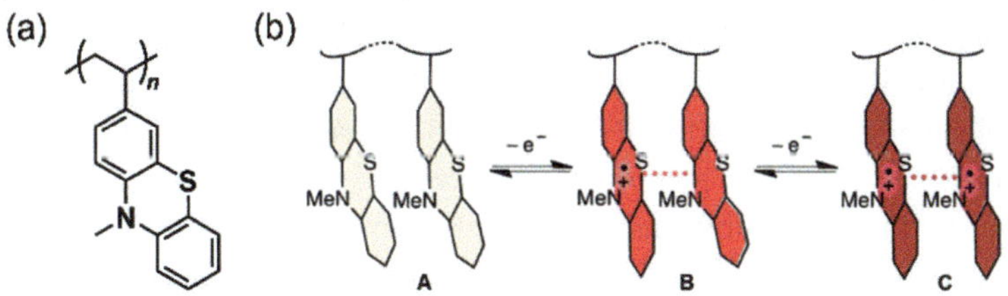

Figure 4.17 (a) Molecular structure of poly(3-vinyl-N-methylphenothiazine). (b) Redox processes of polyvinyl-based redox polymer with a methylpheno-thiazine side group showing the neutral **A**, monoradical **B**, and diradical **C** dimer formation during cycling. Reprinted with permission from Ref. [88]. Copyright 2017 Royal Society of Chemistry.

4.2.2 Redox-Active Conducting Organic Polymers

In 1977, conducting polymers were discovered by H. Shirakawa et al. [98], and in 1981, polyacetylene was used as an electrode-active material in rechargeable batteries [91, 99]. After these discoveries, researchers have been aggressively applying conducting polymers as electrode-active materials by effectively doping and de-doping [92–95]. The fundamental backbones of conducting polymers that have been previously reported can be roughly classified into the following five types (listed in Table 4.6): polyacetylene (PAc) [91], polyparaphenylene (PPP) [92], polyaniline (PAn) [93], polypyrrole (PPy) [94], and polythiophene (PTh) [95].

Among these five types of conducting polymers, PAn, PPy, and PTh are doped and de-doped via electron-release reactions (Table 4.6, entries 27–29), while PAc and PPP are doped and de-doped via both electron release and acceptance reactions (Table 4.6, entries 25 and 26). As shown in Fig. 4.18, the theoretical capacity of each conducting polymer depends on the molecular structure and the associated doping level (0–1), while the doping level controls the redox potential drifts. The theoretical capacities of PAc, PPP, PPy, and PTh are below 150 mAh/g because of low doping levels [91, 92, 94, 95] (Table 4.6, entries 25, 26, 28, and 29). In contrast, PAn has a doping level of 1 and a relatively high theoretical capacity of 295 mAh/g due to the alternating bonding of phenylene and secondary amino groups [93] (Table 4.6, entry 27).

In addition, the representative conducting polymers, poly(3,4-ethylenedioxythiophene) (PEDOT) [96, 101, 102] and polytriphenylamine (PTPAn) [97, 103], have been reported as electrode-active materials (Table 4.6, entries 30 and 31). PEDOT releases electrons at potentials between +2.0 and +4.2 V (versus Li/Li^+) in an organic electrolyte ($LiPF_6$ in EC/DMC) [96] (Table 4.6, entry 30). In addition, PEDOT is very suitable as an electrode-active material because of its high atmospheric stability (when doped) and electrical conductivity [102]. In particular, conductivity can be tuned by adjusting the doping level, the ratio of PEDOT to poly(4-styrenesulfonate) (PSS) as dispersing agent, and its nanomorphology [102].

Table 4.6 Redox-active conducting organic polymers

Entry	Redox-Active Organic Molecule	Redox Potential (V versus Li/Li$^+$)	Redox Capacity (mAh/g)	Electrolyte[a]	Ref.
25	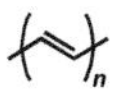 Polyacetylene (PAc)	(+) 0–2.0 (*n*) (+) 3.5–4.0 (*p*)	144	0.5 M $LiClO_4$ in PC	[91]
26	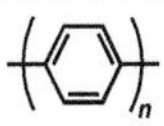 Polyparaphenylene (PPP)	(+) 0–1.0 (*n*) (+) 4.0–4.5 (*p*)	141	1 M $LiPF_6$ in EC/DMC/EMC	[92]
27	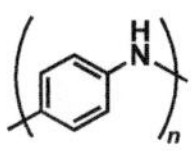 Polyaniline (PAn)	(+) 3.0–4.0	295	1 M $LiClO_4$ in PC	[93]
28	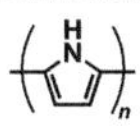 Polypyrrole (PPy)	(+) 3.0–4.0	136	0.1 M KClaq	[94]
29	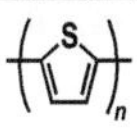 Polythiophene (PTh)	(+) 3.0–4.3	82	1 M $LiPF_6$ in EC/DMC	[95]
30	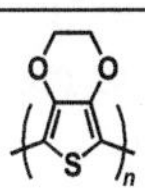 Poly(3,4-ethylenedioxy-thiophene) (**PEDOT**)	(+) 2.0–4.2	45	1 M $LiPF_6$ in EC/DMC	[96]
31	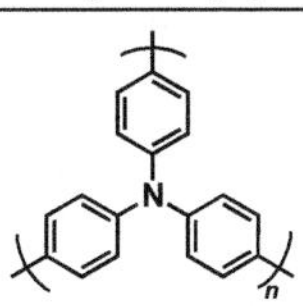 Polytriphenylamine (PTPAn)	+3.7	109	1 M $LiPF_6$ in EC/DMC	[97]

[a] PC: Propylene carbonate, EC: ethylene carbonate, DMC: dimethyl carbonate, EMC: ethyl methyl carbonate.

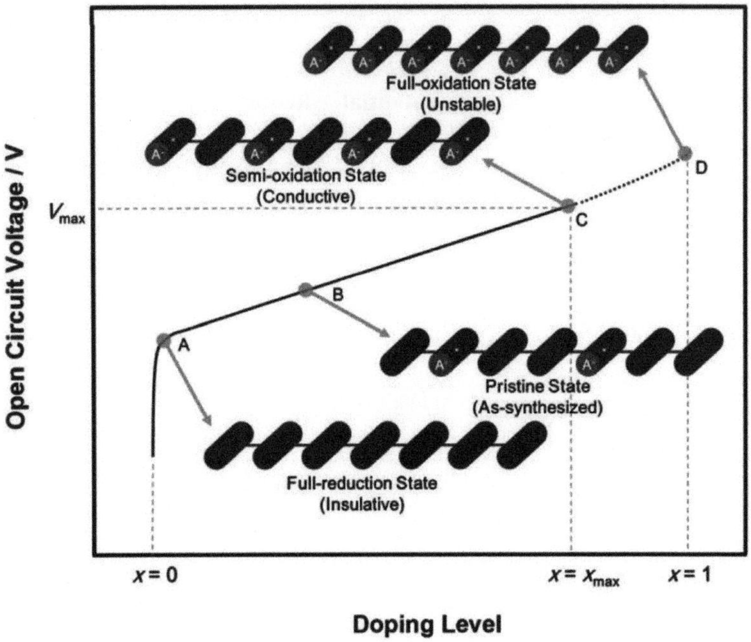

Figure 4.18 Representative electrochemical redox process of an electron-release conducting polymer. Reprinted with permission from Ref. [100]. Copyright 2013 Royal Society of Chemistry.

Polytriphenylamine (PTPAn) releases electrons at potentials between +3.5 and +4.2 V (versus Li/Li^+) in an organic electrolyte ($LiPF_6$ in EC/DMC) [97] (Table 4.6, entry 31). PTPAn has also been investigated as an electrode-active material owing to its excellent morphological and thermal stabilities, as well as charge transport [97, 103].

Since redox-active molecules exhibit negligible electrical conductivity, a conductive additive (e.g., carbon nanotubes) is needed to facilitate an efficient redox reaction. In 2014, a conducting redox polymer containing 1,4-dihydroxybenzene as a redox-active molecule introduced into PPy was reported for the first time [104]. As shown in Fig. 4.19, the conducting redox polymer succeeded in preventing the dissolution of 1,4-hydroxybenzene, which was achieved by introducing this molecule into the conducting polymer while maintaining a high conductivity [104]. Notably, when the number of carbon linkers ($-CH_2-CH_2-$) connecting the conducting polymer backbone and redox-active molecules is two or more, the conductivity of the backbone is not affected by the redox reactions [105]. Therefore, the introduction of a linker is a viable method that facilitates conductivity but does not affect the redox properties of the redox-active molecule or the polymer backbone.

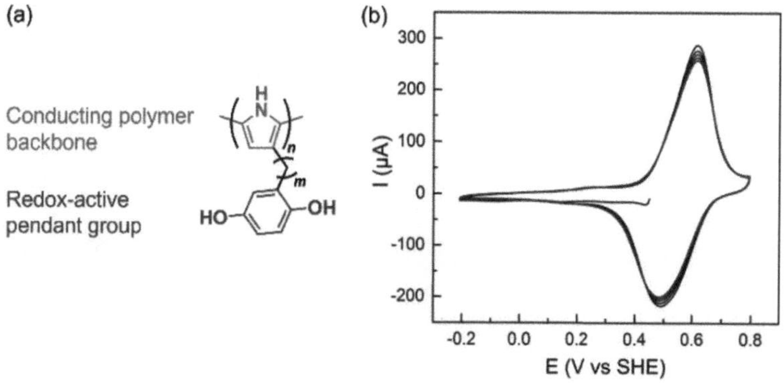

Figure 4.19 (a) Structure of a conducting redox polymer. (b) Cyclic voltammetry (CV) plots of the polymer in aqueous electrolyte buffered at pH 2. First scan in black and subsequent scans in blue [104].

4.3 All-Organic Rechargeable Devices

As shown in Fig. 4.20, a rechargeable battery is composed of two electrodes—a cathode and an anode—and an electrolyte, which is responsible for transporting ions between the electrodes. The cathode-active material releases electrons into the external circuit by undergoing an oxidation reaction during charge. In contrast, the anode-active material accepts electrons from the external circuit by undergoing a reduction reaction during charge. Rechargeable batteries store energy by facilitating reversible redox reactions in the electrode-active materials, a process accompanied by counter-ion compensation from the electrolyte.

For several decades, organic-based rechargeable batteries have attracted attention as a next-generation alternative to metal-based rechargeable batteries [1]. Among organic-based rechargeable batteries, an all-organic rechargeable battery consists of an electrolyte and organic redox-active materials in both electrodes [2, 4]. The electrolyte, a necessary component for improving electrochemical performance, plays an important role in balancing the charge by transporting ions between the cathode and the anode [13]. The type of electrolyte directly affects the properties of a battery (e.g., energy density, cycle life, safety) [13].

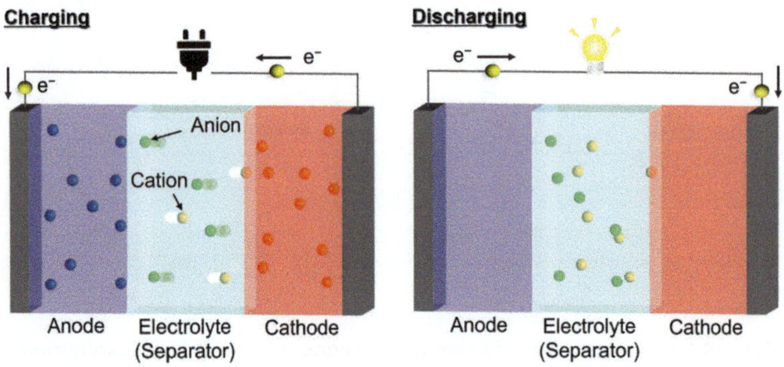

Figure 4.20 Cell configuration of an organic rechargeable battery.

All-organic rechargeable batteries consist of cathodes and anodes that use organic redox-active materials as electrode-active materials, and an electrolyte that transports ions between the cathode and the anode [2, 4, 13] (Fig. 4.20). Section 4.3.1 introduces nonaqueous all-organic rechargeable batteries with organic electrode materials, and Section 4.3.2 introduces aqueous all-organic rechargeable batteries with organic electrode materials. Section 4.3.3 introduces organic redox-flow batteries with redox-active materials. Finally, Section 4.3.4 introduces next-generation recyclable batteries with organic redox materials, focusing on their recyclability or circular economy aspect.

4.3.1 Nonaqueous All-Organic Rechargeable Batteries

All-organic rechargeable batteries originated from the use of conducting polymers as electrode-active materials with charge storage by doping and de-doping. In 1981, MacInnes et al. reported the first all-organic rechargeable battery using PAc, a conducting polymer, as the cathode- and anode-active materials, and an organic electrolyte (0.3 M $LiClO_4$ in THF) [99]. The cathode PAc was doped and de-doped with electron release, and the anode PAc was doped and de-doped with electron acceptance, resulting in an all-organic rechargeable battery with an open-circuit voltage of 2.9 V.

In the same year, the first all-solid-state organic rechargeable battery was reported using PAc in both the cathode- and anode-active materials, and a solid polyethylene oxide (PEO)–NaI complex as the ion carrier (the electrolyte) [106]. The open-circuit voltage was 2.8–3.5 V, obtained at high-temperature conditions (85 °C) to improve the ionic conductivity in the PEO–NaI electrolyte.

In 1983, Kaneto et al. reported, for the first time, the characterization of all-organic rechargeable batteries [107]. The battery using PTh, a conducting polymer, in both the cathode- and anode-active materials exhibited a discharge voltage of 3.1 V, a discharge capacity of 24 mAh/g, and an energy density of 75 Wh/kg in an organic electrolyte (0.2 M $TBABF_4$ in AN).

In 2013, Zhu et al. reported an all-organic rechargeable battery using PPP in both the cathode- and anode-active materials, and an organic electrolyte (1 M $LiPF_6$ in EC/DMC/EMC) [92] (Fig. 4.21a). As shown in Fig. 4.21b, the discharge capacity was 153 mAh/g, which was the highest among all-organic rechargeable batteries that used a conducting polymer as active material in an organic electrolyte at the time, and the discharge voltage was as high as 2.5–4.0 V.

In 2019, Kahsay et al. reported polyisothianaphthene (PITN) as a new conducting polymer [108]. PITN has the smallest band gap of all known conducting polymers—less than 1 eV—compared to 2 eV for PTh, indicating a high conductivity and the ability to both release and accept electrons, resulting in the use of PITN as a cathode- and anode-active material (Fig. 4.22a). As shown in Fig. 4.22b, the all-organic rechargeable battery with PITN in both cathode- and anode-active materials exhibited a discharge capacity of approximately 40 mAh/g. This was the first report that investigated the bipolar function of PITN in batteries containing lithium-based electrolytes.

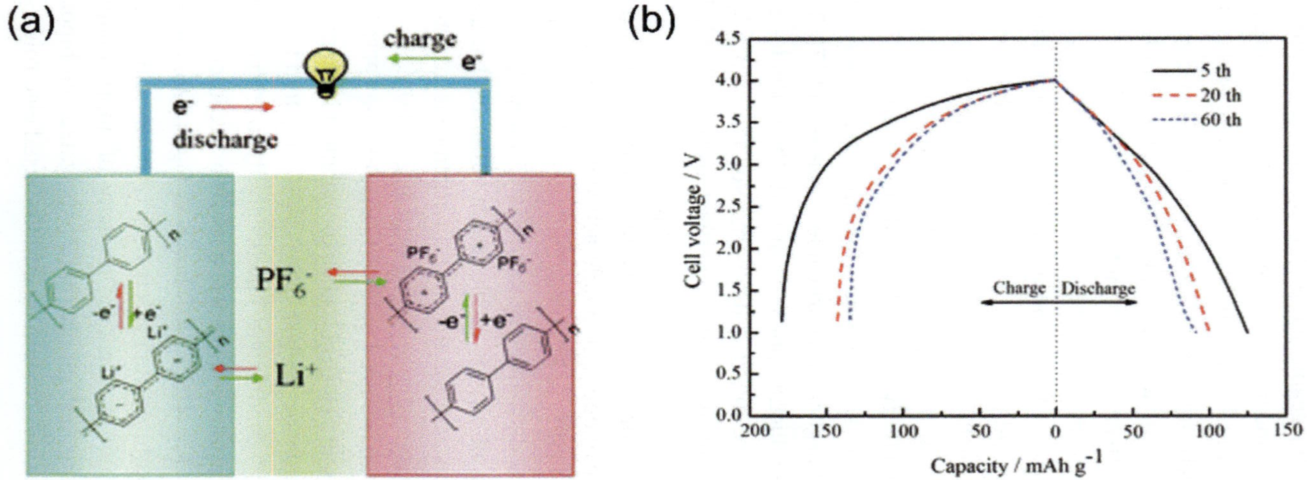

Figure 4.21 (a) Cell configuration of an all-organic rechargeable battery. (b) Charge–discharge profile of coin-type organic batteries using the PPP polymer in both the cathode and anode. The charge–discharge capacities were calculated according to the mass weight of the PPP anode. Charge and discharge currents were set at 40 mA/g. Reprinted with permission from Ref. [92]. Copyright 2013 Royal Society of Chemistry.

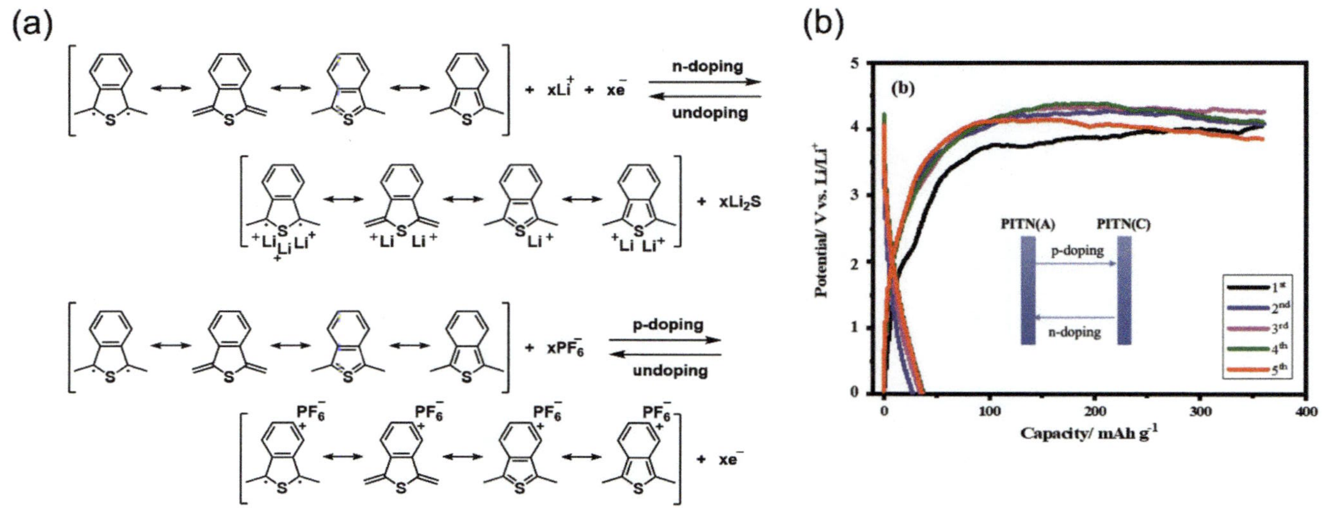

Figure 4.22 (a) Proposed electrochemical reaction mechanism of PITN material in electron-accepting and electron-releasing reactions. (b) Charge–discharge profiles of PITN electrode on electron-doping (PITN/Li; 1.5–4.5 V) and bipolar (PITN/PITN; 0–4 V) modes at 0.1. Reprinted with permission from Ref. [108]. Copyright 2019 Elsevier.

We have described representative examples of all-organic rechargeable batteries that use conducting polymers. Conducting polymers have low capacity, and the shape of the electrode-active material is easily deformed following repeated doping and de-doping and subsequent swelling and shrinkage of the polymer, resulting in poor cycle stability [109]. In addition, the redox potential of the conducting polymer significantly depends on the doping level, and as shown in Fig. 4.23, batteries with conducting polymers tend to have a wide range of nonconstant discharge voltages [109]. Conducting polymers have been investigated for applications in supercapacitors and hybrid battery capacitors rather than rechargeable batteries because of their high conductivity in a wide range of potentials (from (+) 2.0–4.0 V (versus Li/Li^+)), high power density, and fast charge–discharge capability [109].

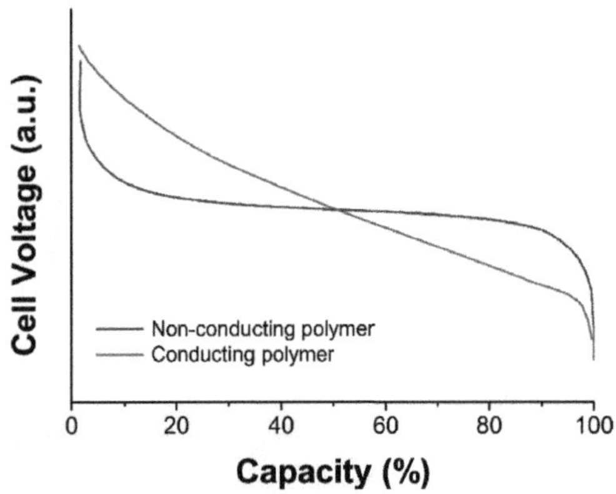

Figure 4.23 Representative discharge voltage profiles of all-polymer batteries based on either conjugated or non-conjugated redox polymers [109].

In contrast to conducting polymers, we will also introduce examples of organic redox-active materials (polymers) that use redox-active molecules as electrodes and will briefly describe their properties. Redox-active molecules are readily soluble in electrolyte solvents (e.g., AN, THF). Polymerization of redox-active molecules, performed by the incorporation of the molecules into an aliphatic polymer backbone, prevents their dissolution into the electrolyte while maintaining their redox properties.

In 2009, Suga et al. reported the first all-organic rechargeable battery based on redox polymers with redox-active molecules introduced into an aliphatic polymer backbone (e.g., polyethylene), which were used in both cathode- and anode-active materials [69] (Fig. 4.24). The rechargeable battery consisted of poly(TEMPO-substituted norbornene) as the cathode-active material, which has TEMPO as the redox-active molecule that releases an electron, and poly(galvinoxylstyrene) as the anode-active material, which has galvinoxylradical as the redox-active molecule that accepts an electron (Fig. 4.24). The all-organic rechargeable batteries with these electrodes and organic electrolyte (0.5 M $TBAClO_4$ and 0.01 M TBAOH in AN) exhibited a constant discharge voltage of 0.66 V, in good agreement with the difference in redox potentials of the respective polymers. The discharge capacity was 28 mAh/g, and in charge–discharge tests, more than 70% of the capacity was retained after 250 cycles. As shown in Fig. 4.24, the redox-active molecules were densely aggregated by polymerization, which facilitated charge transfer between adjacent molecules, resulting in high cycle stability. In addition, the aggregation of redox-active molecules allowed a rapid charging and indicated rate capabilities of 360 C.

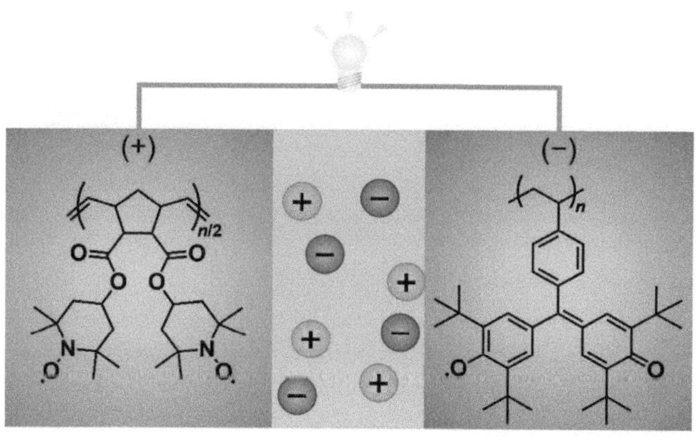

Figure 4.24 Cell configuration of an all-organic rechargeable battery composed of redox polymers.

In 2011, an all-organic rechargeable battery based on poly(nitronylnitroxylstyrene) (PNN) (Fig. 4.25a) was reported by Suga et al. [70]. PNN, a redox polymer exhibiting bipolar redox reactions, was used in both cathode- and anode-active materials, together with an organic electrolyte (0.5 M $TBAClO_4$ and 0.01 M TBAOH in AN). As shown in Figs. 4.25b and 4.25c, this bipolar battery

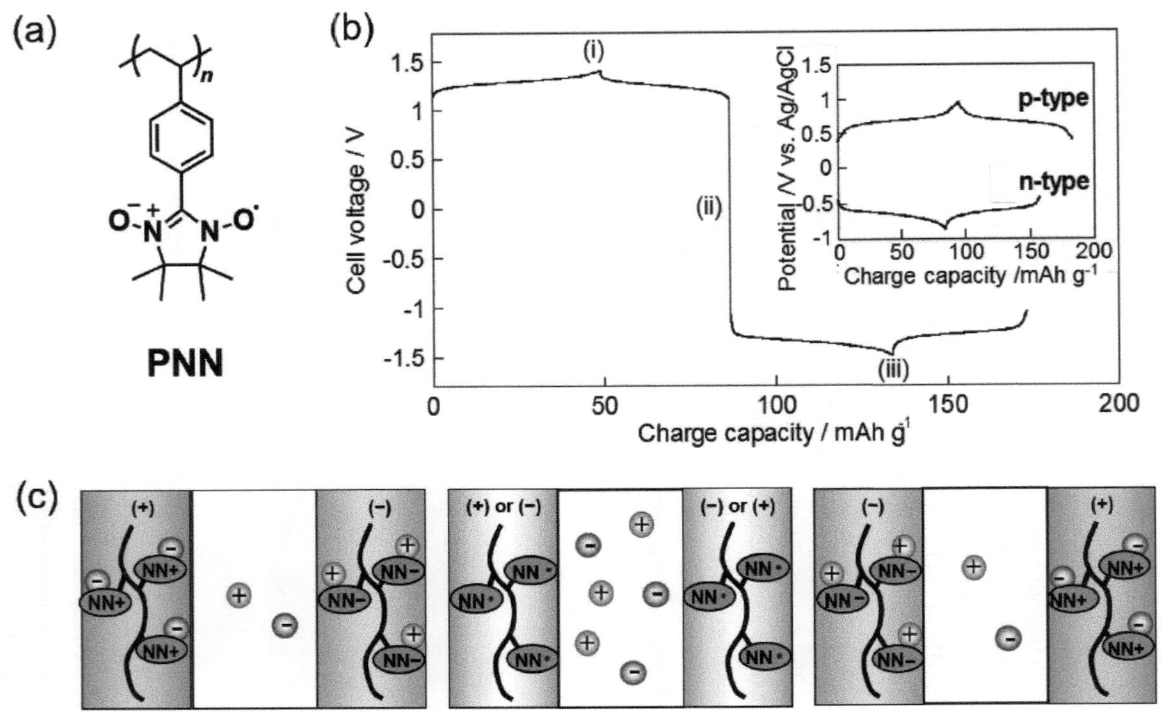

Figure 4.25 (a) Structure of poly(nitronylnitroxylstyrene) (PNN), a cathode- and anode-active material. (b) Charge–discharge curves of a totally organic battery composed of poly(nitronylnitroxylstyrene) in both cathode and anode, performed at a current density of 10 C. Inset: Chronopotentiogram of the half-cell of the polymer electrode. (c) Schematic representation of charged and discharged states.

exhibited a discharge voltage of 1.3 V, in good agreement with the difference in redox potentials of the cathode- and anode-active materials, and a discharge capacity of approximately 42 mAh/g.

In 2018, Oka et al. [58] reported an all-organic rechargeable battery with poly(TEMPO-substituted methacrylate) as the cathode-active material, which undergoes a redox reaction by releasing an electron (Fig. 4.26a), and poly(vinyldibenzothiophenesulfone) as the anode-active material, which undergoes a redox reaction by accepting an electron (Fig. 4.26b), combined with a metal-free organic electrolyte. The discharge voltage of the rechargeable battery was constant at 2.6 V, which was the highest among all-organic rechargeable batteries, and the discharge capacity was about 180 mAh/g (Fig. 4.26c). As shown in Fig. 4.26c, in addition, the high capacity was maintained (>95%) even after 100 charge–discharge cycles.

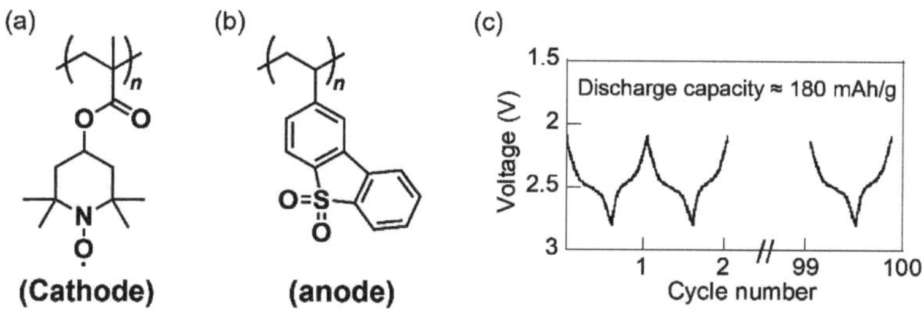

Figure 4.26 (a) Structure of poly(TEMPO-substituted methacrylate), a cathode-active material. (b) Structure of poly(vinyldibenzothiophenesulfo ne), an anode-active material. (c) CV plot at a scan rate of 10 mV/s. Charge and discharge capacities and coulombic efficiency for the first 100 cycles of galvanostatic cycling [58].

In 2022, an all-organic rechargeable battery was reported that contained poly(TEMPO methacrylate$_{0.95}$-co-anthraquinone$_{0.05}$)$_{1610}$ as the cathode-active material, which undergoes a redox reaction by releasing an electron (Fig. 4.27a), and silver terephthalate as the anode-active material, which undergoes a redox reaction by accepting an electron (Fig. 4.27b), combined with an organic electrolyte (3.5 M LiTFSI in TEGDME) [110]. As shown in Fig. 4.27c, the discharge voltage of the rechargeable battery was 2.5–2.8 V, which was the highest among all-organic rechargeable batteries that contain redox polymer or organic salts as the electrode-active materials and a metal-containing electrolyte.

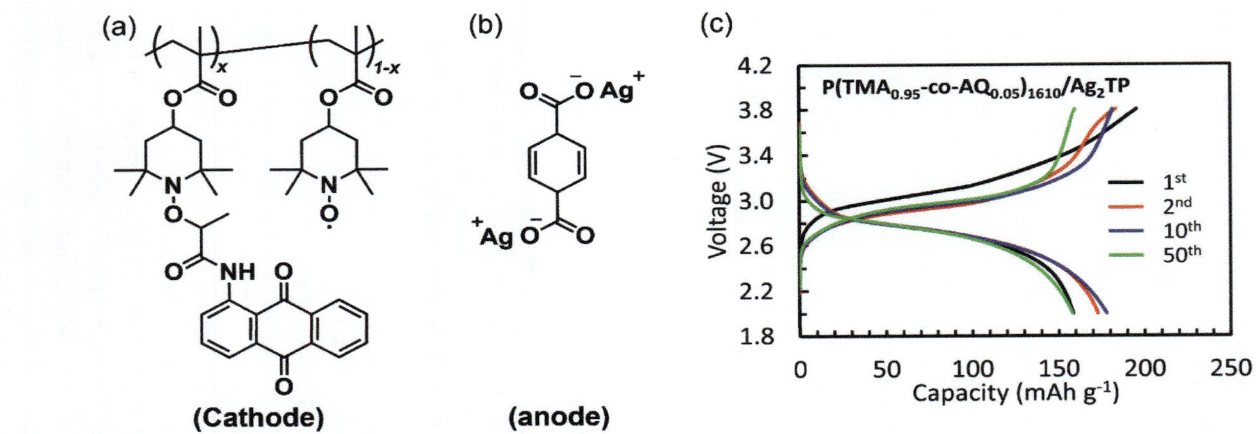

Figure 4.27 (a) Structure of poly(TEMPO methacrylate-co-anthraquinone), a cathode-active material. (b) Structure of silver terephthalate, an anode-active material. (c) Charge–discharge voltage profile of full organic batteries with the constitution of $P(TMA_{0.95}\text{-co-}AQ_{0.05})_{1610}/Ag_2TP$. Reprinted with permission from Ref. [110]. Copyright 2022 Elsevier.

In 2019, Hatakeyama-Sato et al. reported the first all-solid-state organic-polymer-based rechargeable battery that used redox polymers as electrode-active materials with solid imidazolium-substituted polyepichlorohydrin as an ion carrier [111] (Fig. 4.28). The rechargeable battery consisted of a poly(TEMPO-substituted glycidyl ether) as the cathode-active material, which undergoes a redox reaction by releasing an electron, and a poly(viologen) derivative as the anode-active material, which undergoes a redox reaction by accepting an electron. The solid-state organic-polymer-based rechargeable battery having these electrodes and a solid electrolyte was a flexible cell with polymer nanosheets of 2 μm thickness, exhibiting discharge voltages of 1.0 and 1.6 V, and a total discharge capacity of about 80 mAh/g. The rechargeable battery retained approximately 86% of its initial discharge capacity after 25 cycles with rate capabilities of 1 C in charge–discharge cycles.

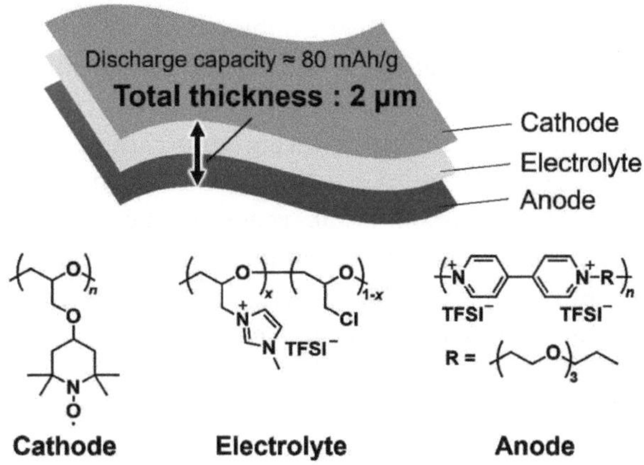

Figure 4.28 Cell configuration of an all-solid-state organic-polymer-based rechargeable battery.

In 2018, Xie et al. reported an all-organic rechargeable battery consisting of a single polymer of poly(2,3-dithiino-1,4-benzoquinone) (PDB) as the cathode- and anode-active materials [112] (Figs. 4.29a and 4.29b). The rechargeable battery used unlithiated PDB as the cathode-active material and superlithiated PDB (details are shown in Section 4.2.1) as the anode-active material, combined with an organic electrolyte (1 M LiTFSI in DOL/DME), resulting in a discharge voltage of 1.0–2.0 V and a discharge capacity of 220 mAh/g. The discharge capacity after 250 cycles was 119 mAh/g, which was very high among all-organic rechargeable batteries.

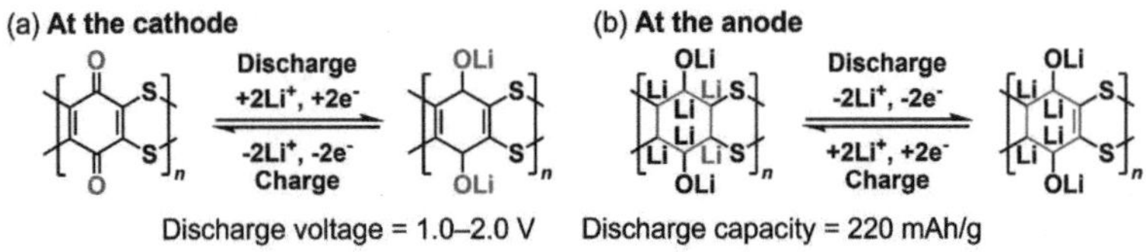

Figure 4.29 (a, b) Proposed redox mechanism of PDB used as active bifunctional electrodes (unlithiated PDB as the cathode and superlithiated PDB as the anode).

Redox polymers, as described earlier, possess a redox capability with redox potentials and reaction kinetics that depend on the type of redox-active molecules. They exhibit relatively high capacities approaching their theoretical values. However, since the backbone and redox-active molecules possess negligible conductivity, a conductive additive (e.g., active carbon and carbon nanotube) is needed for the efficient redox reaction of redox polymers.

In 2017, Emanuelsson et al. reported the first all-organic rechargeable battery having a conductive redox polymer (further details are provided in Section 4.2.2) with a redox-active molecule introduced into the conducting polymer backbone as the active material for both electrodes [113] (Fig. 4.30a). The rechargeable battery consisted of PEDOT-BQ, which was prepared by introducing 1,4-benzoquinone (BQ), a redox-active molecule that releases an electron, into PEDOT, to serve as the cathode-active material; PEDOT-AQ, which was prepared by introducing 9,10-anthraquinone (AQ), a redox-active molecule that accepts an electron, into PEDOT, to serve as the anode-active material; and an organic electrolyte (2-fluoropyridine). As shown in Fig. 4.30b, the rechargeable battery exhibited a discharge voltage of 0.4–0.6 V and a discharge capacity of 110 mAh/g without any conductive additives. This study demonstrates the possibility of fabricating all-organic rechargeable batteries without any conductive additives.

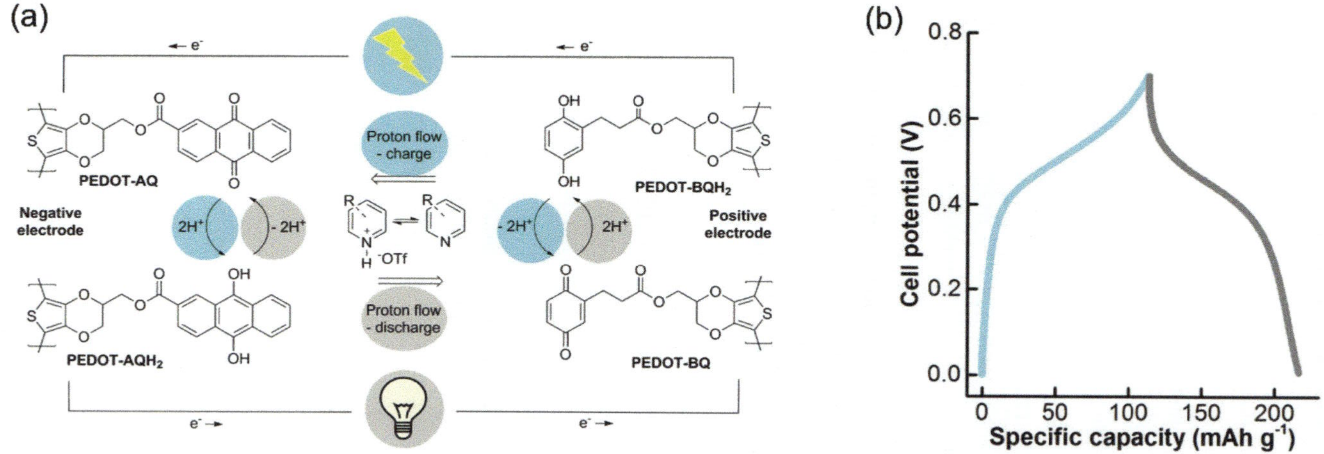

Figure 4.30 Depiction of the conducting redox polymers (electrode materials), the pyridine-based proton donors and acceptors (electrolyte), and the redox reactions that occur during charge (cyan) and discharge (gray). (a) Chemical structures of the anode-active material PEDOT-AQ (top left) and the cathode-active material PEDOT-BQH_2 (top right) in the initial discharged state. The redox reactions during charging are shown in cyan as the AQ form is reduced to AQH_2 (bottom left), while, simultaneously, BQH_2 is oxidized to BQ (bottom right), ending in the charged state. Pyridinium and pyridine electrolytes, seen in the middle, enable the transfer of protons from the cathode to the anode, while the electrons are transported in the external circuit. During discharge, the opposite reactions occur, as shown in gray, resulting in a net flow of protons toward the cathode. (b) Charge and discharge processes shown as the output from a constant-current, two-electrode, galvanostatic, charge–discharge experiment, showing plateaus at cell potentials where the redox reactions occur. Reprinted with permission from Ref. [113]. Copyright 2017 American Chemical Society.

4.3.2 Aqueous All-Organic Rechargeable Batteries

In 1968, Surville et al. reported the first aqueous all-organic rechargeable battery that used PAn as the cathode- and anode-active materials and an aqueous electrolyte (sulfuric acid, pH = 0) [114]. The all-organic rechargeable battery exhibited a discharge capacity of 13 mAh/g, owing to doping and de-doping of PAn with electron release at the cathode and electron acceptance at the anode.

In 2013, an all-organic rechargeable battery that used redox polymers as the electrode-active materials and an aqueous electrolyte was reported for the first time [115]. The rechargeable battery used poly(2,2,6,6-tetramethylpiperidinyloxy-4-yl acrylamide) (PTAm) as the cathode-active material, which undergoes an electron-release redox reaction, and poly(tripyridiniomesitylene) (PTPM), as the anode-active material, which undergoes an electron-acceptance redox reaction (Figs. 4.31a and 4.31b). As shown in Fig. 4.31c, the rechargeable battery with these electrodes and an aqueous electrolyte (0.1 M NaCl(aq)) exhibited two discharge voltages, 1.1 and 1.5 V, depending on the redox state of PTPM. As shown in Figs. 4.31c and 4.31d, the rechargeable battery had a discharge capacity of 165 mAh/g, which was consistent with the theoretical capacity based on a two-electron-accepting redox reaction of the PTPM. As shown in Fig. 4.31e, the rechargeable battery retained 80% of its discharge capacity after 2000 cycles at a rate capability of 60 C in charge–discharge tests with cut-off voltages of 0.3 and 2.0 V.

In 2020, an aqueous all-organic rechargeable battery was reported that was based on conductive redox polymers, PEP(QH_2)E, a PEDOT derivative with hydroquinone as the cathode-active material, and PEP(NQ)E, a PEDOT derivative with 1,4-naphthoquinone, as the anode-active material, and an aqueous electrolyte (0.5 M H_2SO_4) (Figs. 4.32a and 4.32b) [116]. The rechargeable battery exhibited a discharge voltage of 0.4 V and a discharge capacity of 60 mAh/g. The discharge capacity after 500 cycles at 3 C was 51 mAh/g, 85% of the initial capacity.

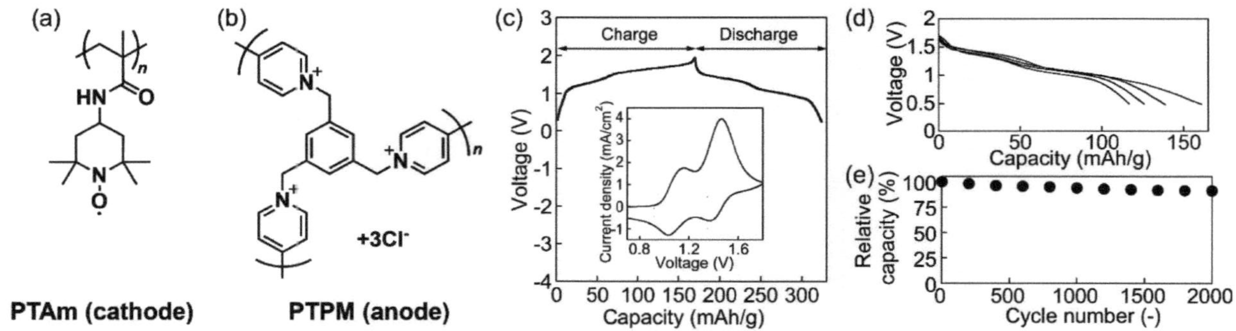

Figure 4.31 (a) Structure of poly(2,2,6,6-tetramethylpiperidinyloxy-4-yl acrylamide), a cathode-active material. (b) Structure of poly(tripyridiniomesitylene), an anode-active material. (c) Charge and discharge curves of a PTPM/PTAm cell with an aqueous NaCl electrolyte at 5 A/g (60 C) (inset shows a cyclic voltammogram of the PTPM/PTAm cell at a scan rate of 50 mV/s). (d) Discharge curves of the PTPM/PTAm cell with the aqueous NaCl electrolyte at discharge rates of 21, 42, 105, and 210 A/g (120–1200 C). (e) Cycle performance of the PTPM/PTAm cell with respect to the discharge capacity. Reprinted with permission from Ref. [115]. Copyright 2013 American Chemical Society.

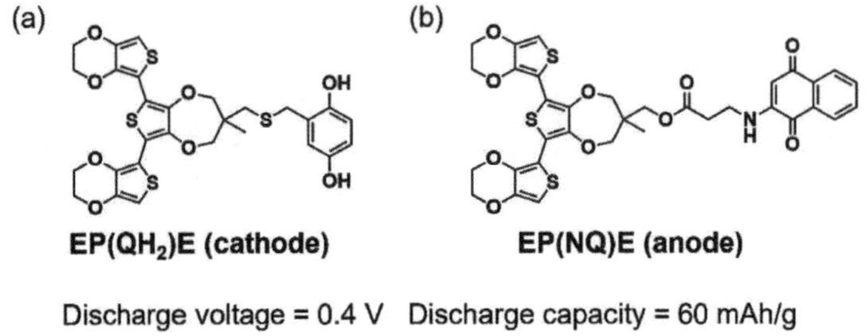

Figure 4.32 (a) Structure of EP(QH$_2$)E, a cathode-active material. (b) Structure of EP(NQ)E, an anode-active material.

Being developed further from aqueous all-organic rechargeable batteries, rechargeable organic–air batteries have been reported using redox polymers as anode-active materials [117, 118]. These batteries were found to exhibit much better cycling performance than metal–air batteries. In addition, an all-solid-state organic rechargeable battery has also been reported using a proton exchange membrane as the electrolyte [119] (Figs. 4.33a and 4.33b), which is expected to undergo additional improvements toward increasing its cyclability and efficiency.

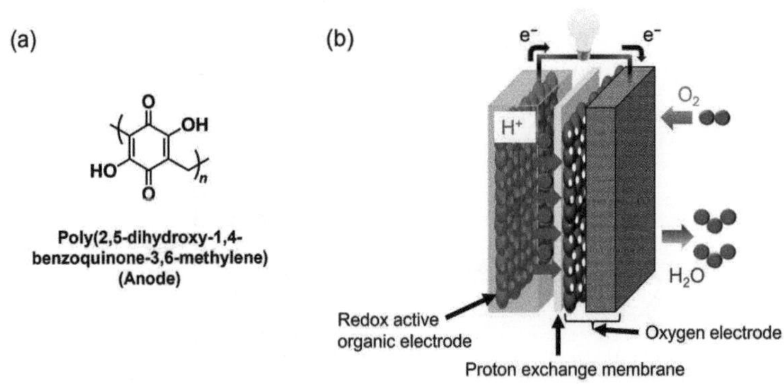

Figure 4.33 (a) Structure of poly(2,5-dihydroxy-1,4-benzoquinone-3,6-methylene), an anode-active material. (b) Cell configuration of an all-solid-state organic rechargeable battery.

4.3.3 All-Organic Redox-Flow Batteries

Organic redox-flow batteries (ORFBs), containing organic redox-active materials, have recently attracted attention owing to their ability to store an ever-increasing quantity of electrical energy in a sustainable way [120]. As shown in Fig. 4.34, ORFBs consist of an electrochemical cell and two tanks in which an electrolyte containing redox-active materials is circulated by a pump. The separator is made of a material that is permeable to cations and anions from the support salt, but not to the redox-active materials. ORFBs consist of organic redox-active materials, which are composed of earth-abundant elements without conventional rare resources such as vanadium. Furthermore, organic redox-active materials for ORFBs have been increasingly reported in recent years since their solubility and redox properties (e.g., redox potential) can be tuned by an appropriate molecular design based on organic synthesis.

In 2011, Li et al. first reported an all-organic redox-flow battery with TEMPO as the cathode-active material, which undergoes a redox reaction by releasing an electron, and *N*-methylphthalimide as the anode-active material, which undergoes a redox reaction by accepting an electron [121] (Figs. 4.35a and 4.35b). As shown in Figs. 4.35c and 4.35d, the redox-flow battery exhibited a discharge voltage of 1.36 V and a coulombic efficiency of about 90% after 20 cycles. In the current study, organic molecules were used as redox-active materials, showing the possibility of fabricating all-organic redox-flow batteries without rare resources such as vanadium.

In 2014, Yang et al. reported the first aqueous organic redox-flow batteries with 1,2-dihydrobenzoquinone-3,5-disulfonic acid (BQH_2DS) as the cathode-active material, which exhibited a redox reaction featuring electron release, and anthraquinone-2-sulfonic acid (AQS) as the anode-active material, which exhibited a redox reaction featuring electron acceptance [122] (Figs. 4.36a and 4.36b). As shown in Fig. 4.36c, the redox-flow battery exhibited a discharge voltage of approximately 0.3–0.5 V, which displayed only minor changes after 12 charge–discharge cycles. In the current study, an aqueous electrolyte was used, which led to the development of nonflammable, safe, all-organic redox-flow batteries.

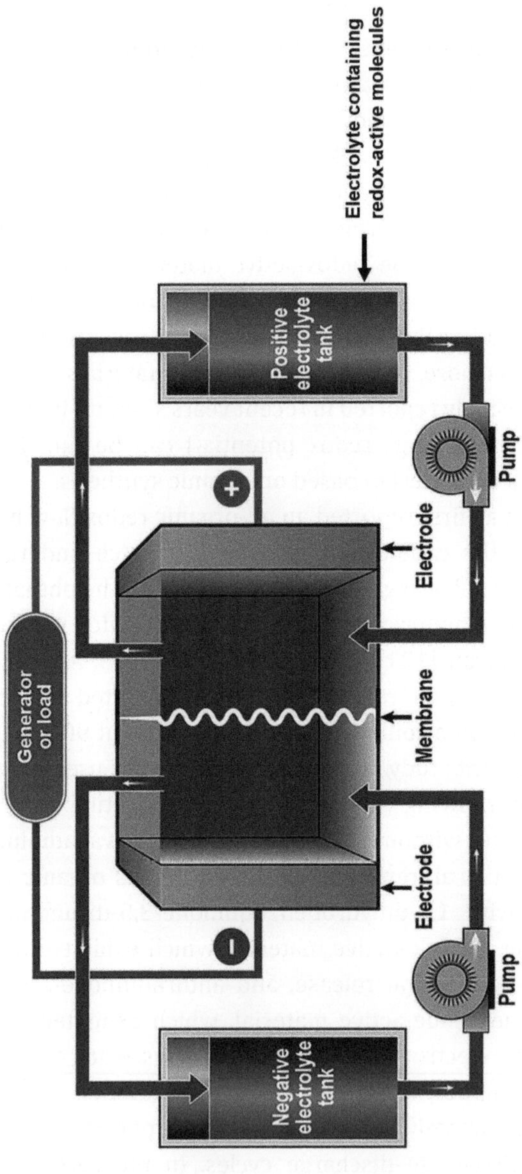

Figure 4.34 Cell configuration of an organic redox-flow battery [120].

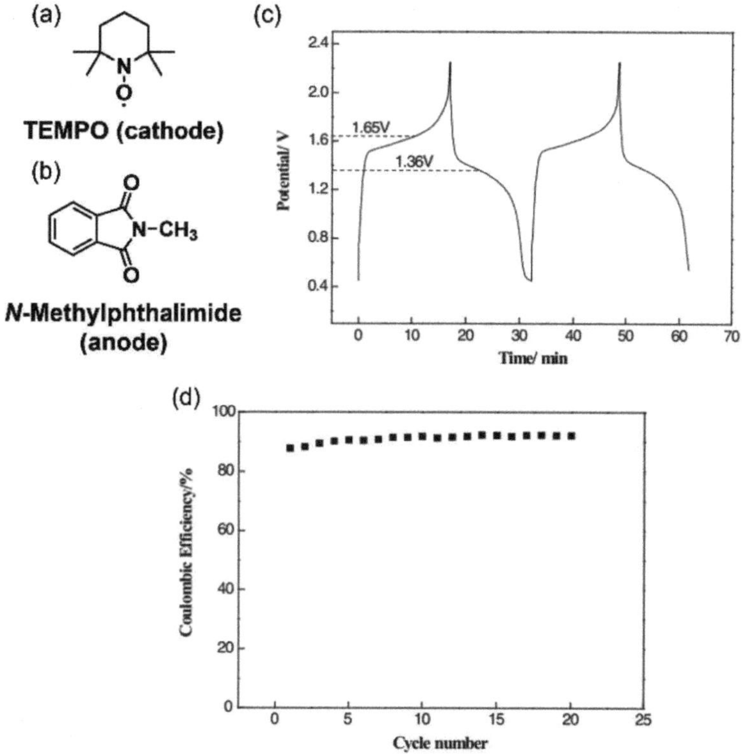

Figure 4.35 (a) Structure of TEMPO, a cathode-active material. (b) Structure of *N*-methylphthalimide, an anode-active material. (c) Charge–discharge curves of an all-organic redox-flow battery using 0.10 M TEMPO/1.00 M $NaClO_4$/acetonitrile (as catholyte) and 0.10 M *N*-Methylphthalimide/1.00 M $NaClO_4$/acetonitrile (as anolyte). (d) Coulombic efficiency of all-organic redox-flow battery for the first 20 cycles. Current density: 0.35 mA/cm^2. Reprinted with permission from Ref. [121]. Copyright 2011 IOPscience.

In 2015, Janoschka et al. reported the first all-organic polymer-based redox-flow battery composed of redox polymer with TEMPO as the cathode-active material, which exhibited a redox reaction featuring electron release, and a redox polymer with viologen as the anode-active material, which exhibited a redox reaction featuring electron acceptance [123]. The redox-flow battery exhibited an open-circuit voltage of 1.1 V. In this concept, polymerization prevented redox-active molecules from aggregating and precipitating, allowing the battery to retain 75% of its initial discharge capacity after 100 charge–discharge cycles. Even after 10,000 charge–discharge cycles in static, unpumped cells, 80% of the initial discharge capacity was retained. To maximize the sustainability of ORFBs, researchers have been actively investigating the use of earth-abundant raw materials and the improvement in ORFBs' lifetime.

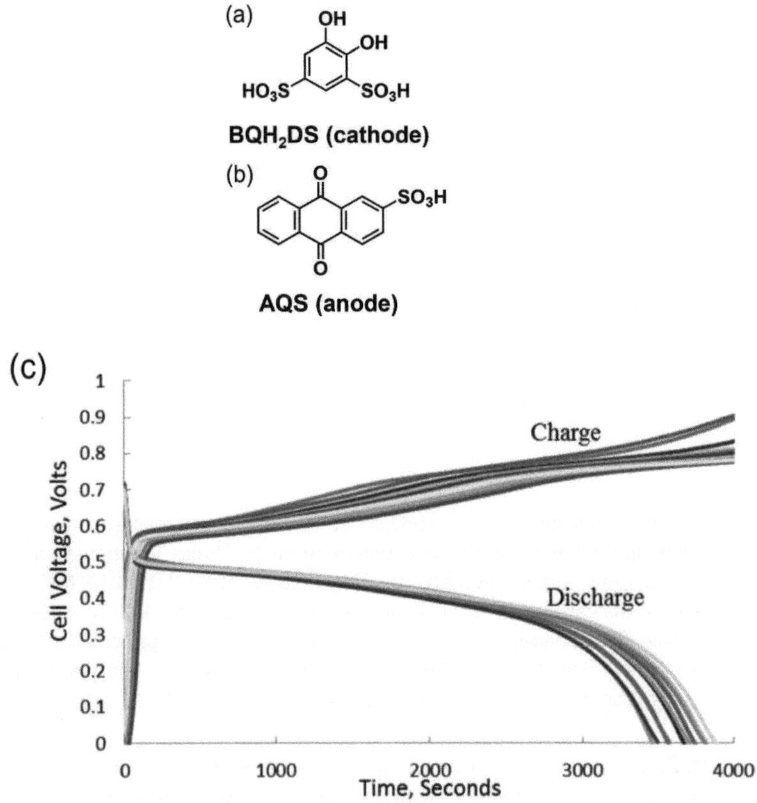

Figure 4.36 (a) Structure of 1,2-dihydrobenzoquinone-3,5-disulfonic acid, a cathode-active material. (b) Structure of anthraquinone-2-sulfonic acid, an anode-active material. (c) Twelve charge–discharge curves with a 25 cm^2 redox-flow cell; 0.2 M BQH_2DS and 0.2 M AQS, 1 M sulfuric acid, charge and discharge at 200 mA, flow rate of 0.5 L/min [122].

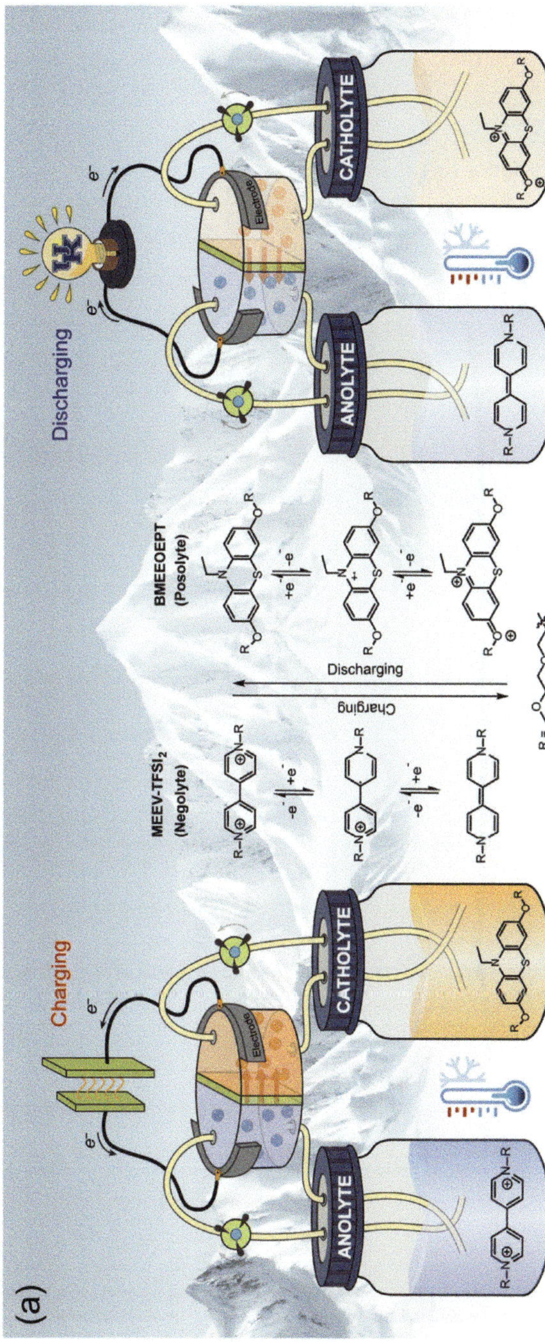

(a)
Charging
Discharging
ANOLYTE
CATHOLYTE
Electrode
MEEV-TFSI2
(Negolyte)
BMEEOEPT
(Posolyte)
Charging
Discharging
R =

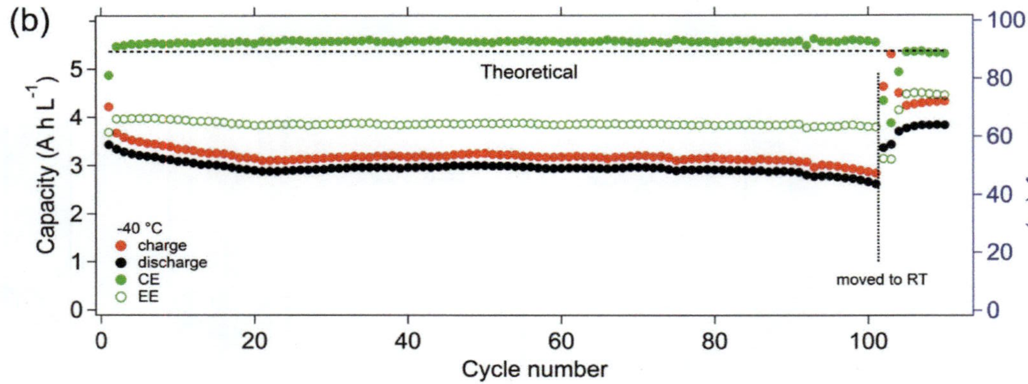

Figure 4.37 (a) Illustrative design of an ultra-low temperature (–40 °C) redox-flow battery during charge and discharge. (b) Long-term cycling of the –40 °C cell. The current rate was 15 mA/cm. Reprinted with permission from Ref. [124]. Copyright 2022 Royal Society of Chemistry.

In 2022, Liang et al. reported the first all-organic redox-flow battery usable at ultra-low temperatures (−40 °C) for use in cold environments, composed of glycolated phenothiazine as the cathode-active material, which exhibited redox reaction featuring electron release, and glycolated viologen as the anode-active material, which exhibited redox reaction featuring electron acceptance [124] (Fig. 4.37a). As shown in Fig. 4.37b, the redox-flow battery exhibited a discharge voltage of approximately 0.6–0.8 V at −40 °C and retained 78% of its initial discharge capacity after 100 charge–discharge cycles.

4.3.4 Next-Generation All-Organic and Sustainable Batteries

As previously noted, there has been active research and development in the field of rechargeable batteries to improve their electric energy storage capacity. A significant portion of this research has been concentrated on optimizing the post-use decomposition of all-organic rechargeable batteries.

In 2021, Nguyen et al. reported an on-demand degradable all-organic rechargeable battery having a polymer with a polypeptide backbone comprising TEMPO as the cathode-active material, which exhibited a redox reaction featuring electron release, and a polymer with viologen as the anode-active material, which exhibited a redox reaction featuring electron acceptance [126] (Fig. 4.38). The all-organic rechargeable battery exhibited discharge voltages of 1.1 and 1.6 V and a discharge capacity of 25 mAh/g. Each polypeptide could be completely degraded to the backbone and redox-active molecules by exposing them to 1 M $HCl_{(aq)}$ at 110 °C.

A redox polymer with the polypeptide backbone was also applied to an all-organic redox-flow battery [127] (Figs. 4.39a and 4.39b). As shown in Fig. 4.39c, the redox-flow battery had a discharge voltage of 1.1 V and a discharge capacity of 530 mAh/L. As shown in Fig. 4.39c, after 500 cycles, the discharge capacity was 390 mAh/L, retaining 74% of the initial discharge capacity.

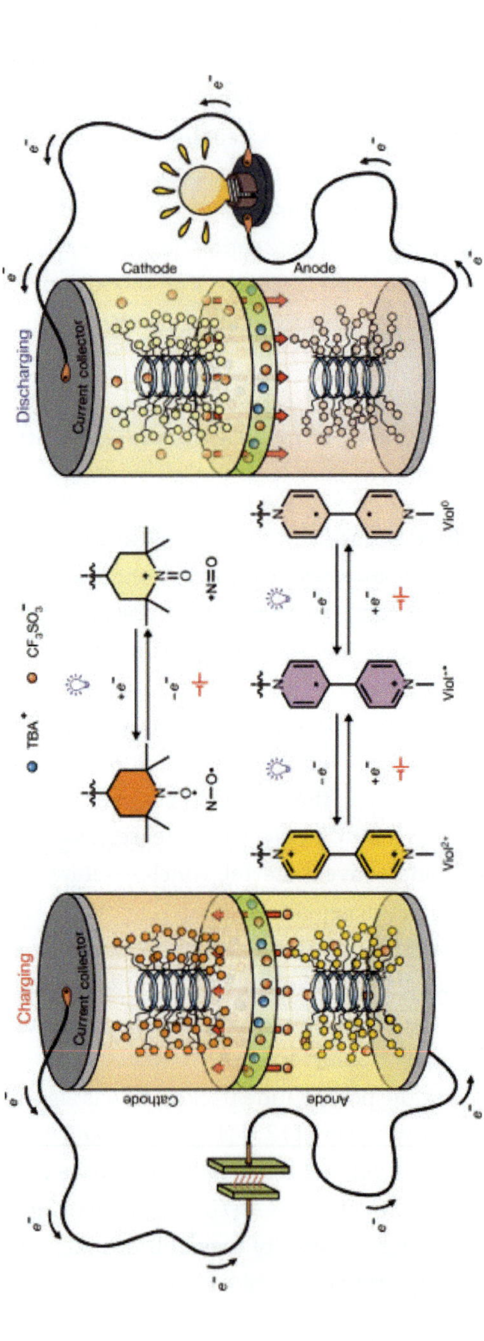

Figure 4.38 Schematics of a polypeptide-based organic radical battery, and the reactions that occur during charging and discharging are shown on the left and right, respectively [125].

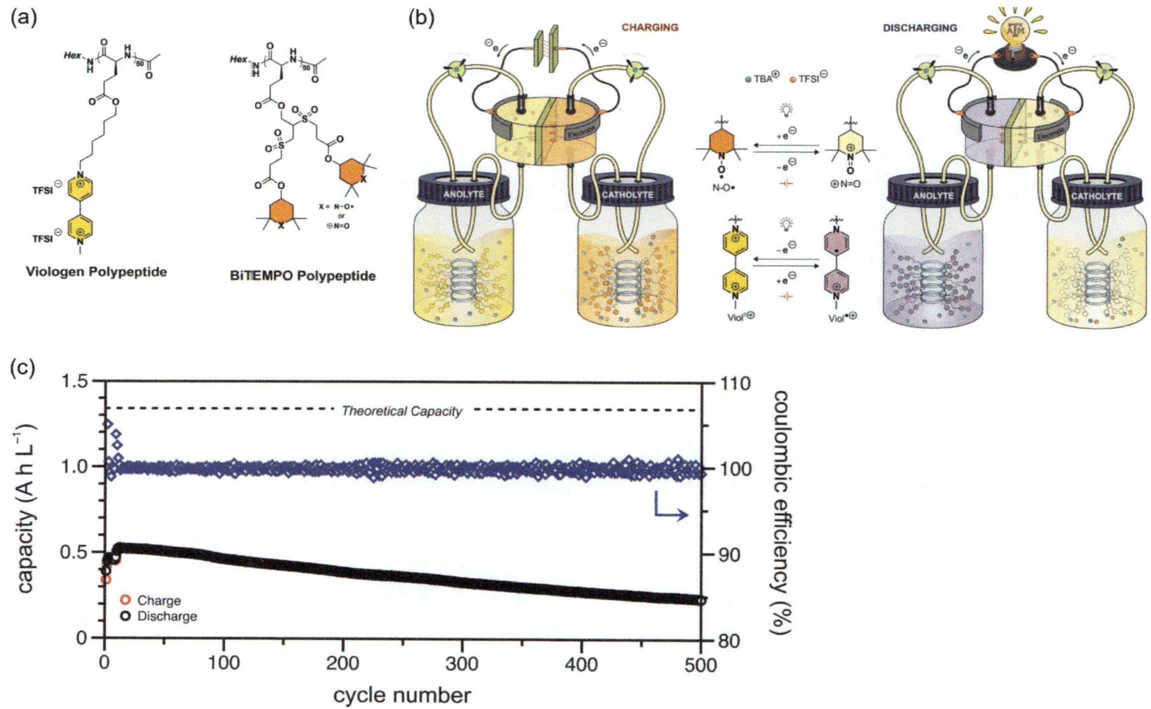

Figure 4.39 (a) Structures of redox-active polypeptides, a cathode- and anode-active material. (b) Illustrative design of polypeptide-based redox-flow battery during charge and discharge. (c) Long-term cycling of the polypeptide flow cell. The current density was 10 mA/cm. Each cell consisted of 7.5 mL of 50 mM (by repeat unit) viologen polypeptide anolyte and 25 mM (by repeat unit) biTEMPO polypeptide catholyte in 0.5 M TEATFSI in AN. Reprinted with permission from Ref. [127]. Copyright 2022 Royal Society of Chemistry.

In Section 4.3, all-organic rechargeable batteries containing organic redox-active materials as electrode-active materials were introduced. Numerous all-organic rechargeable batteries with excellent performance (e.g., voltage, capacity, and stability) have been reported based on the design of the electrode-active materials in these batteries. In the continuous studies, eco-friendly all-organic rechargeable batteries will be fabricated by synthesizing electrode-active materials from earth-abundant raw materials with low environmental burden. Subsequently, when decomposing and separating a defunct battery into its raw material components become viable, realizing sustainable energy storage will also become possible (Fig. 4.40).

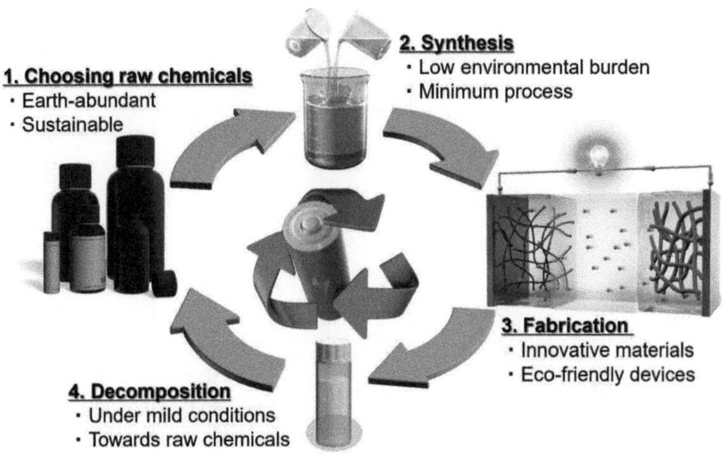

Figure 4.40 Sustainable and environmentally friendly recycling method for organic redox-active materials.

4.4 Conclusion

In this chapter, research history and the latest trends in all-organic rechargeable batteries that use organic redox-active materials as electrode-active materials were introduced. Organic redox-active materials are composed of earth-abundant building blocks (C, H, N, O, and S) and should be a promising alternative to conventional transition-metal-based electrodes. All-organic rechargeable batteries demonstrate excellent performance (e.g., voltage, capacity, and stability) based on the design of their electrode-active materials. Since all-organic rechargeable batteries do not need metal-based materials, the batteries are potentially lightweight and flexible. Very recently, eco-friendly all-organic rechargeable batteries, which focus on decomposition after use, have also appeared. In the continuous work, the development of organic electrode-active materials will, in turn, accelerate the evolution of all-organic rechargeable batteries, which should ultimately lead to their implementation as sustainable energy storage devices.

Acknowledgments

This work was partially supported by the Grants-in-Aids for Scientific Research (22K14732, 23H03827, 23K17945) from MEXT, Japan. K.O. also acknowledges the support from LNest Grant from Nipponham, FUSO Innovative Technology Fund, Masuyakinen Basic Research Foundation, the Shorai Foundation for Science and Technology, Yamada Science Foundation, JACI Prize for Encouraging Young Researcher, TEPCO Memorial Foundation, Amano Institute of Technology, Sugiyama Houkoukai, and Kenjiro Takayanagi Foundation. R.A. acknowledges the support from JST SPRING, Grant number JPMJSP2138. R.A. also acknowledges the support from Kato Foundation for Promotion of Science (KS-3514).

References

1. Poizot, P., Gaubicher, J., Renault, S., Dubois, L., Liang, Y., and Yao, Y. (2020). Opportunities and challenges for organic electrodes in electrochemical energy storage, *Chem. Rev.*, **120**, pp. 6490–6557.
2. Kim, J., Kim, Y., Yoo, J., Kwon, G., Ko, Y., and Kang, K. (2023). Organic batteries for a greener rechargeable world, *Nat. Rev. Mater.*, **8**, pp. 54–70.
3. Li, J., Fleetwood, J., Hawley, W. B., and Kays, W. (2022). From materials to cell: State-of-the-art and prospective technologies for lithium-ion battery electrode processing, *Chem. Rev.*, **122**, pp. 903–956.
4. Oka, K. and Nishide, H. (2020) *Radical Polymers for Rechargeable Batteries* 34, eds. Casado, N. and Mecerreyes, D., Chapter 4 "Redox Polymers for Energy and Nanomedicine," (The Royal Society of Chemistry, UK), pp. 137–165.

5. Yanamandra, K., Pinisetty, D., and Gupta, N. (2023). Impact of carbon additives on lead-acid battery electrodes: A review, *Renew. Sust. Energ. Rev.*, **173**, pp. 113078.
6. Lopes, P. P. and Stamenkovic, V. R. (2020). Past, present, and future of lead–acid batteries, *Science*, **369**, pp. 923–924.
7. Manthiram, A. (2020). A reflection on lithium-ion battery cathode chemistry, *Nat. Commun.*, **11**, pp. 1550.
8. Yoshino, A. (2022). The lithium-ion battery: Two breakthroughs in development and two reasons for the Nobel Prize, *Bull. Chem. Soc. Jpn.*, **95**, pp. 195–197.
9. Goodenough, J. B. and Kim, Y. (2010). Challenges for rechargeable Li batteries, *Chem. Mater.*, **22**, pp. 587–603.
10. Whittingham, M. S. (2004). Lithium batteries and cathode materials, *Chem. Rev.*, **104**, pp. 4271–4302.
11. Yoshino, A. (2012). The birth of the lithium-ion battery, *Angew. Chem. Int. Ed.*, **51**, pp. 5798–5800.
12. Roy, K., Banerjee, A., and Ogale, S. (2022). Search for new anode materials for high performance Li-ion batteries, *ACS Appl. Mater. Interfaces*, **14**, pp. 20326–20348.
13. Li, M., Hicks, R. P., Chen, Z., Luo, C., Guo, J., Wang, C., and Xu, Y. (2023). Electrolytes in organic batteries, *Chem. Rev.*, **123**, pp. 1712–1773.
14. Hawkins, T. R., Singh, B., Majeau-Bettez, G., and Strømman, A. H. (2013). Comparative environmental life cycle assessment of conventional and electric vehicles, *J. Ind. Ecol.*, **17**, pp. 53–64.
15. Ambrose, H. and Kendall, A. (2016). Effects of battery chemistry and performance on the life cycle greenhouse gas intensity of electric mobility, *Transportation Research Part D: Transport and Environment*, **47**, pp. 182–194.
16. Månberger, A. and Stenqvist, B. (2018). Global metal flows in the renewable energy transition: Exploring the effects of substitutes, technological mix and development, *Energy Policy*, **119**, pp. 226–241.
17. Chen, Q., Lv, Y., Yuan, Z., Li, X., Yu, G., Yang, Z., and Xu, T. (2022). Organic electrolytes for pH-neutral aqueous organic redox flow batteries, *Adv. Funct. Mater.*, **32**, pp. 2108777.
18. Leguizamon, S., Díaz-Orellana, K. P., Velez, J., Thies, M. C., and Roberts, M. E. (2015). High charge-capacity polymer electrodes comprising alkali lignin from the Kraft process, *J. Mater. Chem. A*, **3**, pp. 11330–11339.

19. Miroshnikov, M., Divya, K. P., Babu, G., Meiyazhagan, A., Reddy Arava, L. M., Ajayan, P. M., and John, G. (2016). Power from nature: Designing green battery materials from electroactive quinone derivatives and organic polymers, *J. Mater. Chem. A*, **4**, pp. 12370–12386.
20. Lu, Y., Hou, X., Miao, L., Li, L., Shi, R., Liu, L., and Chen, J. (2019). Cyclohexanehexone with ultrahigh capacity as cathode materials for lithium-ion batteries, *Angew. Chem. Int. Ed.*, **58**, pp. 7020–7024.
21. Li, M., Case, J., and Minteer, S. D. (2021). Bipolar redox-active molecules in non-aqueous organic redox flow batteries: Status and challenges, *ChemElectroChem*, **8**, pp. 1215–1232.
22. Williams, D. L., Byrne, J. J., and Driscoll, J. S. (1969). A high energy density lithium/dichloroisocyanuric acid battery system, *J. Electrochem. Soc.*, **116**, pp. 2.
23. Visco, S. J. and DeJonghe, L. C. (1988). Ionic conductivity of organosulfur melts for advanced storage electrodes, *J. Electrochem. Soc.*, **135**, pp. 2905.
24. Nakahara, K., Iwasa, S., Satoh, M., Morioka, Y., Iriyama, J., Suguro, M., and Hasegawa, E. (2002). Rechargeable batteries with organic radical cathodes, *Chem. Phys. Lett.*, **359**, pp. 351–354.
25. Ding, Y., Li, Y., and Yu, G. (2016). Exploring bio-inspired quinone-based organic redox flow batteries: A combined experimental and computational study, *Chem*, **1**, pp. 790–801.
26. Huynh, M. T., Anson, C. W., Cavell, A. C., Stahl, S. S., and Hammes-Schiffer, S. (2016). Quinone $1e^-$ and $2e^-/2H^+$ reduction potentials: Identification and analysis of deviations from systematic scaling relationships, *J. Am. Chem. Soc.*, **138**, pp. 15903–15910.
27. Armand, M., Grugeon, S., Vezin, H., Laruelle, S., Ribière, P., Poizot, P., and Tarascon, J. M. (2009). Conjugated dicarboxylate anodes for Li-ion batteries, *Nat. Mater.*, **8**, pp. 120–125.
28. Vadehra, G. S., Maloney, R. P., Garcia-Garibay, M. A., and Dunn, B. (2014). Naphthalene diimide based materials with adjustable redox potentials: Evaluation for organic lithium-ion batteries, *Chem. Mater.*, **26**, pp. 7151–7157.
29. Deng, W., Shen, Y., Qian, J., Cao, Y., and Yang, H. (2015). A perylene diimide crystal with high capacity and stable cyclability for Na-ion batteries, *ACS Appl. Mater. Interfaces*, **7**, pp. 21095–21099.
30. Lee, S., Hong, J., Jung, S.-K., Ku, K., Kwon, G., Seong, W. M., Kim, H., Yoon, G., Kang, I., Hong, K., Jang, H. W., and Kang, K. (2019). Charge-transfer complexes for high-power organic rechargeable batteries, *Energy Stor. Mater.*, **20**, pp. 462–469.

31. Liang, Y., Jing, Y., Gheytani, S., Lee, K.-Y., Liu, P., Facchetti, A., and Yao, Y. (2017). Universal quinone electrodes for long cycle life aqueous rechargeable batteries, *Nat. Mater.*, **16**, pp. 841–848.

32. Oka, K., Murao, S., Kataoka, M., Nishide, H., and Oyaizu, K. (2021). Hydrophilic anthraquinone-substituted polymer: Its environmentally friendly preparation and efficient charge/proton-storage capability for polymer–air secondary batteries, *Macromolecules*, **54**, pp. 4854–4859.

33. Tong, L., Jing, Y., Gordon, R. G., and Aziz, M. J. (2019). Symmetric all-quinone aqueous battery, *ACS Appl. Energy Mater.*, **2**, pp. 4016–4021.

34. Jouhara, A., Dupré, N., Gaillot, A.-C., Guyomard, D., Dolhem, F., and Poizot, P. (2018). Raising the redox potential in carboxyphenolate-based positive organic materials via cation substitution, *Nat. Commun.*, **9**, pp. 4401.

35. Lakraychi, A. E., Fahsi, K., Aymard, L., Poizot, P., Dolhem, F., and Bonnet, J. P. (2017). Carboxylic and sulfonic N-substituted naphthalene diimide salts as highly stable non-polymeric organic electrodes for lithium batteries, *Electrochem. Commun.*, **76**, pp. 47–50.

36. Gheytani, S., Liang, Y., Wu, F., Jing, Y., Dong, H., Rao, K. K., Chi, X., Fang, F., and Yao, Y. (2017). An aqueous Ca-ion battery, *Adv. Sci.*, **4**, pp. 1700465.

37. Hernández, G., Casado, N., Zamarayeva, A. M., Duey, J. K., Armand, M., Arias, A. C., and Mecerreyes, D. (2018). Perylene polyimide-polyether anodes for aqueous all-organic polymer batteries, *ACS Appl. Energy Mater.*, **1**, pp. 7199–7205.

38. Cobb, S. J., Ayres, Z. J., Newton, M. E., and Macpherson, J. V. (2019). Deconvoluting surface-bound quinone proton coupled electron transfer in unbuffered solutions: Toward a universal voltammetric pH electrode, *J. Am. Chem. Soc.*, **141**, pp. 1035–1044.

39. Singh, V., Kim, S., Kang, J., and Byon, H. R. (2019). Aqueous organic redox flow batteries, *Nano Res.*, **12**, pp. 1988–2001.

40. Han, X., Qing, G., Sun, J., and Sun, T. (2012). How many lithium ions can be inserted onto fused C6 aromatic ring systems? *Angew. Chem. Int. Ed.*, **51**, pp. 5147–5151.

41. Han, X., Yi, F., Sun, T., and Sun, J. (2012). Synthesis and electrochemical performance of Li and Ni 1,4,5,8-naphthalenetetracarboxylates as anodes for Li-ion batteries, *Electrochem. Commun.*, **25**, pp. 136–139.

42. Lee, M., Hong, J., Lee, B., Ku, K., Lee, S., Park, C. B., and Kang, K. (2017). Multi-electron redox phenazine for ready-to-charge organic batteries, *Green Chem.*, **19**, pp. 2980–2985.

43. Bard, A. J., Faulkner, L. R., and White, H. S. (2001) *Electrochemical Methods: Fundamentals and Applications*, 2nd Ed. (John Wiley & Sons, USA).

44. Zhang, L., Qian, Y., Feng, R., Ding, Y., Zu, X., Zhang, C., Guo, X., Wang, W., and Yu, G. (2020). Reversible redox chemistry in azobenzene-based organic molecules for high-capacity and long-life nonaqueous redox flow batteries, *Nat. Commun.*, **11**, pp. 3843.

45. Wang, X., Tang, W., and Loh, K. P. (2021). Para-substituted triphenylamine as a catholyte for zinc–organic aqueous redox flow batteries, *ACS Appl. Energy Mater.*, **4**, pp. 3612–3621.

46. Huang, L., Chen, Y., Liu, Y., Wu, T., Li, H., Ye, J., Dai, G., Zhang, X., and Zhao, Y. (2020). π-Extended dihydrophenazine-based polymeric cathode material for high-performance organic batteries, *ACS Sustain. Chem. Eng.*, **8**, pp. 17868–17875.

47. Peterson, B. M., Gannett, C. N., Melecio-Zambrano, L., Fors, B. P., and Abruña, H. (2021). Effect of structural ordering on the charge storage mechanism of p-type organic electrode materials, *ACS Appl. Mater. Interfaces*, **13**, pp. 7135–7141.

48. Peng, Z., Yi, X., Liu, Z., Shang, J., and Wang, D. (2016). Triphenylamine-based metal–organic frameworks as cathode materials in lithium-ion batteries with coexistence of redox active sites, high working voltage, and high rate stability, *ACS Appl. Mater. Interfaces*, **8**, pp. 14578–14585.

49. Yamamoto, K., Suemasa, D., Masuda, K., Aita, K., and Endo, T. (2018). Hyperbranched triphenylamine polymer for ultrafast battery cathode, *ACS Appl. Mater. Interfaces*, **10**, pp. 6346–6353.

50. Shimizu, T., Tanifuji, N., and Yoshikawa, H. (2022). Azo compounds as active materials of energy storage systems, *Angew. Chem. Int. Ed.*, **61**, pp. e202206093.

51. Hong, J., Lee, M., Lee, B., Seo, D.-H., Park, C. B., and Kang, K. (2014). Biologically inspired pteridine redox centres for rechargeable batteries, *Nat. Commun.*, **5**, pp. 5335.

52. Fujihara, Y., Kobayashi, H., Takaishi, S., Tomai, T., Yamashita, M., and Honma, I. (2020). Electrical conductivity-relay between organic charge-transfer and radical salts toward conductive additive-free rechargeable battery, *ACS Appl. Mater. Interfaces*, **12**, pp. 25748–25755.

53. Hanyu, Y. and Honma, I. (2012). Rechargeable quasi-solid state lithium battery with organic crystalline cathode, *Sci. Rep.*, **2**, pp. 453.

54. Häupler, B., Burges, R., Janoschka, T., Jähnert, T., Wild, A., and Schubert, U. S. (2014). PolyTCAQ in organic batteries: Enhanced capacity at

constant cell potential using two-electron-redox-reactions, *J. Mater. Chem. A*, **2**, pp. 8999–9001.

55. Hatakeyama-Sato, K., Sadakuni, K., Kitagawa, K., and Oyaizu, K. (2023). Thianthrene polymers as 4 V-class organic mediators for redox targeting reaction with $LiMn_2O_4$ in flow batteries, *Sci. Rep.*, **13**, pp. 5711.
56. Naskar, K., Dey, A., Maity, S., Ray, P. P., Ghosh, P., and Sinha, C. (2020). Biporous Cd(II) coordination polymer via in situ disulfide bond formation: Self-healing and application to photosensitive optoelectronic device, *Inorganic Chemistry*, **59**, pp. 5518–5528.
57. Chen, S., Wang, D., Zhao, Y., and Wang, D. (2018). Superior performance of a lithium–sulfur battery enabled by a dimethyl trisulfide containing electrolyte, *Small Methods*, **2**, pp. 1800038.
58. Oka, K., Kato, R., Oyaizu, K., and Nishide, H. (2018). Poly(vinyldibenz othiophenesulfone): Its redox capability at very negative potential toward an all-organic rechargeable device with high-energy density, *Adv. Funct. Mater.*, **28**, pp. 1805858.
59. Speer, M. E., Kolek, M., Jassoy, J. J., Heine, J., Winter, M., Bieker, P. M., and Esser, B. (2015). Thianthrene-functionalized polynorbornenes as high-voltage materials for organic cathode-based dual-ion batteries, *Chem. Commun.*, **51**, pp. 15261–15264.
60. Tinker, L. A. and Bard, A. J. (1979). Electrochemistry in liquid sulfur dioxide. 1. Oxidation of thianthrene, phenothiazine, and 9,10-diphenylanthracene, *J. Am. Chem. Soc.*, **101**, pp. 2316–2319.
61. Wang, M. and Tang, Y. (2018). A review on the features and progress of dual-ion batteries, *Adv. Energy Mater.*, **8**, pp. 1703320.
62. Lee, S., Kwon, G., Ku, K., Yoon, K., Jung, S.-K., Lim, H.-D., and Kang, K. (2018). Recent progress in organic electrodes for Li and Na rechargeable batteries, *Adv. Mater.*, **30**, pp. 1704682.
63. Li, M., Yan, G., Zou, P., Ji, H., Wang, H., Hu, Z., Yang, Z., Feng, Y., Ben, H., and Zhang, X. (2022). Dynamic disulfide bonds contained covalent organic framework modified separator as efficient inhibit polysulfide shuttling in Li–S batteries, *ACS Sustain. Chem. Eng.*, **10**, pp. 13638–13649.
64. Zhou, J., Qian, T., Xu, N., Wang, M., Ni, X., Liu, X., Shen, X., and Yan, C. (2017). Selenium-doped cathodes for lithium–organosulfur batteries with greatly improved volumetric capacity and coulombic efficiency, *Adv. Mater.*, **29**, pp. 1701294.
65. Visco, S. J., Liu, M., Armand, M. B., and de Jonghe, L. C. (1990). Solid redox polymerization electrodes and their use in all-solid-

state batteries, *Mol. Cryst. Liq. Cryst. Inc. Nonlinear Opt. Lett.*, **190**, pp. 185–195.

66. Shadike, Z., Tan, S., Wang, Q.-C., Lin, R., Hu, E., Qu, D., and Yang, X.-Q. (2021). Review on organosulfur materials for rechargeable lithium batteries, *Mater. Horiz.*, **8**, pp. 471–500.
67. Kwon, G., Lee, S., Hwang, J., Shim, H.-S., Lee, B., Lee, M. H., Ko, Y., Jung, S.-K., Ku, K., Hong, J., and Kang, K. (2018). Multi-redox molecule for high-energy redox flow batteries, *Joule*, **2**, pp. 1771–1782.
68. Pisoschi, A. M., Cheregi, M. C., and Danet, A. F. (2009). Total antioxidant capacity of some commercial fruit juices: Electrochemical and spectrophotometrical approaches, *Molecules*, **14**, pp. 480–493.
69. Suga, T., Ohshiro, H., Sugita, S., Oyaizu, K., and Nishide, H. (2009). Emerging N-type redox-active radical polymer for a totally organic polymer-based rechargeable battery, *Adv. Mater.*, **21**, pp. 1627–1630.
70. Suga, T., Sugita, S., Ohshiro, H., Oyaizu, K., and Nishide, H. (2011). p- and n-type bipolar redox-active radical polymer: Toward totally organic polymer-based rechargeable devices with variable configuration, *Adv. Mater.*, **23**, pp. 751–754.
71. Nishide, H. (2022). Organic redox polymers as electrochemical energy materials, *Green Chem.*, **24**, pp. 4650–4679.
72. Oyaizu, K. and Nishide, H. (2009). Radical polymers for organic electronic devices: A radical departure from conjugated polymers? *Adv. Mater.*, **21**, pp. 2339–2344.
73. Takeo, S., Yong-Jin, P., Kenichi, O., and Hiroyuki, N. (2004). Electron-transfer kinetics of nitroxide radicals as an electrode-active material, *Bull. Chem. Soc. Jpn.*, **77**, pp. 2203–2204.
74. Bai, W.-L., Zhang, Z., Chen, X., Wei, X., Zhang, Q., Xu, Z.-X., Liu, Y.-S., Chang, B., Wang, K.-X., and Chen, J.-S. (2020). Boosting the electrochemical performance of $Li–O_2$ batteries with DPPH redox mediator and graphene-luteolin-protected lithium anode, *Energy Stor. Mater.*, **31**, pp. 373–381.
75. Nakanishi, I., Yamashita, S., Shimokawa, T., Kamibayashi, M., Sekine-Suzuki, E., Ueno, M., Ogawa, Y., Ozawa, T., and Matsumoto, K.-I. (2018). Analysis of redox states of protic and aprotic solutions irradiated by low linear energy transfer carbon-ion beams using a 2,2-diphenyl-1-picrylhydrazyl radical, *Org. Biomol. Chem*, **16**, pp. 1272–1276.
76. DaRooge, M. A. and Mahoney, L. R. (1967). Reaction of 2,4,6-tri-tert-butylphenoxy radical with unhindered phenols, *J. Org. Chem.*, **32**, pp. 1–6.

77. Liu, T., Wei, X., Nie, Z., Sprenkle, V., and Wang, W. (2016). A total organic aqueous redox flow battery employing a low cost and sustainable methyl viologen anolyte and 4-HO-TEMPO catholyte, *Adv. Energy Mater.*, **6**, pp. 1501449.

78. Otteny, F., Perner, V., Wassy, D., Kolek, M., Bieker, P., Winter, M., and Esser, B. (2020). Poly(vinylphenoxazine) as fast-charging cathode material for organic batteries, *ACS Sustain. Chem. Eng.*, **8**, pp. 238–247.

79. Godet-Bar, T., Leprêtre, J. C., Le Bacq, O., Sanchez, J. Y., Deronzier, A., and Pasturel, A. (2015). Electrochemical and ab initio investigations to design a new phenothiazine based organic redox polymeric material for metal-ion battery cathodes, *Phys. Chem. Chem. Phys.*, **17**, pp. 25283–25296.

80. Chen, L., Bridges, C. R., Gao, G., Baumgartner, T., and He, X. (2019). Poly(5-vinylbenzothiadiazole) for high-performance lithium-ion batteries, *ACS Appl. Energy Mater.*, **2**, pp. 7315–7320.

81. Gélinas, B., Das, D., and Rochefort, D. (2017). Air-stable, self-bleaching electrochromic device based on viologen- and ferrocene-containing triflimide redox ionic liquids, *ACS Appl. Mater. Interfaces*, **9**, pp. 28726–28736.

82. Yao, M., Sano, H., Ando, H., and Kiyobayashi, T. (2015). Molecular ion battery: A rechargeable system without using any elemental ions as a charge carrier, *Sci. Rep.*, **5**, pp. 10962.

83. Madasamy, K., Velayutham, D., Suryanarayanan, V., Kathiresan, M., and Ho, K.-C. (2019). Viologen-based electrochromic materials and devices, *J. Mater. Chem. C*, **7**, pp. 4622–4637.

84. Shah, K. W., Wang, S.-X., Soo, D. X. Y., and Xu, J. (2019). Viologen-based electrochromic materials: From small molecules, polymers and composites to their applications, *Polymers*, **11**, pp. 1839.

85. Ding, J., Zheng, C., Wang, L., Lu, C., Zhang, B., Chen, Y., Li, M., Zhai, G., and Zhuang, X. (2019). Viologen-inspired functional materials: Synthetic strategies and applications, *J. Mater. Chem. A*, **7**, pp. 23337–23360.

86. Lee, K., Serdiuk, I. E., Kwon, G., Min, D. J., Kang, K., Park, S. Y., and Kwon, J. E. (2020). Phenoxazine as a high-voltage p-type redox center for organic battery cathode materials: Small structural reorganization for faster charging and narrow operating voltage, *Energy Environ. Sci.*, **13**, pp. 4142–4156.

87. Poon, J.-F., Farmer, L. A., Haidasz, E. A., and Pratt, D. A. (2021). Temperature-dependence of radical-trapping activity of phenoxazine, phenothiazine and their aza-analogues clarifies the way forward for new antioxidant design, *Chem. Sci.*, **12**, pp. 11065–11079.

88. Kolek, M., Otteny, F., Schmidt, P., Mück-Lichtenfeld, C., Einholz, C., Becking, J., Schleicher, E., Winter, M., Bieker, P., and Esser, B. (2017). Ultra-high cycling stability of poly(vinylphenothiazine) as a battery cathode material resulting from π–π interactions, *Energy Environ. Sci.*, **10**, pp. 2334–2341.

89. Kolek, M., Otteny, F., Becking, J., Winter, M., Esser, B., and Bieker, P. (2018). Mechanism of charge/discharge of poly(vinylphenothiazine)-based Li–organic batteries, *Chem. Mater.*, **30**, pp. 6307–6317.

90. Makarova, M. V., Akkuratov, A. V., Sideltsev, M. E., Stevenson, K. J., and Romadina, E. I. (2022). Novel ethylene glycol substituted benzoxadiazole and benzothiadiazole as anolytes for nonaqueous organic redox flow batteries, *ChemElectroChem*, **9**, pp. e202200483.

91. Nigrey, P. J., MacInnes, D., Nairns, D. P., MacDiarmid, A. G., and Heeger, A. J. (1981). Lightweight rechargeable storage batteries using polyacetylene, $(CH)_x$ as the cathode-active material, *J. Electrochem. Soc.*, **128**, pp. 1651.

92. Zhu, L. M., Lei, A. W., Cao, Y. L., Ai, X. P., and Yang, H. X. (2013). An all-organic rechargeable battery using bipolar polyparaphenylene as a redox-active cathode and anode, *Chem. Commun.*, **49**, pp. 567–569.

93. MacDiarmid, A. G., Yang, L. S., Huang, W. S., and Humphrey, B. D. (1987). Polyaniline: Electrochemistry and application to rechargeable batteries, *Synth. Met.*, **18**, pp. 393–398.

94. Nie, X., Xiao, T., and Liu, Z. (2019). A redox-generated biomimetic membrane potential across polypyrrole films, *Chem. Commun.*, **55**, pp. 10023–10026.

95. Liu, L., Tian, F., Wang, X., Yang, Z., Zhou, M., and Wang, X. (2012). Porous polythiophene as a cathode material for lithium batteries with high capacity and good cycling stability, *React. Funct. Polym.*, **72**, pp. 45–49.

96. Moni, P., Lau, J., Mohr, A. C., Lin, T. C., Tolbert, S. H., Dunn, B., and Gleason, K. K. (2018). Growth temperature and electrochemical performance in vapor-deposited poly(3,4-ethylenedioxythiophene) thin films for high-rate electrochemical energy storage, *ACS Appl. Energy Mater.*, **1**, pp. 7093–7105.

97. Xu, T., Xiong, J., Du, X., Zhang, Y., Song, S., Xiong, C., and Dong, L. (2018). Polytriphenylamine derivative and carbon nanotubes as cathode materials for high-performance polymer-based batteries, *J. Phys. Chem. C*, **122**, pp. 20057–20063.

98. Shirakawa, H., Louis, E. J., MacDiarmid, A. G., Chiang, C. K., and Heeger, A. J. (1977). Synthesis of electrically conducting organic polymers:

Halogen derivatives of polyacetylene, (CH), *J. Chem. Soc., Chem. Commun.*, pp. 578–580.

99. MacInnes, D., Druy, M. A., Nigrey, P. J., Nairns, D. P., MacDiarmid, A. G., and Heeger, A. J. (1981). Organic batteries: Reversible n- and p-type electrochemical doping of polyacetylene, (CH), *J. Chem. Soc., Chem. Commun.*, pp. 317–319.

100. Song, Z. and Zhou, H. (2013). Towards sustainable and versatile energy storage devices: An overview of organic electrode materials, *Energy Environ. Sci.*, **6**, pp. 2280–2301.

101. Groenendaal, L., Jonas, F., Freitag, D., Pielartzik, H., and Reynolds, J. R. (2000). Poly(3,4-ethylenedioxythiophene) and its derivatives: Past, present, and future, *Adv. Mater.*, **12**, pp. 481–494.

102. Jiang, Y., Liu, T., and Zhou, Y. (2020). Recent advances of synthesis, properties, film fabrication methods, modifications of poly(3,4-ethylenedioxythiophene), and applications in solution-processed photovoltaics, *Adv. Funct. Mater.*, **30**, pp. 2006213.

103. Su, C., He, H., Xu, L., Zhao, K., Zheng, C., and Zhang, C. (2017). A mesoporous conjugated polymer based on a high free radical density polytriphenylamine derivative: Its preparation and electrochemical performance as a cathode material for Li-ion batteries, *J. Mater. Chem. A*, **5**, pp. 2701–2709.

104. Karlsson, C., Huang, H., Strømme, M., Gogoll, A., and Sjödin, M. (2014). Probing polymer–pendant interactions in the conducting redox polymer poly(pyrrol-3-ylhydroquinone), *J. Phys. Chem. C*, **118**, pp. 23499–23508.

105. Karlsson, C., Huang, H., Strømme, M., Gogoll, A., and Sjödin, M. (2015). Impact of linker in polypyrrole/quinone conducting redox polymers, *RSC Adv.*, **5**, pp. 11309–11316.

106. Chiang, C. K. (1981). An all-polymeric solid state battery, *Polymer*, **22**, pp. 1454–1456.

107. Kaneto, K., Yoshino, K., and Inuishi, Y. (1983). Characteristics of polythiophene battery, *Jpn. J. Appl. Phys.*, **22**, pp. L567.

108. Kahsay, B. A., Ramar, A., Wang, F.-M., Yeh, N.-H., Lin, P.-L., Luo, Z.-J., Chan, T.-S., and Su, C.-H. (2019). Investigating an all-organic battery using polyisothianaphthene as a redox-active bipolar electrode material, *J. Power Sources*, **428**, pp. 115–123.

109. Goujon, N., Casado, N., Patil, N., Marcilla, R., and Mecerreyes, D. (2021). Organic batteries based on just redox polymers, *Prog. Polym. Sci.*, **122**, pp. 101449.

110. Jiang, S., Li, W., Xie, Y., Yan, X., Zhang, K., and Jia, Z. (2022). An all-organic battery with 2.8 V output voltage, *Chem. Eng. J.*, **434**, pp. 134651.

111. Hatakeyama-Sato, K., Tezuka, T., Ichinoi, R., Matsumono, S., Sadakuni, K., and Oyaizu, K. (2020). Metal-free, solid-state, paperlike rechargeable batteries consisting of redox-active polyethers, *ChemSusChem*, **13**, pp. 2443–2448.

112. Xie, J., Wang, Z., Xu, Z. J., and Zhang, Q. (2018). Toward a high-performance all-plastic full battery with a'single organic polymer as both cathode and anode, *Adv. Energy Mater.*, **8**, pp. 1703509.

113. Emanuelsson, R., Sterby, M., Strømme, M., and Sjödin, M. (2017). An all-organic proton battery, *J. Am. Chem. Soc.*, **139**, pp. 4828–4834.

114. De Surville, R., Jozefowicz, M., Yu, L. T., Pepichon, J., and Buvet, R. (1968). Electrochemical chains using protolytic organic semiconductors, *Electrochim. Acta*, **13**, pp. 1451–1458.

115. Sano, N., Tomita, W., Hara, S., Min, C.-M., Lee, J.-S., Oyaizu, K., and Nishide, H. (2013). Polyviologen hydrogel with high-rate capability for anodes toward an aqueous electrolyte-type and organic-based rechargeable device, *ACS Appl. Mater. Interfaces*, **5**, pp. 1355–1361.

116. Strietzel, C., Sterby, M., Huang, H., Strømme, M., Emanuelsson, R., and Sjödin, M. (2020). An aqueous conducting redox-polymer-based proton battery that can withstand rapid constant-voltage charging and sub-zero temperatures, *Angew. Chem. Int. Ed.*, **59**, pp. 9631–9638.

117. Choi, W., Harada, D., Oyaizu, K., and Nishide, H. (2011). Aqueous electrochemistry of poly(vinylanthraquinone) for anode-active materials in high-density and rechargeable polymer/air batteries, *J. Am. Chem. Soc.*, **133**, pp. 19839–19843.

118. Oka, K., Furukawa, S., Murao, S., Oka, T., Nishide, H., and Oyaizu, K. (2020). Poly(dihydroxybenzoquinone): Its high-density and robust charge storage capability in rechargeable acidic polymer–air batteries, *Chem. Commun.*, **56**, pp. 4055–4058.

119. Yonenaga, M., Kaiwa, Y., Oka, K., Oyaizu, K., and Miyatake, K. (2023). All-solid-state rechargeable air batteries using dihydroxybenzoquinone and its polymer as the negative electrode, *Angew. Chem. Int. Ed.*, **62**, pp. e202304366.

120. Wei, X., Pan, W., Duan, W., Hollas, A., Yang, Z., Li, B., Nie, Z., Liu, J., Reed, D., Wang, W., and Sprenkle, V. (2017). Materials and systems for organic redox flow batteries: Status and challenges, *ACS Energy Letters*, **2**, pp. 2187–2204.

121. Li, Z., Li, S., Liu, S., Huang, K., Fang, D., Wang, F., and Peng, S. (2011). Electrochemical properties of an all-organic redox flow battery using 2,2,6,6-tetramethyl-1-piperidinyloxy and N-methylphthalimide, *Electrochem. Solid-State Lett.*, **14**, pp. A171.

122. Yang, B., Hoober-Burkhardt, L., Wang, F., Surya Prakash, G. K., and Narayanan, S. R. (2014). An inexpensive aqueous flow battery for large-scale electrical energy storage based on water-soluble organic redox couples, *J. Electrochem. Soc.*, **161**, pp. A1371.

123. Janoschka, T., Martin, N., Martin, U., Friebe, C., Morgenstern, S., Hiller, H., Hager, M. D., and Schubert, U. S. (2015). An aqueous, polymer-based redox-flow battery using non-corrosive, safe, and low-cost materials, *Nature*, **527**, pp. 78–81.

124. Liang, Z., Jha, R. K., Suduwella, T. M., Attanayake, N. H., Wang, Y., Zhang, W., Cao, C., Kaur, A. P., Landon, J., and Odom, S. A. (2022). A prototype of high-performance two-electron non-aqueous organic redox flow battery operated at −40 °C, *J. Mater. Chem. A*, **10**, pp. 24685–24693.

125. Tao, H., Mengqi, L., Jun, X., Hao, L., and Guoxiu, W. (2021). Recent research on emerging organic electrode materials for energy storage, *Energy Materials*, **1**, pp. 100009.

126. Nguyen, T. P., Easley, A. D., Kang, N., Khan, S., Lim, S.-M., Rezenom, Y. H., Wang, S., Tran, D. K., Fan, J., Letteri, R. A., He, X., Su, L., Yu, C.-H., Lutkenhaus, J. L., and Wooley, K. L. (2021). Polypeptide organic radical batteries, *Nature*, **593**, pp. 61–66.

127. Liang, Z., Nguyen, T. P., Attanayake, N. H., Easley, A. D., Lutkenhaus, J. L., Wooley, K. L., and Odom, S. A. (2022). Metal-free polypeptide redox flow batteries, *Mater. Adv.*, **3**, pp. 6558–6565.

Chapter 5

Regenerative Fuel Cells

J. Rubio-Garcia,[a] Andres Parra-Puerto,[b] and Barun Kumar Chakrabarti[c]
[a]*Schlumberger Cambridge Research, Madingley Rd, Cambridge CB3 0EL, United Kingdom*
[b]*Universitat d'Alacant, Departamento de Química Física, Carretera de San Vicente del Raspeig s/n, 03690, San Vicente del Raspeig – Alicante, Spain*
[c]*Sabancı Üniversitesi Nanoteknoloji Araştırma ve Uygulama Merkezi (SUNUM), Orta Mahalle Üniversite Caddesi No:27, 34956, Tuzla – İstanbul, Turkey*
jrubio@slb.com

Electrochemical flow reactors are device architectures that present benefits for medium to large electrical energy storage (i.e., from 2 to 10 h) due to their ability to store the energy in active species dispersed in a fluid (gas or liquid). Those species can be placed in reservoirs isolated from the electrodes, and this enables the independent scaling of power and capacity, which is not possible for traditional batteries such as lithium-ion or lead–acid. Depending on the nature of those fluids and each half-cell reaction, the electrochemical flow reactors can be called redox-flow batteries (RFBs, liquid–liquid), hybrid redox-flow batteries (HRFBs, solid–liquid), or regenerative fuel cells (RFCs, gas–liquid).

Batteries: The Future of Energy Storage
Edited by Subashani Maniam

ISBN 978-981-5129-38-0 (Hardcover), 978-1-003-51288-2 (eBook)
www.jennystanford.com

State-of-the-art RFBs such as the all-vanadium RFB (VRFB) are limited by the cost of the electrolyte, which accounts for up to 54% of the total cost of a system (8 h capacity) [1]. Decreasing the electrolyte volume by replacing the negative side by hydrogen reactions (or by oxygen reactions on the positive side) in RFC configuration contributes to lower the capital cost. Other advantages of RFCs are fast kinetics of the hydrogen reaction when catalyzed on metal surfaces, as well as the opportunity to compress the gas, thus achieving a higher system energy density. In alignment with this, the RFC configuration enables a drop in operational costs, for instance those associated with rebalancing the electrolyte and conditioning with recovery of the electrolyte to a known state-of-heath. Electrolyte state-of-health is indeed a major problem for liquid–liquid electrochemical flow reactions because of the transport of solvated species across the membrane, which is enhanced due to electroosmotic transport and hydraulic convection. This leads to electrolyte imbalance associated with water crossover and, depending on the nature of the redox couple employed for each half-cell reaction, it might lead to irreversible capacity degradation. In contrast, for the gas–liquid configuration of an RFC, the liquid that has crossed to the gas side can be easily pumped back to the liquid reservoir.

This chapter will cover in-depth explanations of the RFC system and its key performance indicators with comparisons to other technologies. A short overview of emerging RFC chemistries is presented.

5.1 Introduction to RFCs

RFCs are efficient renewable energy converters that can be operated in a fully reversible manner to ensure on-demand power delivery (Fig. 5.1). RFC systems employing oxygen reactions at the positive side, such as the V/O_2 or the quinone-derivative/O_2 RFCs, have been reported in the literature [2–4]. However, the RFC configuration employing hydrogen evolution reaction and hydrogen oxidation reaction (HER/HOR) as negative electrode reactions (charge and discharge, respectively) has received significantly more attention due to the fast and reversible nature of HOR/HER. The hydrogen-liquid

RFC embodiment offers high specific energy density (theoretically 3660 Wh/kg), high charge/discharge efficiency (>70%), and long-life capability (~10,000 h) [5]. Other features linked to RFCs are their fast response, reliability, light weight, and eco-friendliness as they utilize abundant water as a fuel carrier [6]. The initial scope for the application of hydrogen-based energy storage devices such as the RFC or the H_2-Ni battery was aerospace applications such as automotive, aircraft, spacecraft, transportation, and portable electronics. In this context, the versatility of the RFC design allows for applicability in space habitats such as Mars or moon surface missions where several tens of kilowatts of electrical power are required [7]. More importantly, unlike the bulkiness of secondary batteries when considering energy storage for long periods, RFCs have smaller footprint and extended longevity [8, 9].

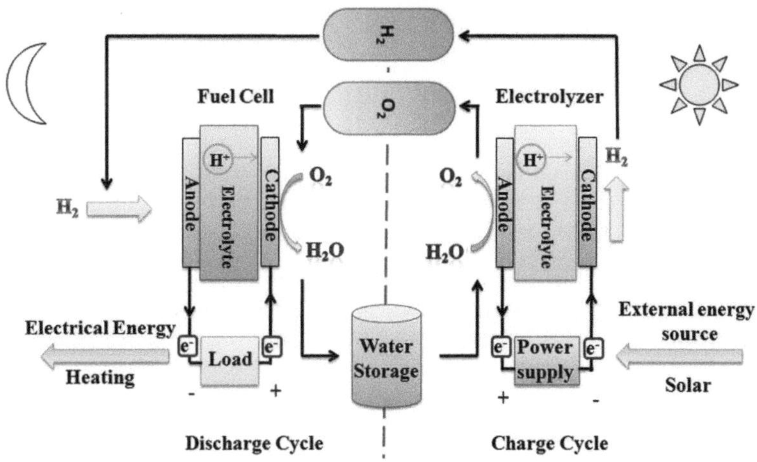

Figure 5.1 Schematic diagram of conventional discrete RFC system combined with solar energy sources. Reprinted with permission from Ref. [10]. Copyright (2021) Elsevier.

The VRFBs have been proposed as a key competitor to lithium-ion batteries for grid support and integration of clean energy sources. RFCs can compete with VRFBs and even surpass their performances for the integration of intermittent and fluctuating clean energy sources such as wind or solar [10]. RFCs present fast response, long lifetime, low degradation rate, long cycle life, and high energy efficiency. They also present a modular character, which simplifies scaling up to hundreds of megawatt-hours. When compared to commercial VRFBs or emerging RFB chemistries, the RFC configuration can operate at higher current densities while retaining high electrolyte utilization, efficiency, and capacity retention over time. This means that, for a given power rate, the RFC stack will be smaller than the RFB counterpart, which leads to not only lower levelized cost of energy storage but also lower capital investment, in line with the US Department of Energy's target of less than USD 100 kW/h for broad market penetration in the long term [11]. RFCs also present lower maintenance and operational costs [12]. This chapter aims at providing a realistic view on the status of RFCs as well as emerging chemistries.

5.1.1 RFCs Historically

Historically, optimization of RFC cell components and the discovery of new liquid electrolyte formulations for the positive half-cell side has been the focus of research activities. Table 5.1 shows a historical overview of the developments in the RFC technology. It aims to emphasize the importance of developing novel concepts rather than providing a comprehensive review on practical scale devices [13].

The pioneering example of the halogen-RFC (H_2/Br_2) was patented in 1964. The performance of this early-stage system was rather poor with peak power close to 100 mW/cm^2 because of the lack of optimization strategies. Significant progress was quickly achieved, leading to a system that operated for more than 10,000 h. Almost in parallel, a solid oxide electrolyte version of the RFC operating at 500 °C was presented by replacing Br_2 with Cl_2. Further halogen-based RFC improvements led to implementation beyond aerospace applications because of reaching extremely high peak power (>1500 mW/cm^2). More recently, the scientific community was pushed to investigate new redox couples for the liquid phase to eradicate issues with toxic halogen fumes and the stability of both the hydrogen catalyst and the membrane. Those issues are linked to the highly corrosive environment and liquid crossing to the gas side. In terms of examples of promising chemistries with the ability to replace halogen-based electrolytes, it needs to be pointed out that the first RFC utilizing organic redox couples was reported in 2013. Organic redox couples are desirable due to their eco-friendliness, geographic availability, and low cost. Although the performance of organic RFCs was limited by the solubility of the quinone-active materials, a major improvement was enabled with the implementation of functionalized organic species such as 1,2-dihydrobenzoquinone-3,5-disulfonic acid (BQDS) [14].

New metal-based RFCs were investigated as well; for example, the H_2/Fe RFC delivered a peak power density of 257 mW/cm^2. The system presents a relatively low cell voltage (0.77 V), which limits further improvements. In alignment with this, the H_2/Ce RFC presents very high cell voltage (around 1.74 V), but the occurrence of oxygen evolution reaction as a competitive reaction during battery charge decreases coulombic efficiency. To merge the benefits of the all-vanadium RFB and RFC architecture, Yufit et al. presented the H_2/V RFC [13]. Their proof-of-concept was ameliorated using HCl as the supporting electrolyte that pushed the cell voltage up when compared to the standard H_2SO_4 supporting electrolyte. Indeed, this H_2/V RFC version showed better performance than the all-vanadium RFB counterpart, which has benefited from decades of optimization. Finally, the H_2/Mn RFC, which was patented in 2017, delivered a high energy density of 30 Wh/L and exceptionally high-power densities greater than 1.4 W/cm^2. In their proof-of-concept article, Rubio-Garcia et al. presented some preliminary economic analysis based on real cell performance data, suggesting a lower cost per kilowatt-hour than any of the competing technologies [15].

Table 5.1 Historical development of RFCs along with power density enhancement

Year	RFC Detail	Efficiency	Cycles/ Life	Peak Power Density/ mW/cm^2
1964	H_2/Br_2 patented for the first time [16]	—	—	50
1966	Second patent on H_2/Br_2 [17]	—	—	30
1969	First scientific paper on H_2/Br_2 published [18]	—	—	100

(*Continued*)

Table 5.1 (*Continued*)

Year	RFC Detail	Efficiency	Cycles/ Life	Peak Power Density/ mW/cm^2
1974	First report on H_2/Cl_2 (500 °C) [19]	0.9 (I)[a]	100 h	120
1980	Detailed electrochemical and mass transport studies on H_2/Cl_2 [20]	0.9 (En)[b]	Several	60
1984	Lifetime study on H_2/Br_2 [21]	0.5 (En)	10,000 h	150
1994	Study on H_2/Cl_2 for space application [22]	0.95 (V)[c]	1000 h	590
2003	H_2/Cl_2 [23]	0.43 (V)	120 h	500
2006	Simulation of a single cell H_2/Cl_2 [24]	—	Few	580
2006	Highest power density for H_2/Br_2 [25]	0.9 (En)	Few	1500
2011	Demonstration of a V/O_2 five-cell stack [26]	0.41 (En)	>100 h	82
2012	Low precious metal content catalyst H_2/Cl_2 [27]	0.9 (V)	—	1100
2012	First patent on air/PS and H_2/V granted [28]	—	—	2
2013	Membrane-less laminar flow H_2/Br_2 [29]	0.92 (V)	Few	795
2013	First single-cell study on H_2/V reported [30]	0.6 (En)	10 h	114
2013	First H_2/benzoquinone (H_2/BQ) reported [31]	—	—	150
2015	First dithionite/air reported [32]	0.12 (En)	45 h	2
2016	Co-CNT electrodes for Br-side in H_2/Br_2 [33]	0.8 (V)	Few	640
2016	Improvements in V/O_2 performance using better membranes [34]	0.45 (En)	Few	>85

Year	RFC Detail	Efficiency	Cycles/ Life	Peak Power Density/ mW/cm^2
2016	H_2/Ce performance enhancements reported [35]	0.9 (En)	<10 h	900
2017	First unit cell model on H_2/V [36]	0.8 (V)	25 h	—
2017	H_2/Mn patent [37]	0.72 (En)	Several	120
2017	Co-CNT electrodes for V side in H_2/V [38]	—	—	610
2017	Experimental results on Air/PS reported [39]	0.74 (V)	30	5
2018	H_2/Mn experimental results reported [15]	0.82 (En)	160	1410
2019	Crossover tolerant catalysts for Br-side in H_2/Br_2 [40]	0.72 (En)	100	550
2019	Application of rGO-modified electrodes for H_2/V [41]	0.92 (En)	10	452
2020	Enhanced H_2/BQ performance reported [14]	>0.6 (En)	200	122
2020	H_2-V using HCl as electrolyte, enhanced V concentration and cell potential [42]	>0.8 (En)	Few	—
2021	Electrospun carbonized mats as electrode catalysts for H_2/V [43]	>0.9 (En)	500	645
2022	Dual-membrane approach for air/PS [44]	>0.4 (En)	80	5.8
2023	Electrospun carbonized mats as electrode catalysts for H_2-Mn [45]	>0.8 (En)	100	570

[a] (I) is current efficiency; [b] (En) is energy efficiency; [c] (V) is voltage efficiency.

Note: The list highlights the most important/breakthrough innovation/milestones. Adapted from Chakrabarti et al. [13].

A testimony to the attractiveness of the RFC architecture is supported by Fig. 5.2 and illustrated by the number of peer-reviewed articles recorded by Scopus on May 28, 2023. In the last 5 years, the increment in articles devoted to RFC has increased exponentially. Interestingly, in terms of keywords, studies devoted to energy, chemical engineering, and engineering account for more than 50% of the research, which might suggest the versatility of RFCs and their orientation to applied research.

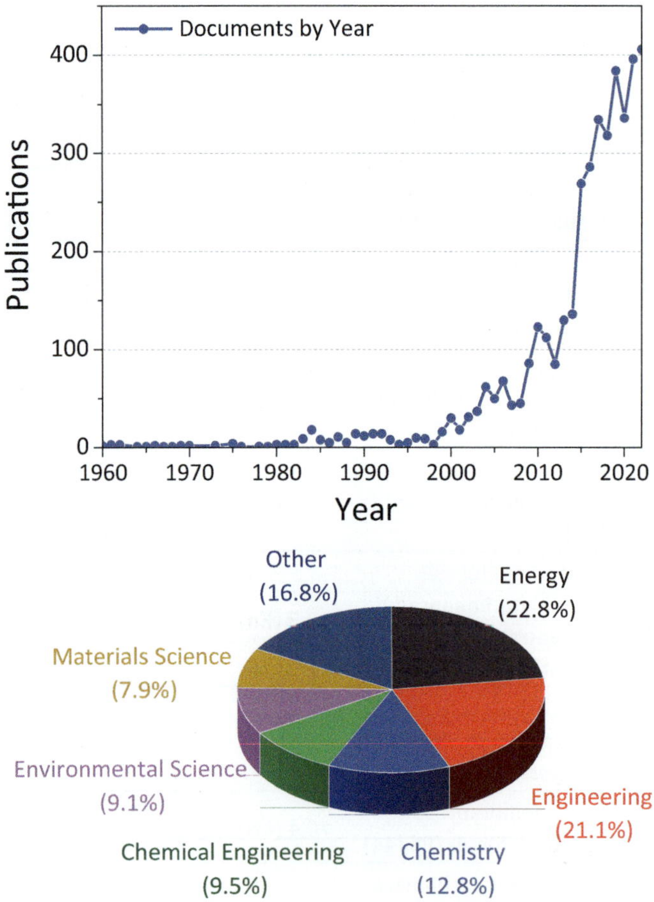

Figure 5.2 Statistics showing the exponential increase in RFC publications from the inception of the technology until May 28, 2023 (data from Scopus).

5.1.2 RFC Single Cell

Figure 5.3 illustrates the assembly of an H_2/V RFC single cell. During charge, the system is operated electrolytically, and during discharge, the polarity is reversed so that the cell operates in the galvanic mode. Current can flow throughout the circuit via proton transport across an ion-selective membrane. During charging, two protons reach the surface of a catalyst where they are combined with two electrons coming from the external circuit to generate hydrogen. The counterreaction takes place on a porous carbon catalyst and relies on the oxidation of VO^{2+} dissolved in the liquid phase to VO_2^+. The opposite reactions occur during RFC discharge. Unlike liquid–liquid RFBs, in the RFC configuration, the gas can be compressed and stored in an external pressurized vessel to reduce volume footprint. The material employed for the construction of the external tanks will depend on gas pressure and/or chemical stability with respect to the liquid electrolyte. Some companies have tested interesting concepts such as storage of fluids in a single reservoir [46].

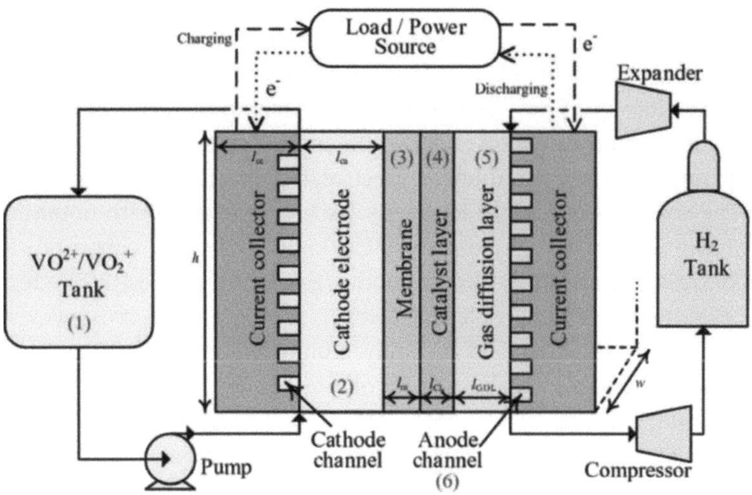

Figure 5.3 A simplified model for a unit cell of a hydrogen/vanadium system. Reprinted with permission from Ref. [36]. Copyright (2017) IOP science.

Typically, perfluorinated polymers such as Nafion are employed as cell separators. Although the polymer is functionalized with negatively charged sulfonated groups to enable proton transport, other cationic species can diffuse to the other reservoir. For example, negatively charged and neutral species can permeate through the polymer pores, which leads to issues such as flooding of the negative electrode, catalyst poisoning, and, depending on how the system is operated along with its liquid formulation, metal catalyst dissolution.

The rate-limiting step in RFCs is normally associated with the kinetics of the redox couple employed at the positive side (cathode during discharge). Since most of the positive redox couples undergo out-of-sphere reactions, they can be catalyzed on carbon-based surfaces. As a result, the performance of RFCs is dominated by the electrochemical surface of the carbon catalyst, its wetting properties, and reactant distribution within its porous structure. The combination of the negative electrode, membrane, and porous carbon electrode is known as the membrane electrode assembly (MEA). The MEA is sandwiched between two bipolar plates with careful control of electrode compression to reduce through-plane resistance. Besides that, the bipolar plates provide mechanical support and current collection. In some cases, the plates may include a flow field design to favor homogenous mass transport and reactant distribution. Those plates are often made of either graphite or chemically stable plastics that contain conductive fillers.

Scaling-up of the system relies on stacking the cell depicted in Fig. 5.3 with connection in series. In an RFB stack, the introduction of two ionically conductive electrolytes in the electrochemical flow reactor can lead to longevity issues associated with potential distribution across the entire stack (known as shunt currents). In contrast, the use of a nonconductive gas phase in the RFC system helps mitigate this problem. However, the RFC liquid crossover to the gas phase should be carefully monitored as it can electrically connect different cells and promote shunt-current-linked cell degradation. The RFC is integrated in an energy storage system by including elements such as AC/DC converters, control systems, thermal management, pumps, or blowers.

5.2 Reaction Thermodynamics and Kinetics

The energy in RFCs is stored via a conversion step, as chemical energy in the electrolyte by means of an electrochemical process. This system is quite particular, as it is a mixture of two types of electrochemical energy storage and conversion systems, a secondary battery (where charge and discharge processes occur), and a fuel cell (where a fuel and an oxidant are supplied to the cell and converted to energy). In these systems, two terms are of importance: energy efficiency (EE, the relationship between the energy recovered/energy feed) and coulombic efficiency (CE, the relationship between the discharge/charge amount). These figures of merit are influenced not only by the thermodynamics and kinetics of the respective redox reactions but also by mass transport phenomena and electrical resistance.

$$\mathrm{EE} = \frac{E_{\mathrm{out}}}{E_{\mathrm{in}}} \qquad \mathrm{CE} = \frac{C_{\mathrm{discharge}}}{C_{\mathrm{Charge}}}$$

5.2.1 Redox Reaction: Thermodynamics

The equilibrium potential, E_{eq} (V versus reference), is effectively a measure of the chemical potential of an electron and is a key thermodynamic parameter for an electrochemical reaction. Similar to a Gibbs free energy model in solution thermodynamics, the equilibrium potential is influenced by the solution temperature, T (K), and the solution composition according to the Nernst equation, provided here for a concerted electron transfer reaction, as shown in Eq. (5.1).

$$\mathrm{Ox} + ne^- \rightleftarrows \mathrm{Red}$$

$$E_{\mathrm{eq}} = E^0 - \frac{RT}{n\mathrm{F}} \ln \frac{\alpha_{\mathrm{Red}}}{\alpha_{\mathrm{Ox}}} \tag{5.1}$$

R (8.314 J/molK) is the universal gas constant, n is the number of electrons transferred in the redox reaction, F (96485 C/mol) is the Faraday constant, α_{Ox} and α_{Red} (dimensionless) are the activities of the oxidized and reduced species, respectively, and E^0 (V versus reference) is the standard redox potential. In many cases, a more experimentally accessible form of the Nernst equation is used (Eq. (5.2)) where the more general formal redox potential, $E^{0'}$ (V versus reference), is used in place of the standard redox potential to capture nonideal species activity and deviations from standard conditions.

$$E_{\mathrm{eq}} = E^{0'} - \frac{RT}{n\mathrm{F}} \ln \frac{C_{\mathrm{Red}}}{C_{\mathrm{Ox}}} \tag{5.2}$$

C_{Ox} and C_{Red} (mol/m^3) are the concentrations of the oxidized and reduced species, respectively. Within the context of RFB performance, the difference in equilibrium potentials between half-cell reactions defines the open-circuit voltage of the full cell and is thus directly proportional to the maximum available energy in the system, making the standard, or formal, redox potential an essential quantity in the evaluation of novel redox electrolytes.

The equilibrium potentials cannot be measured, as it is for a single point potential. The only way possible is by measuring the potential difference between two points, in this case the electrodes, because the half-cell potentials are measured by means of a second half-cell relative to the standard hydrogen electrode (SHE).

If we consider a generic redox cell reaction in a galvanic cell, this can be written as

$$\mathrm{aOx_1 + bRed_2 \leftrightarrows aRed_1 + bOx_2} \tag{5.3}$$

To keep the reaction simple, the electric charges carried by the Ox and Red species in Eq. (5.3) are not mentioned. In the cell reaction, the electrons are stoichiometrically compensated, and the reversible resistance of the reaction tends to be infinite, thereby constituting a reversible process. For these conditions, the electrochemical Gibbs free energy ($\tilde{\mathrm{G}}$) can be written as

$$-\Delta\tilde{G} = nFE_{\mathrm{cell}} \tag{5.4}$$

$$E_{\mathrm{cell}} = E^0_{\mathrm{cell}} + \frac{RT}{nF}\ln\frac{\alpha^b_{\mathrm{Ox_2}}\alpha^a_{\mathrm{Red_1}}}{\alpha^b_{\mathrm{Red_2}}\alpha^a_{\mathrm{Ox_1}}} \tag{5.5}$$

E_{cell} is the cell potential (or cell voltage), the so-called electromotive force (emf) of the cell, i.e., the potential difference, measured at zero current between the two half-cells under equilibrium conditions; E^0_{cell} is the relevant standard cell potential. Breaking apart Eq. (5.5) into the corresponding half-cell reactions and assuming a reversible process, we obtain

$$E_{\mathrm{eq,1}} = E^0_1 + \frac{RT}{n\mathrm{F}}\ln\frac{\alpha^a_{\mathrm{Red_1}}}{\alpha^a_{\mathrm{Ox_1}}} \tag{5.6}$$

$$E_{\mathrm{eq,2}} = E^0_2 + \frac{RT}{n\mathrm{F}}\ln\frac{\alpha^b_{\mathrm{Ox_2}}}{\alpha^b_{\mathrm{Red_2}}} \tag{5.7}$$

As mentioned earlier, we can only measure the potential difference between two electrodes, obtaining the E_{cell}; considering this, we can express the following relationships:

$$E_{cell} = E_{eq,2} - E_{eq,1} \tag{5.8}$$

$$E^0_{cell} = E^0_{Ox_1,Red_1} - E^0_{Ox_2,Red_2} \tag{5.9}$$

Summarizing Eqs. (5.4) to (5.9) into a scheme is possible and depicted in Fig. 5.4 [47].

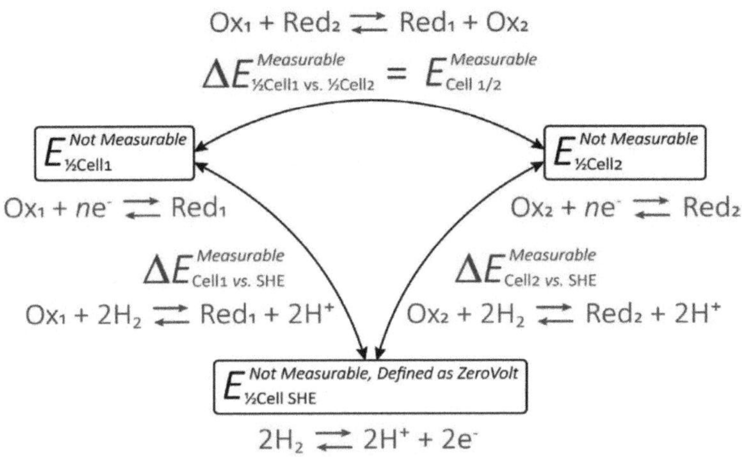

Figure 5.4 Relationship of the physical meaning for the individual half-cell redox reactions.

The reduction half-cell reaction, Fig. 5.4 (left side), happens in the half-cell 1, where the electrons are being transferred from the electrode to the Ox_1 species in solution, thereby obtaining a positive charge in the electrode (working electrode, cathode). The opposite happens in the half-cell 2 (counter electrode, anode). Consequently, the potential difference between both half-cells will give a positive E_{cell} (emf) value. A positive E_{cell} implies the occurrence of a spontaneous redox reaction (Eq. (5.4)). The electrons "travel" counterclockwise in the external subcircuit, while in the internal subcircuit, anions migrate inside each half-cell and from the left to the right half-cell, while cations follow the opposite direction [47].

As an example, the equilibrium potential for the most studied RFC (H_2/Br_2), $E_{eq,Br_2/Br^-}$, and $E_{eq,H^+/H_2}$ of the $Br_{2(aq)}/Br^-$ – H^+/H_2 electrochemical half-cell couples are defined by the Nernst equation as

Positive side

$$E_{eq,Br_2/Br^-} = E^o_{Br_2/Br^-} + \frac{RT}{2F}\ln\left(\frac{\alpha_{Br_2}}{\alpha^2_{Br^-}}\right)$$

Negative side

$$E_{eq,H^+/H_2} = E^o_{H^+/H_2} + \frac{RT}{2F}\ln\left(\frac{\alpha^2_{H^+}}{\alpha_{H_2}}\right)$$

where $E^o_{Br_2/Br^-}$ and $E^o_{H^+/H_2}$ are the standard potentials; T is the temperature in K; *R* is the universal gas constant; *n* is the number of electrons involved, 2 in both cases; *F* is the Faraday constant; and α is the activity of species [48].

5.2.2 Redox Reaction: Kinetics

5.2.2.1 Gas-side kinetics

To understand the gas-side kinetics for the different RFC configurations explained in Sections 5.5 and 5.6, we need to introduce two known reactions in traditional hydrogen fuel cells.

Hydrogen Oxidation Reaction/Hydrogen Evolution Reaction (HOR/HER) [49]

Reaction	$2H^+ + 2e^- \rightleftarrows H_2$
Mechanism	
Tafel	$H_2 + 2^* \rightleftarrows 2H_{ad}$
Heyrovsky	$H_2 + * \rightleftarrows H_{ad} + H^+ + e^-$
Volmer	$H_{ad} \rightleftarrows * + H^+ + e^-$

* denotes a site on the surface

Oxygen Reduction Reaction/Oxygen Evolution Reaction (ORR/OER) [50, 51]

Reaction	$O_2 + 4H^+ + 4e^- \rightleftarrows 2H_2O$
Mechanism	
Dissociative	**Associative**
$O_2 + 2^* \rightarrow 2O^*$	$O_2 + {}^* \rightarrow O_2^*$
$O^* + H^+ + e^- \rightarrow HO^*$	$O_2^* + (H^+ + e^-) \rightarrow HO_2^*$
$HO^* + H^+ \rightarrow H_2O + {}^*$	$HO_2^* + (H^+ + e^-) \rightarrow H_2O + O^* \text{or}\, H_2O_2$
	$O^* + (H^+ + e^-) \rightarrow HO^*$ $H_2O_2 + 2H^+ + 2e^- \rightarrow 2H_2O$
	$HO^+ + (H^+ + e^-) \rightarrow H_2O + {}^*$

Reaction	$H_2O \rightleftarrows O_2 + 4H^+ + 4e^-$
Mechanism	
$H_2O + {}^* \rightarrow HO^* + H^+ + e^-$ $HO^* \rightarrow O^* + H^+ + e^-$ $O^* + H_2O \rightarrow HO_2^* + H^+ + e^-$ $HO_2^* \rightarrow {}^* + O_2 + H^+ + e^-$	

Both reactions, either involving H_2 or O_2, have been studied for years predominantly on precious metal surfaces as catalysts; the most used one is platinum (Pt) as it usually has the best activity in acid media. In terms of reaction rate-limiting steps, the H_2 reaction mechanism is simpler compared with the O_2 one. HOR and HER can be considered reversible as they require lower overpotential, making the use of hydrogen a preference in RFCs [52].

In contrast, for the ORR/OER reactions, the reversibility drops drastically, with a considerable increase in the overpotential [51]. While oxygen redox reactions are attractive owing to their relatively high potential (1.23 V versus SHE), their sluggish kinetics and considerable activation barrier are not ideal for fully reversible battery cells or in an RFC setup [53] even though some examples can be found (Section 5.5) with reasonable performance. In addition to this, the precious catalyst loading required for the O_2 reactions is at least eight times (approximately 0.40 mgPt/cm^2) that of the H_2 side (anode) (approximately 0.05 mgPt/cm^2). The dependence on expensive catalysts can be alleviated with anion-exchange membranes (AEMs), which implies cell operation at a high pH. However, the alkaline electrolyte leads to issues such as the presence of CO_2 and formation of solid deposits, or the decrease of nearly two orders of magnitude in the HER/HOR kinetics [54]. This might hinder the full commercial exploitation of AEM-based devices in the near term [52].

An aspect to consider about these reactions is the kinetic parameters such as rate or exchange current density j_o ($\mathrm{Acm}^{-2}_{\mathrm{Metal}}$) at a certain temperature/pressure that related with the corresponding activation energy, E_a (kJ/mol), and the anodic and cathodic transfer coefficients, α. These parameters can be obtained by investigating the behavior of the material in a simplified experimental setup (rather than a full RFC). For instance, one can study these reactions with the well-established rotating disk electrode (RDE) technique.

The Butler–Volmer equation expresses the relationship between the electrode potential and the current density (j) in a general form [55]. The equation is fitted to the current density ($\mathrm{A\,cm}^{-2}_{\mathrm{Metal}}$) versus HOR/HER overpotential (η) using semilogarithmic plots (Eq. (5.10)):

$$j = j_0 \times \left(e^{\frac{\alpha_a F\eta}{RT}} - e^{\frac{\alpha_c F\eta}{RT}} \right) \tag{5.10}$$

α_a and α_c are the anodic (HOR) and cathodic (HER) transfer coefficients, F is Faraday's constant, and R is the universal gas constant. At a high overpotential, the Butler–Volmer equation can often be expressed as the Tafel equation, which was first empirically demonstrated. Interestingly, the Tafel slopes (mV/dec) contribute to elucidate the rate-limiting step of the reaction. To obtain the anodic and cathodic Tafel slopes (b in V/dec), these can be calculated using Eq. (5.11), using the respective anodic and cathodic transfer coefficients (Fig. 5.5).

$$b_{a/c} = \frac{2.303RT}{F\alpha_{a/c}} \tag{5.11}$$

The slope value can be an indication of the rate-determining step, which, in the case of the H_2 reactions, could be a Tafel, Volmer, or

Heyrovsky step or a combination of them. In acidic media, standard values of the kinetic parameters for Pt are agreed to be in the range of $j_o \sim 1\ \text{mA cm}_{\text{Pt}}^{-2}$ with a Tafel slope of 30 mV/decade [52].

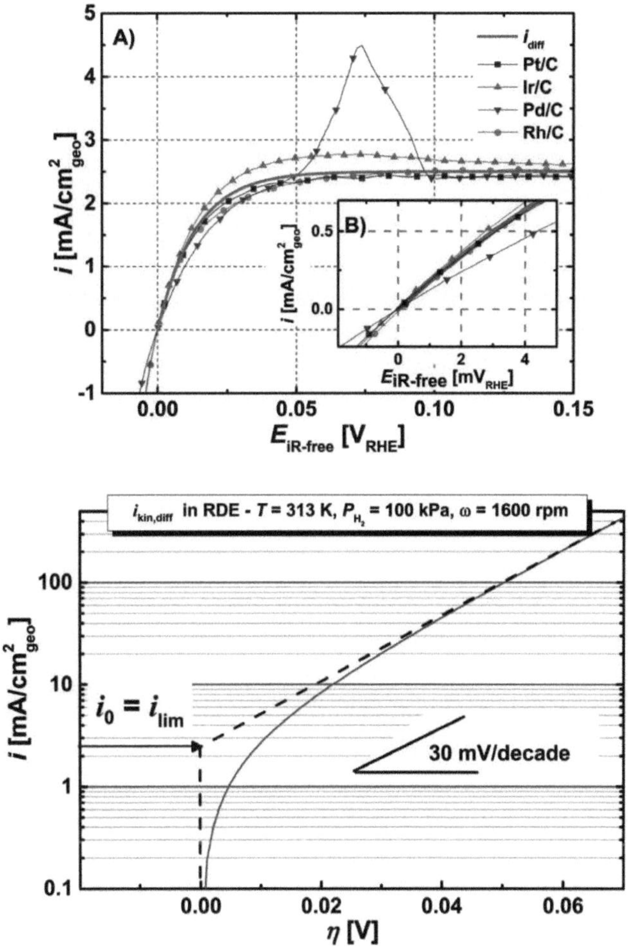

Figure 5.5 Top: HOR experiments using different noble metals and the theoretical diffusion j ($j_{\text{lim}} = 2.4\ \text{mA}_{\text{geo}}^{-2}$). Bottom: theoretical Tafel representation of HOR currents extracted from diffusion limited voltammograms recorded with RDE (assumed $j_{\text{lim}} = 2.4\ \text{mA cm}_{\text{geo}}^{-2}$). The dashed lines represent construction lines for the visual determination of i_0 and the Tafel slope. Reprinted with permission from Ref. [52]. Copyright (2014) IOP Science (The Electrochemical Society).

Finally, E_a (kJ/mol) can be obtained from the electrochemical data, but to achieve that, j_0 must be fitted at different temperatures, using Eq. (5.12):

$$\frac{\partial \log j_0}{\partial(1/T)} = \frac{E_a}{2.303R} \tag{5.12}$$

The RDE technique is, however, not adequate for extremely fast reactions, for example HOR/HER on Pt electrodes in acid (Fig. 5.5) [49, 56]. Likewise, the overpotential range in which the reaction is purely kinetically controlled is rather small for the RDE technique. The mass transport limiting current is quickly achieved for processes in which the concentration of the reactant is small, for instance due to limited gas solubility (HOR or ORR).

In the case of HOR, the voltammogram obtained fits to the Nernstian diffusion overpotential, η_{diff}. This term describes the potential change due to the H_2 surface concentration (Eq. (5.13)):

$$\eta_{diff} = -\frac{RT}{2F}\ln\left(1-\frac{i}{i_{lim}}\right) \tag{5.13}$$

where i and i_{lim} are the measured and diffusion-limiting currents, respectively. Something that needs to be considered is the kinetic currents that must be plotted versus the potential with the iR-drop (ΔE_Ω) and η_{diff} corrections [52]. After the corrections, we can calculate the maximum HOR current (i_{diff}) under pure diffusion conditions, Eq. (5.14):

$$i_{diff} = i_{lim}\left(1-e^{-\frac{2F}{RT}\eta}\right) \tag{5.14}$$

Since the kinetics of the HER/HOR are very fast, which can be considered quasi-reversible (Nernst equation) in the j range for the RDE, the kinetic currents (i_{kin}) can be obtained (or estimated) using the Koutecký–Levich correction for a first-order reaction (Eq. (5.15)) [52]:

$$i_{kin,diff} = i_{diff}\left(\frac{i_{lim}}{i_{lim}-i}\right) \tag{5.15}$$

$$i_{kin,\,diff} = i_{lim}\left(e^{\frac{2F}{RT}\eta}-1\right) \tag{5.16}$$

Equation (5.16) (combination of Eqs. (5.14) and (5.15)) represents the pseudo-kinetic HOR current, obtained from the voltammograms, which closely follow η_{diff}.

It should be noted that the development of new experimental protocols to overcome mass transport limitations imposed by RDE techniques is a subject of research in the field of electrocatalysis. For example, Zalitis and Kucernak et al. demonstrated a novel approach to extend the kinetic current potential range by increasing the gas feed in what the authors called a floating electrode configuration [57, 58]. This technique allows establishment of a closer catalyst performance to a real-life operation regime in an RFC.

5.2.2.2 Liquid-side kinetics

The electrochemical kinetics in the liquid side play a critical role in influencing RFC performance, notably overpotential and cell power density. Thus, determining the kinetic parameters for the employed redox-active species is essential [59]. A good RFC requires favorable kinetic characteristics to enable usability over hundreds or thousands of cycles and to deliver a desirable potential window and current density. One should determine the reversibility of the redox reaction, kinetics, and charge-transfer mechanisms of a redox couple before integrating within an RFC. This important information will enable rational improvements in the design of the RFC. However, to achieve that, the redox couple is often studied with a dilute electrolyte formulation that utilizes high concentration of a supporting electrolyte to ensure the reaction is diffusion-controlled. Likewise, the implementation of a reference electrode facilitates the experimental observation of potential shifts as well as the identification of side reactions. This can be achieved, for example, *via* cyclic voltammetry experiments.

A "chemically reversible" redox reaction is essential for a good cycle life in RFCs, along with the "electron transfer reversibility," which indicates the electron transfer rate between the electrodes and the redox species. In most of the electrochemical literature, the term "reversible" indicates a fast electron transfer reaction that maintains thermodynamic equilibrium at the electrode/solution interface, whereas "irreversible" indicates a slow reaction that can seriously limit power output in RFCs. "Quasi-reversible" refers to intermediate rates between these two extremes. Figure 5.6 shows

Step 1 **Carry out CV of redox couple**

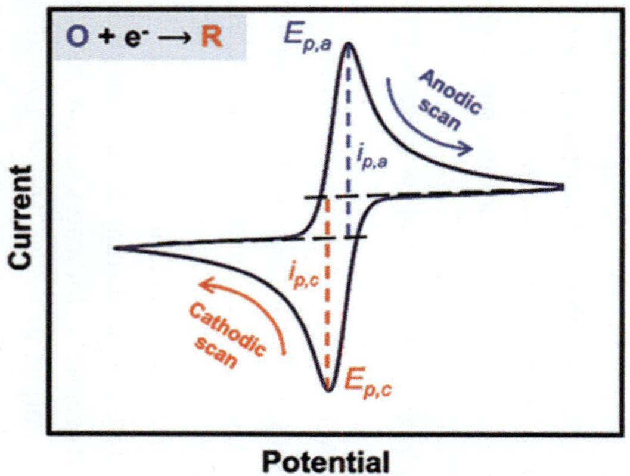

Run the CV at a reasonable scan rate, *e.g.*, 50 mV/s.

Step 2 **Define type of electron transfer mechanism**

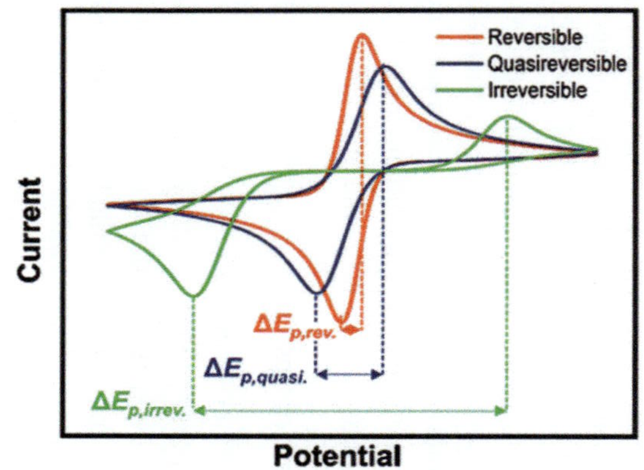

Determine the peak separation, ΔE_p. ΔE_p = anodic peak potential ($E_{p,a}$) - cathodic peak potential ($E_{p,c}$).

Reversible if $\Delta E_{p,rev.}$ = (57/n) mV where n = number of e^-

Quasireversible if $\Delta E_{p,rev.} < \Delta E_{p,quasi.} \lesssim$ (200/n) mV *

Irreversible if $\Delta E_{p,irrev.} >$ (200/n) mV *

Step 3a Run CVs at different scan rates

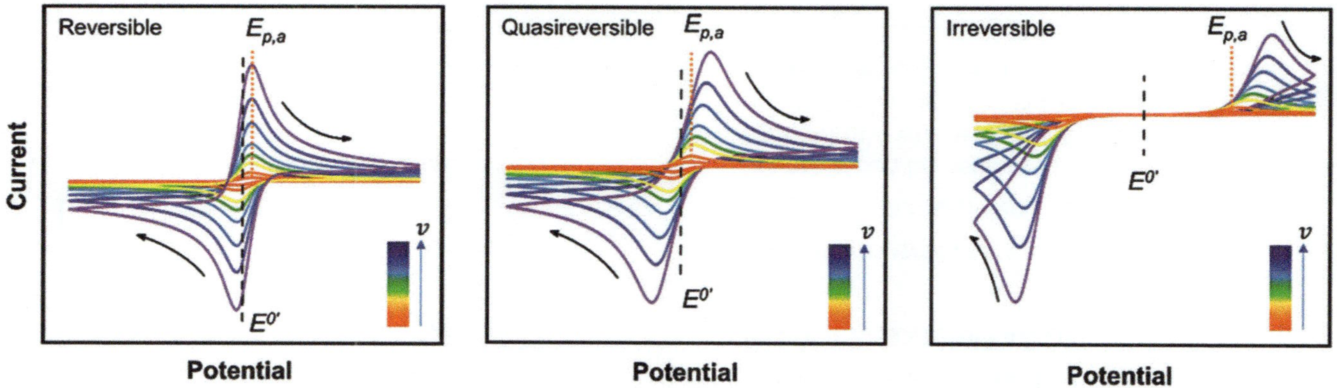

Correct for uncompensated resistance.[14,15] (Detailed directions in Supplementary Information.)
If $E_{p,a}$ does not shift with scan rate (v), system is reversible.

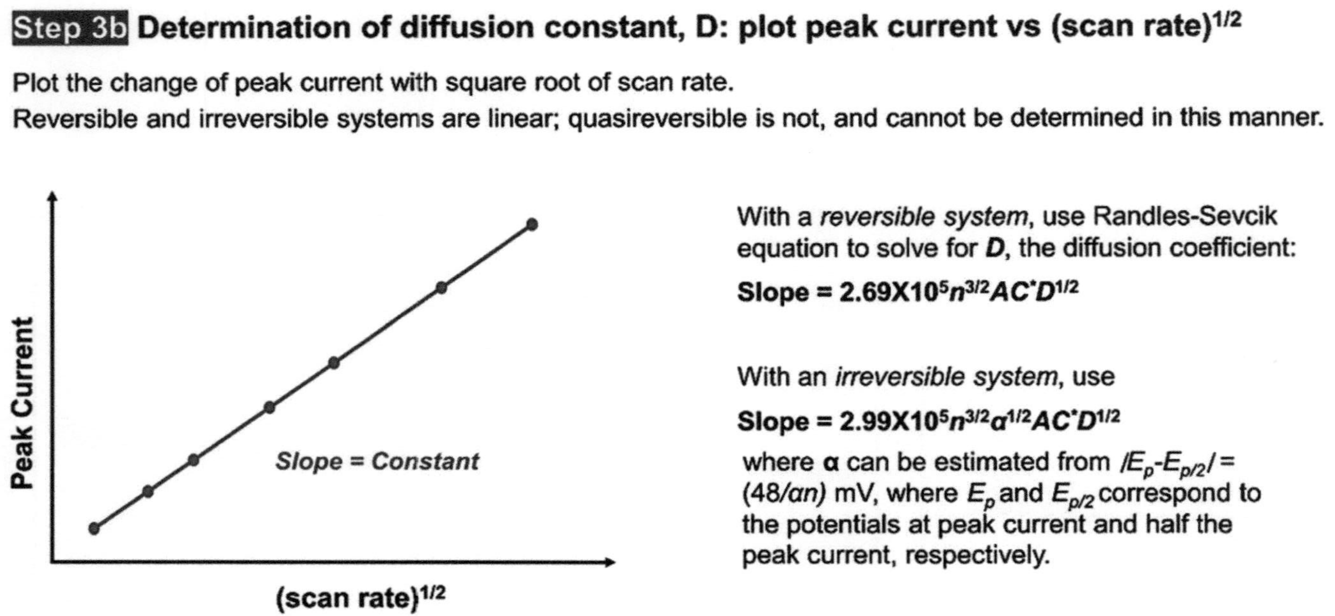

Figure 5.6 First set of cyclic voltammetry (CV) experiments to define the kinetic regimes (reversible, quasi-reversible, and irreversible) for an electron transfer (ET) mechanism (steps 1–3a). CVs for reversible, quasi-reversible, and irreversible systems were simulated using Bio-Logic EC-Lab software. Parameters: $k^0 = 0.5$ cm/s (reversible, red), 0.001 cm/s (quasi-reversible, blue), and 1×10^{-6} cm/s (irreversible, green); $\upsilon = 50$ mV/s; diffusion coefficient ($D_O = D_R$) = 1×10^{-5} cm^2/s. *The final potential range for the quasi-reversible system in step 2 [(200/n) mV] was chosen based on values obtained from the Nicholson equation. Step 3b illustrates the representative linear relationship between the peak currents and the square root of scan rates $\upsilon_{1/2}$ for reversible and irreversible systems. Reprinted with permission from Ref. [59]. Copyright (2020) ACS.

the standard experiments that should be performed to determine the reversibility of the redox reaction (reversible, quasi-reversible, or irreversible) and how to calculate the associated diffusion coefficient of the redox-active species in solution.

The diffusion coefficient is an important parameter to determine, as species with a slow diffusion will impart mass transport overpotential. Another important parameter to consider is the standard rate constant (k^0); this constant can quantify the kinetics of a reaction [60]. For low values of k^0, the reaction has sluggish kinetics requiring larger overpotential (more energy), which is not the case for RFCs. In the case of reversible reactions, the value of k^0 is high (>10^{-1} cm/s); it is lower for quasi-reversible reactions (10^{-1} to 10^{-5} cm/s), and even less for irreversible reactions (<10^{-5} cm/s) [59].

To calculate k^0, it is required to use the Nicholson procedure, in which the value is calculated from the peak-to-peak potential separation (ΔE) at different scan rates (ν), following Eq. (5.17):

$$k^0 = \psi \frac{\left[\pi D_{\mathrm{Ox}} v (nF / RT)\right]^{1/2}}{\left(D_{\mathrm{Ox}} / D_{\mathrm{red}}\right)^{\alpha/2}} \tag{5.17}$$

$$\psi = \frac{\left(-0.6288 + 0.0021 \Delta E_{\mathrm{p}}\right)}{1 - 0.017 \Delta E_{\mathrm{p}}} \tag{5.18}$$

Here, ψ is the kinetic parameter; D_{Ox} and D_{red} are the diffusion coefficients of the oxidized and reduced species, respectively; ν is the scan rate; and α is the transfer coefficient. Following the Nicholson procedure, ΔE_{p} is a function of ψ, which is independent of α, in the range of $0.3 < \alpha < 0.7$. The calculation involves the measurement of ΔE_{p} at different ν values, and finally determining the values for ψ [59]. However, obtaining the values of k^0 for a reversible system is complex due to the limitation of the standard electrodes in terms of size with diameters of a few millimeters. This only allows scan rates of over 100 V/s because of the ohmic potential error, as well as reaching large currents associated with the capacitance. To avoid this issue, Mirkin and Bard developed a method using micro- and nanoelectrodes to determine the value of k^0. This is possible as the micro- and nanoelectrodes improve the mass transport rate and minimize the ohmic drop in the solution, thereby reducing capacitive currents and the adsorption reaction. In their method, k^0 is determined directly from two easily accessible experimental parameters, ($E_{1/4}$ - $E_{1/2}$) and ($E_{1/2}$ - $E_{3/4}$), where $E_{1/2}$ is the half-wave potential and $E_{1/4}$ and $E_{3/4}$ are quartile wave potentials [60].

Another well-established method to obtain the k^0 value of the redox couple and the diffusion coefficient is RDE (mentioned for the gas side) [61]. In the case of kinetically limited reactions, the charge-transfer rate limits the current, classified as irreversible reactions [62]. This can be shown via the Koutecký–Levich equation:

$$\frac{1}{j}=\frac{1}{j_k}+\frac{1}{j_{\text{lim}}}=\frac{1}{j_k}+\frac{1}{0.62nFD_0^{2/3}\omega^{-1/2}v^{-1/6}C_0}$$

Plotting the $1/j$ versus $1/\omega^{1/2}$, we can obtain the j_k from the intercept with $1/\omega^{1/2} = 0$ at a particular potential. To calculate k^0, we need to plot the log(i_k) at different potentials versus the overpotential (E - E^0), as shown in Eq. (5.19).

$$j_k = nFk_fC_0 = nFk^0C_0\exp\left(\frac{-\alpha nF\left(E-E^0\right)}{RT}\right) \tag{5.19}$$

5.3 Key Performance Indicators for RFC

5.3.1 Introduction

A polarization curve is an experiment often employed to characterize a polymer electrolyte membrane fuel cell (PEMFC), which is applicable as well in the context of RFC performance evaluation (Fig. 5.7). It relies on holding a constant current until the cell reaches steady state and the cell voltage remains constant. Current is then subsequently and gradually increased in steps, and the voltage drop is monitored at each step. The obtained curve allows direct visualization of different losses that dictate the behavior of the reactor as a full cell. The polarization of an RFC under load is commonly described by Eq. (5.20) where E_{cell} is the RFB cell voltage, IR is the voltage drop associated to the ohmic resistance, η^{act} is the activation polarization of both anode and cathode, and η_+^{c}, η_-^{c} are the concentration polarizations of the positive and negative half-cells, respectively. For example, in a fuel cell (and RFC) system, Eq. (5.20) can be summarized as Eq. (5.21):

$$E_{\text{cell}} = E_{\text{OCV}} - IR - \eta^{\text{act}} - \eta_{+}^{\text{c}} - \eta_{-}^{\text{c}} \quad (5.20)$$

$$E_{\text{Fuel Cell}} = E_{\text{OCV}} - IR - \eta_{\text{reaction}} - \eta_{\text{transport}} \quad (5.21)$$

The cell is not normally operated under standard conditions, which pushes the potential away from the thermodynamic one. Likewise, the Nernst equation is not fulfilled because the solution does not contain species that are the products of the redox reaction. This means that the open-circuit voltage (OCV) might not be constant over time. However, as soon as the RFC is forced to deliver power, voltage losses are expected to overcome the activation energy for both redox reactions. As the operational current density increases, the cell voltage gradually drops by a combination of reaction kinetics and ohmic losses. In case the RFC utilizes hydrogen reactions in the negative side, both the activation energy and overpotential associated with HOR in the RFC are significantly lower than those linked to the soluble redox couple ($\eta_{+}^{\text{c}} \gg \eta_{-}^{\text{c}}$). As the rate of the reaction increases with current output, the depletion of the active species in the liquid side leads to reactant starvation and mass transport overpotential.

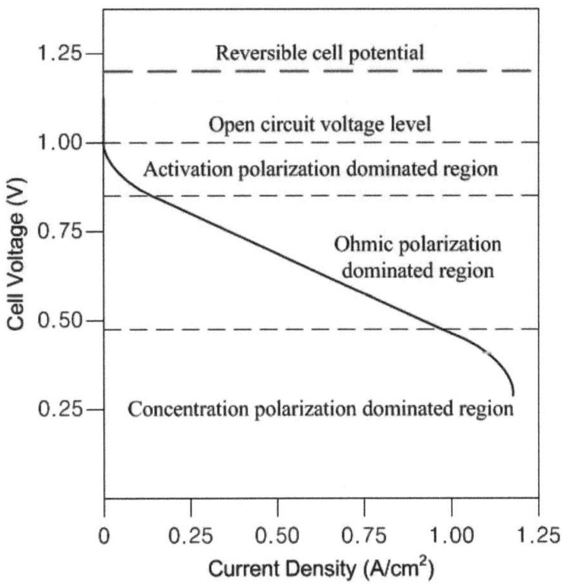

Figure 5.7 Voltage breakdown curves for PEMFCs (and RFCs). Reprinted with permission from Ref. [63]. Copyright (2021) IOP science (The Electrochemical Society).

It is worth mentioning that, although it provides valuable information on the performance of the RFC, polarization curve experiments are not suitable to deconvolute each polarization (ohmic, activation, and concentration) at each current density. There have been efforts to build arrays of reference electrodes on a single cell or stack as a strategy to map potential distribution across the cell [64–67]. Other works employed techniques such as electrochemical impedance spectroscopy (EIS), which involves an equivalent circuit to determine different resistive elements [68]. However, the ability to implement experimental methods to decouple these polarization elements remains a challenge. For example, the protocol for polarization curve measurements has not been standardized for the RFC. The measurement is strongly influenced by variables, including flow rate, cell assembly protocol, or electrolyte recirculation [69]. The latter is particularly relevant as the stoichiometry of the system does not remain constant if the RFC is operated in a closed loop. In alignment with this, a polarization curve might not adequately describe the operational regime for the RFC. For RFCs employing highly reversible but expensive electrolytes such as the H_2/V RFC, the system is operated at a relatively low current density when compared to a PEMFC to ensure high electrolyte utilization. The next subsection is devoted to the main key performance indicators (KPIs) and testing protocols commonly used to evaluate RFCs.

5.3.2 RFC Fundamental KPIs

The volumetric capacity of the liquid electrolyte, whether negative or positive (anode and cathode during RFC discharge), is a crucial factor in determining the charge capacity per unit volume of the electrolyte in an RFC (for the liquid-side half-cell only). This parameter is particularly important for large-scale applications as it helps project the required footprint of the liquid-side RFC tank [70]. To ensure accurate and unbiased comparisons, it is essential to clearly state which electrolyte volume (positive or negative depending upon whether the RFC is hydrogen- or oxygen-based) is considered when reporting and comparing volumetric capacity of the liquid-side half-cell. Neglecting the electrolyte volume and gas pressure can lead to inaccurate assessments of the cell's energy storage capacity. By including information from both half-cells, a more comprehensive evaluation of the RFC system's energy density can be achieved. This transparency helps avoid misleading or biased comparisons between different RFC systems [70].

Table 5.2 summarizes some relevant performance metrics. These factors substantially affect overall RFC performance and should be controlled and considered collectively when making comparisons. A very good summary of the parameters that must be considered for RFB was done by Yao et al. [70]. The volumetric cell capacity of the liquid ($Q_{v,+/-}$, Eq. (5.27)) is determined by the combination of the volumetric capacity of the liquid electrolyte ($Q_{v,+/-}$, Eq. (5.28)). This value represents the total charge capacity per unit volume of the liquid-side RFC half-cell [70].

Table 5.2 Cell attributes and performance metrics of RFBs (relevant to the liquid-side half-cell of any RFC)

	Definition		Remarks
	Cell Attributes		
[a]$V_{cell,OCV}$ at 50% SOC	$V_{cell,OCV}(V) = V_+ + V_-$	(5.22)	The electromotive force of the RFB can be evaluated at 50% SOC.
Theoretical electrolyte capacity[b]	$Q_t = n \times [C] \times v \times \frac{F}{3600} (\mathrm{A\,h})$	(5.23)	Theoretical capacity stored in each volume of electrolyte.
Current density[c]	$j_a = I / A (\mathrm{mA\,cm_{geo}^{-2}})$	(5.24)	Current applied per area
SOC	$SOC = Q_{Charge} / Q_{Total}$	(5.25)	Charged capacity stored over theoretical capacity
Solid areal capacity	$Q_{a(solid)} - \frac{Q_{rev}}{A} (\mathrm{mA\,h\,cm_{geo}^{-2}})$	(5.26)	Achieved reversible capacity stored per membrane area.
	Performance Metrics		
Volumetric capacity	$Q_{v,+/-} = \frac{Q_{Discharge,+/-}}{v_{+/-}} (\mathrm{A\,h\,L_{pos./neg.}^{-1}})$	(5.27)	Discharge capacity stored per unit volume of electrolyte (positive, negative, or both)
	$Q_{v,cell} = \frac{Q_{Discharge}}{v_+ + v_-} = \frac{Q_{v+} \times Q_{v-}}{Q_{v+} + Q_{v-}} (\mathrm{A\,h\,L_{pos.+neg.}^{-1}})$	(5.28)	

	Definition	Remarks
Capacity decay rate	The slope of capacity-time or capacity-cycle number plots divided by the initial capacity (% per day or % per cycle)	Capacity decay in % over a total test duration time (% per day) or cycle number (% per cycle). The total duration or cycle number should be reported along with the decay rate.
Coulombic efficiency	$CE = Q_{Discharge} / Q_{Charge}$ (5.29)	The ratio of discharge capacity over the charge capacity
Voltage efficiency	$VE = V_{Discharge} / V_{Charge}$ (5.30)	The ratio of the discharge voltage over the charge voltage
Volumetric energy density[d]	$E_{v,cell} = \frac{\int dQ_{discharge} \times dV_{cell}}{v_{+} + v_{-}} \left(W\,h\,L^{-1}_{pos.+neg.}\right)$ (5.31)	Energy output per unit volume of total electrolyte (both half-reactions)
Energy efficiency	$EE = E_{v,\,cell\,discharge} / E_{v,\,cell\,charge}$ (5.32)	The ratio of discharge energy over charge energy at a certain current density
Power density	$P\left(mW\,cm^{-2}_{geo}\right) = \left(I \times V_{cell}\right) / A$ (5.33)	Power output per membrane area at a certain SOC

[a,d] V_{cell} is the cell voltage at the corresponding current (V), V_{+} for positive and V_{-} for negative.
[b] $[C]$ is the concentration (mol/L), n is the number of electrons per mole, v is the volume (L), and F is the Faraday constant.
[c] I is the current (mA) and A is the active area (cm^{-2}_{geo}, membrane area).
Note: Reproduced with permission from Ref. [70]. Copyright (2021) Elsevier.

5.3.3 Self-Discharge and Cell Degradation

Lifetime evaluation of RFCs can be performed by not only noting their CE but also measuring the voltaic and energy efficiencies. The voltaic efficiency of an RFC system is influenced by various factors, including the contributions to the overpotential. Additionally, the discharge capacity of the RFC upon cycling can indicate how well the system performs upon stressful charge and discharge conditions at high operating current densities. For example, the H_2/V RFC has been cycled efficiently at 150 mA/cm^2 in a recent investigation and displayed exceptional performance in terms of cycle life, as shown in Fig. 5.8 [43].

The capacity fade rate can be expressed as a percentage per day or per cycle, depending on the preferred unit of measurement for the RFC system in question. The relative importance of these metrics varies based on the most critical degradation mechanism of the specific RFC system under consideration [71].

One of the challenges faced by vanadium-based RFB systems (VRFBs) is electrolyte imbalance. While VRFBs have the advantage of avoiding cross-contamination between the negative and positive sides, they can experience imbalances in the electrolyte composition over time. The primary cause of electrolyte imbalance in VRFBs is the asymmetric water crossover that occurs during the charge and discharge operations. Water molecules can pass through the ion-selective membrane, leading to a difference in water content between the negative and positive compartments. To address this issue, periodic electrolyte rebalancing procedures are required during long-term usage. These procedures involve adjusting the water content or adding water to rebalance the electrolyte concentrations in both compartments. However, these procedures can be costly and time-consuming. Efforts have been made to develop strategies to mitigate the electrolyte imbalance in VRFB systems. These include optimizing the design of the ion-selective membrane to minimize water crossover, developing advanced membrane materials with improved selectivity, and exploring alternative electrolyte compositions [72].

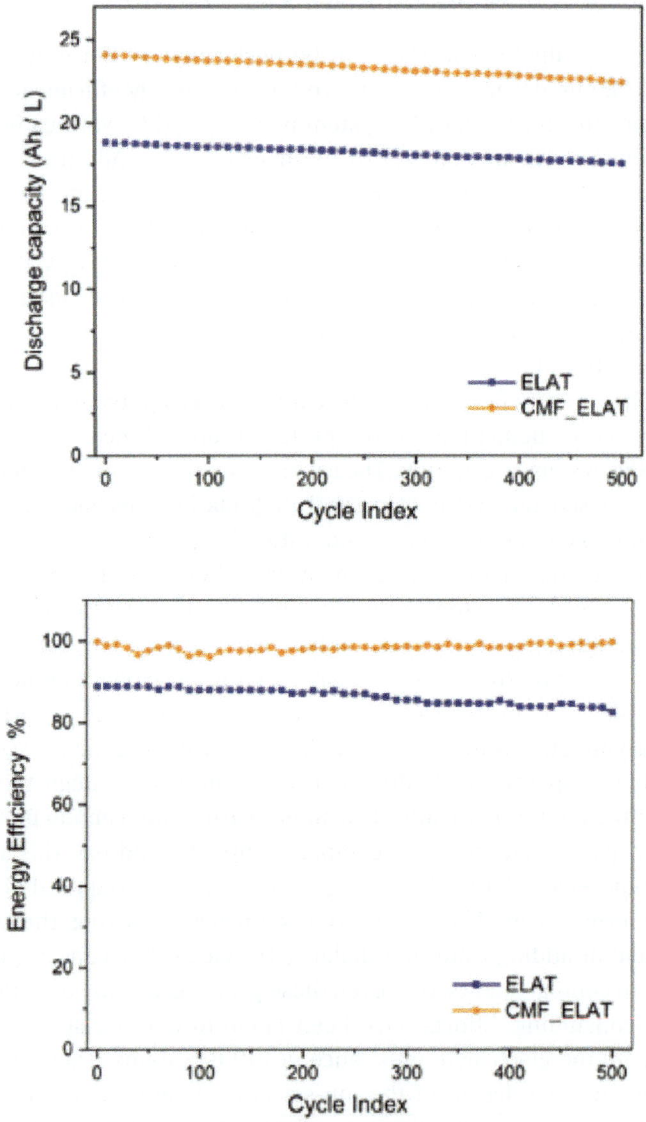

Figure 5.8 Discharge capacity utilization and energy efficiency of an H_2/V RFC upon continuous cycling at 150 mA/cm^2. This shows that the new electrode modification steps can enhance the RFC lifetime significantly. ELAT refers to a commercially sourced carbon cloth electrode, while CMF refers to a carbonized electrospun metal fabric applied as an electrode catalyst for the vanadium-based half-cell. Reprinted with permission from Ref. [43]. Copyright (2021) ACS.

RFC systems, especially for the case of the liquid/air systems, also face similar or even worse situations than VRFBs. For these liquid/air RFC systems, depending on the chemistry being employed, e.g., vanadium/air, crossover of electrolytes, as well as oxygen, can not only cause slow capacity fade but also major failure by the oxidation of V(II) ions in the presence of oxygen (that permeated through the membrane), as discussed in the following section. In contrast, hydrogen-based RFC systems, except for H_2/Br_2 RFC (that faces severe issues due to bromide crossover through ion-exchange membranes that result in HOR/HER catalyst poisoning) are less likely to face capacity fade and contamination problems if the hydrogen gas being supplied is pure. For such hydrogen-based RFC systems, as hydrogen/vanadium or hydrogen/manganese, crossover liquid to the negative half-cell is simply collected in an overflow vessel and recycled back to the positive half-cell in periodic steps to reduce capacity losses in the systems. In the case of halogen-based RFCs, liquid crossover can lead to irreversible degradation of the system. The platinum catalyst on the negative side undergoes corrosion reactions and poisoning by ion adsorption in the presence of bromide. To mitigate this, strategies such as replacing Pt by corrosion-resistant electrodes such as Rh_xS_y and operation under hydrogen-pressurized conditions have been reported (also discussed in Section 5.6). Likewise, extensive work has been done to find a balance between membrane proton conductivity and its role as a selective barrier.

5.3.4 Considerations for Cell Stacking

A typical RFC stack comprises several components that work together to enable RFC operation. These components can be categorized into the electrolyte system, hydrogen (or oxygen) system, bidirectional inverters and system controls, and containment or housing arrangements, as described in the following paragraphs [46]:

- **Electrolyte system:** This system includes the electrolyte storage tank, which holds the electrolytes used in the RFC. It also consists of relevant pumps for circulating the electrolytes, a heat-exchanging unit to control the temperature of the electrolyte, tubing, and couplings for fluid transport within the system.

- **Hydrogen (or oxygen) system:** The hydrogen (or oxygen) system is responsible for the supply and management of hydrogen gas (or oxygen in some cases). It includes components such as the hydrogen storage tank, Venturi ejectors (used for hydrogen recirculation), a liquid separator (to capture and store crossover liquid), a gas washer (to remove impurities from hydrogen gas), safety and vent valves for controlling gas flow and pressure, and tubing and couplings for gas transport.
- **Bidirectional inverters and system controls:** These components are used for power conversion and control within the RFC system. Bidirectional AC/DC inverters are employed to convert AC power from the grid to DC power during the charging mode, and vice versa during the discharging mode. Sensors are used to measure various parameters such as voltages, currents, state of charge, electrolyte flow rate, electrolyte pressure drop, temperature, fluid levels, hydrogen (or oxygen) pressure, hydrogen flow rate, and hydrogen level and to detect the liquid electrolyte concentration. Printed circuit boards and programming are employed for controlling and monitoring the system.
- **Containment or housing arrangements:** The RFC system is housed in a container, typically a standard 40 ft (36.7 m) container (taken from an example of an H_2/Br_2 RFC), which serves as the main housing for the stack, electrolyte system, power conversion electronics, and control system. The container provides a compact and enclosed space for the RFC components. However, the hydrogen storage tank is positioned outside the container due to safety considerations.

These various components and systems work together to facilitate the operation of the RFC stack, enabling the flow of electrolytes, control of power conversion, and monitoring of system parameters. The containerized design helps to provide a standardized and compact solution for deploying RFC systems [46].

5.3.5 Safety and Environmental Considerations

The inclusion of hydrogen pressure tanks in systems is a safety concern. The tank should be certified for pressures above 10 bar, which are in alignment with the operation of the system. This also might avoid the integration of additional gas compression. Liquid crossover and its impact on the development of shunt currents upon reaching the gas side can be mitigated with liquid recirculation ejectors. Likewise, a gas washer might be required to remove impurities in the gas phase and protect the integrity of the catalyst. Another safety consideration, which is crucial especially in the case of H_2/Br_2 RFCs, is the corrosive and hazardous character of bromine in the gas phase. Indeed, H_2/Br_2 RFCs often implement complexation agents to decrease the bromine vapor pressure, which might mitigate this safety concern. However, those chemicals have strong influence on the cost of the electrolyte and the chemical stability of the membrane [73, 74]. Here are some points related to safety concerns and measures in H_2/Br_2 RFCs [75]:

- **Heat management:** To prevent the release of bromine to the environment, reliable heat management systems are required. Controlling the temperature within the RFC stack is crucial to ensure that the bromine remains in its liquid state and does not evaporate. Effective cooling and thermal regulation mechanisms are implemented to maintain safe operating temperatures.
- **Complexing agents:** One approach to lower bromine vapor pressures is to dissolve complexing agents in the positive electrolyte. These agents help reduce the tendency of bromine to evaporate. However, the use of complexing agents can introduce challenges such as the formation of a viscous layer on the proton exchange membrane (PEM). This layer may hinder the transport of ions, which could affect the performance and efficiency of the RFC.
- **Hydrogen gas safety:** Hydrogen gas, which is involved in the H_2/Br_2 RFC system, also requires careful safety measures. Hydrogen is a highly flammable gas, and its handling, storage, and containment must follow strict safety protocols. Special precautions are taken to prevent leaks, ensure proper ventilation, and minimize the risk of hydrogen-related hazards.

- **Sealing technology:** Sealing technology plays a vital role in the cell assembly of H_2/Br_2 RFCs. It ensures gas and liquid tightness at elevated pressures, temperatures, and mechanical stresses inside the cell. The seals and gaskets used in the system are carefully selected to be compatible with the specific chemistries and conditions present in the RFC. They must be inert to the various species involved and maintain their integrity throughout the operational lifespan of the RFC.

Overall, safety considerations in H_2/Br_2 RFCs involve effective heat management, careful selection and use of complexing agents, proper handling and containment of hydrogen gas, and reliable sealing technology. By implementing robust safety measures, it is possible to mitigate risks and ensure the safe and reliable operation of the RFC system [75].

The implementation of safety measures in H_2/Br_2 RFCs is crucial to prevent hazards and ensure the overall safety of the system. Here are some additional points related to safety considerations [46]:

- **ATEX zone:** ATEX (atmosphere explosible) zones are designated areas where the presence of flammable gases, such as hydrogen, requires special precautions to prevent explosions. Electrical components used within the H_2/Br_2 RFC system are designed to meet the requirements of ATEX standards, ensuring they are safe for operation in potentially hazardous environments.
- **Safety valves:** Safety valves are installed within the system to prevent pressure buildup above a certain threshold, typically set at 10 bar. These valves release excess pressure to maintain safe operating conditions and avoid potential damage or rupture of the system components.
- **Vent valve:** The vent valve plays a role in purging hydrogen into the air, primarily to remove excess liquid crossover. It helps prevent the accumulation of hydrogen gas within the system, reducing the risk of flammable concentrations and maintaining a safe environment.
- **Tubing and couplings:** The use of polyvinylidene fluoride (PVDF) for tubing and couplings provides high chemical resistance. PVDF is known for its ability to withstand the

corrosive properties of the electrolyte, ensuring the integrity and longevity of the system components.

- **Neutralizing agent:** To enhance electrolyte safety, a neutralizing agent such as calcium carbonate may be present. In the event of a major electrolyte tank rupture, the neutralizing agent can help mitigate the consequences by neutralizing the electrolyte, reducing the risk of harmful chemical reactions or spills.
- **Safety factors in design:** The design and deployment of storage tanks consider safety factors. By incorporating a safety factor of 2.5 and a production factor of 3, the tanks are designed to withstand higher pressures and loads than expected under normal operating conditions, ensuring an additional margin of safety [76].

These safety measures aim to minimize risks associated with hydrogen and electrolyte handling, pressure buildup, and potential system failures. By implementing such precautions and adhering to safety standards, the H_2/Br_2 RFC and other RFC systems can operate safely and reliably in various applications.

5.4 Liquid/Air RFCs

5.4.1 Vanadium/Air

The first report of the vanadium/air redox-flow battery (VARFB) can be traced back to 1992 when it was initially patented [77, 78]. In 1997, Menictas and Skyllas-Kazacos expanded on this technology [26, 79]. The concept behind this redox pair is to replace the positive side (VO^{2+} / VO_2^+) of the standard VRFB by the O_2/H_2O redox couple. Implementing this system helps reduce the volume of the positive tank significantly [2]. Additionally, removing the positive electrolyte tank doubles the energy density compared to the standard VRFB [80]. Moreover, this system allows for higher concentrations of vanadium, with levels reaching close to 4 M [81]. The reactions occurring in a typical VARFB are shown below [80].

Positive side:	$O_2 + 4H^+ + 4e^- \rightleftarrows 2H_2O$	$E^0 = 1.23$ V vs. SHE
Negative side:	$4V^{2+} \rightleftarrows 4V^{3+} + 4e^-$	$E^0 = -0.26$ V vs. SHE
Overall reaction:	$4V^{2+} + O_2 + 4H^+ \rightleftarrows 4V^{3+} + 2H_2O$	$E^0 = 1.49$ V

One of the major issues of the VARFB is the crossover from the negative to the positive side (Fig. 5.9, first process), which leads to a reduction in the CE and cycle life due to an irreversible capacity loss [2]. In addition to this, the CE can also be reduced due to the oxygen permeation (Fig. 5.9, second process), and, therefore, V^{2+} is oxidized to V^{3+} [2].

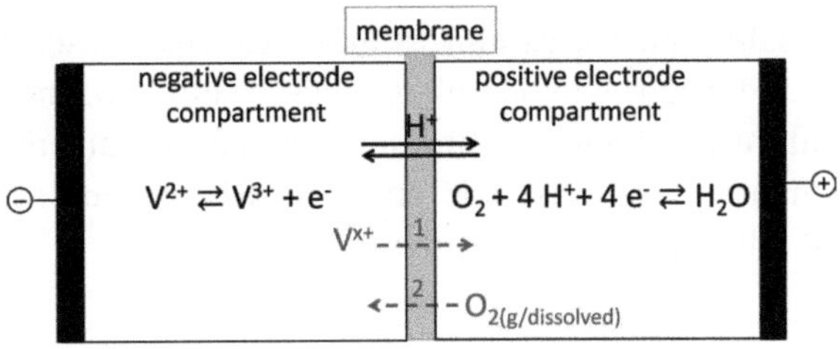

Figure 5.9 Schematic of VARFB, the main reactions (solid lines), and undesired transfer processes through the membrane (numbered, dashed lines). Reprinted with permission from Ref. [2]. Copyright (2016) Elsevier.

The technology of VARFBs is still in its early stages [4]. One of the primary challenges faced by VARFBs is capacity decay and catalyst poisoning [82]. To mitigate capacity decay due to the crossover of vanadium ions, the selection of an appropriate membrane is crucial [83]. Catalyst poisoning, on the other hand, is mainly caused by the contact of vanadium ions, particularly V^{2+} ions, with the catalyst. Currently, most VARFB systems utilize cation-exchange membranes (CEMs), which are also used in VRFBs [83]. Examples of such membranes include Nafion, which is the most widely applied CEM [84], multilayer CEMs that add additional layers to Nafion [2], sulfonated poly(ether either ketone) membranes (SPEEK) with lower resistance and slower self-discharge rates [85], and other composite membranes [84]. The development of suitable membranes is crucial as it is a critical component of VARFB systems and may even be a solution to the oxygen permeation issue [83]. In a recent study, the performances of VARFBs with three different commercial CEMs (Nafion, CMS, and CMX) were investigated and compared using a proposed dynamic model. This research aimed to assess and understand the impact of different CEM choices on the performance of VARFB systems [83].

While there have been numerous models developed for VRFBs, there are still limited models explicitly focused on VARFBs [80]. Austing et al. [2] utilized Nernst–Planck equations to investigate the charge and discharge processes of VARFBs, specifically examining voltage and concentration changes. Their study aimed to understand the fundamental behavior of VARFBs based on mathematical modeling. Barelli et al. employed a three-dimensional computational fluid dynamics (CFD) model to analyze the impact of airflow patterns on VARFBs when applied in electric vehicles (EVs) [4]. They investigated the thermal runaway risks associated with the utilization of VARFBs for charging EVs. However, their study did not specifically address the efficiency, capacity, or long-term performance of VARFBs [83].

Due to the more complex structure of VARFBs, constructing an experimental platform for them can be challenging compared to VRFBs. Therefore, developing a mathematical model presents an attractive approach to reduce costs and time, compared to experimental studies [80]. In a previous research study, the performance of a VARFB was compared, considering the impacts of ion diffusion, bulk electrolyte transfer, and side reactions [86]. Building upon this research, a more recent study expands its scope to include the effects of mass balance, oxygen transport, and operating conditions [83].

Furthermore, this model allows for the comparison of VARFB performance using different membranes. Various membrane-manufacturing technologies, such as radiation grafting [87], supercritical fluid techniques [88], and coating methods [89], can be evaluated and their impact on membrane properties can be assessed without the need for additional experiments [83]. Ultimately, a solution to the problem of oxygen permeation through the membrane can also be explored using such models.

5.4.2 Polysulfide/Air

The polysulfide redox couple has garnered significant attention among various redox chemistries due to its low cost, environmentally friendly characteristics, and high energy density attributed to the high aqueous solubility of polysulfide [90, 91]. This chemistry has been extensively studied in organic-solvent-based systems such as lithium- and sodium-sulfur batteries [92]. Additionally, the aqueous polysulfide redox couple has been adopted in several RFB systems, including polysulfide-bromine, polysulfide-iodide, polysulfide-ferrocyanide, and polysulfide-air (PSA) [13, 93–98]. Among these, the polysulfide/air RFB (PSARFB) stands out for its lowest chemical cost of storage, thanks to the utilization of the inexpensive oxygen redox couple [44].

According to the analysis presented in the article by Xia et al. [44], an all-alkaline PSARFB system was developed in 2011. This system employed an alkaline-based oxygen (air) redox couple and a polysulfide redox couple, with an AEM separating the two [13]. The discharge process involved the oxidation of shorter-chain-length polysulfides to longer-chain-length polysulfides in the negative side, while O_2 was reduced to OH^- in the positive side, and vice versa. The reactions occurring in both half-cells and the overall cell reaction can be summarized as follows:

Positive side:	$O_2 + 2H_2O + 4e^- \rightleftarrows 4OH^-$	$E^0 = 0.401$ V vs. SHE
Negative side:	$2S_2^{2-} \rightleftarrows S_4^{2-} + 2e^-$	$E^0 = -0.476$ V vs. SHE
Overall reaction:	$4S_2^{2-} + O_2 + 2H_2O \rightleftarrows 2S_4^{2-} + 4OH^-$	$E^0 = 0.877$ V

Based on these reactions, the alkaline PSARFB exhibited a standard equilibrium potential of 0.88 V [44]. The PSARFB for large-scale energy storage shows great potential but also faces scientific challenges. A major challenge is the development of low-cost electrocatalysts for the polysulfide oxidation/reduction reactions (PSOR/PSRR) and the OER/ORR. The complex chemistry of aqueous polysulfide solutions makes it difficult to fully understand the reaction mechanisms of PSOR/PSRR [99]. Alkaline polysulfide solutions contain various sulfur species, including polysulfide ions with chain lengths between 2 and 5 (S_x^{2-}, x = 2–5), bisulfide ions (HS^-), hydroxide ions (OH^-), and alkali-metal cations (e.g., Na^+, K^+) [100, 101]. These sulfur species exist in dynamic equilibria within the aqueous polysulfide solutions [99]. Transition metal sulfides such as nickel sulfides (NiS_x) and cobalt sulfides (CoS/CoS_2) have shown promise as catalysts for polysulfide reactions [97–99].

In PSA systems, the aqueous polysulfide electrolyte must be alkaline to suppress hydrogen sulfide formation, while the air electrolyte can be alkaline or acidic [99]. However, using an acidic air electrolyte requires expensive platinum-group metal (PGM) catalysts and limits the use of polymer-based CEM due to electrolyte neutralization issues in the long term. In contrast, using an alkaline air electrolyte avoids the neutralization problem and allows for the use of nonprecious metals and metal oxides as the air electrode [102]. However, this comes at the cost of a lower OCV compared to an acidic positive-electrolyte-based system [99]. To overcome these challenges, further research and development are needed to explore cost-effective catalysts for polysulfide and oxygen reactions, understand the complex polysulfide chemistry, and optimize the choice of electrolytes and membrane materials for the improved performance of alkaline PSARFBs.

One of the key challenges in PSARFBs is the crossover of polysulfides through membranes, which leads to self-discharge and rapid capacity decay. Conventional alkaline PSARFBs typically use a single-AEM or a single-CEM as the membrane separator, as shown in Fig. 5.10a,b [95]. However, the high permeability of polysulfide ions through AEMs can result in the crossover of sulfur species toward the air-side half-cell. This leads to the loss of active materials in the negative side and the poisoning of the sulfur-intolerant air electrode [99].

While CEMs such as Nafion show good rejection of polysulfides through Donnan exclusion, they also exhibit poor hydroxide conductivity, limiting the OER at the air electrode [103]. Efforts have been made to synthesize new membranes with size-selective functions to effectively block aqueous polysulfide ions [99]. However, effectively addressing the crossover of polysulfides remains a challenging task [103]. As a result, stable alkaline PSARFB systems with long-term performance have been difficult to achieve until

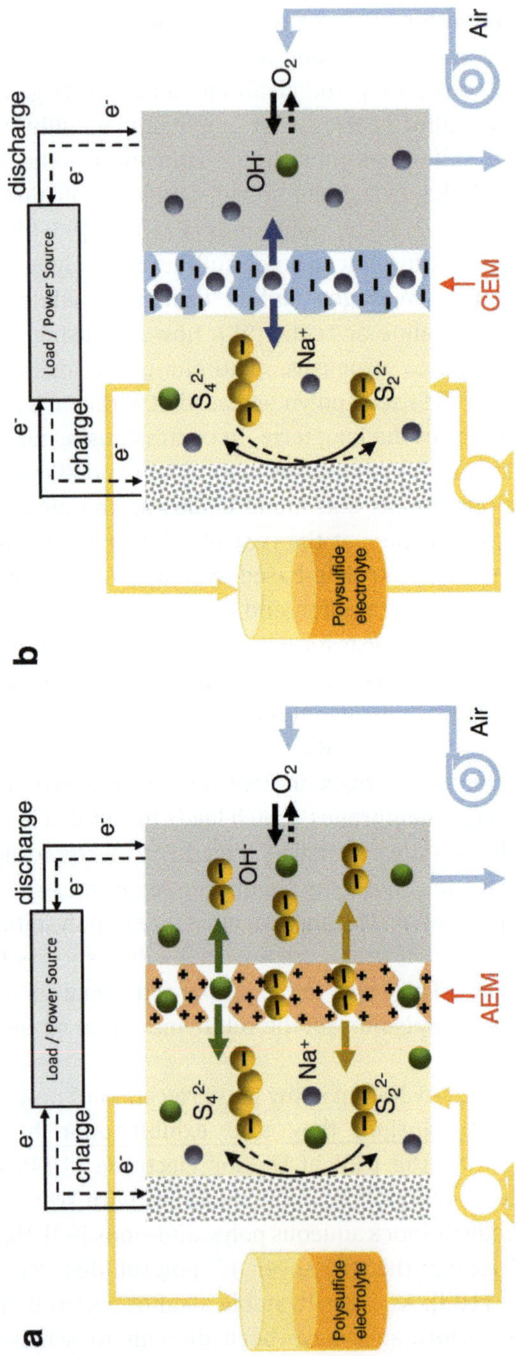
a
b
discharge
charge
Load / Power Source
e⁻
O2
Air
OH⁻
Na⁺
S4 2-
S2 2-
AEM
CEM
Polysulfide electrolyte

c

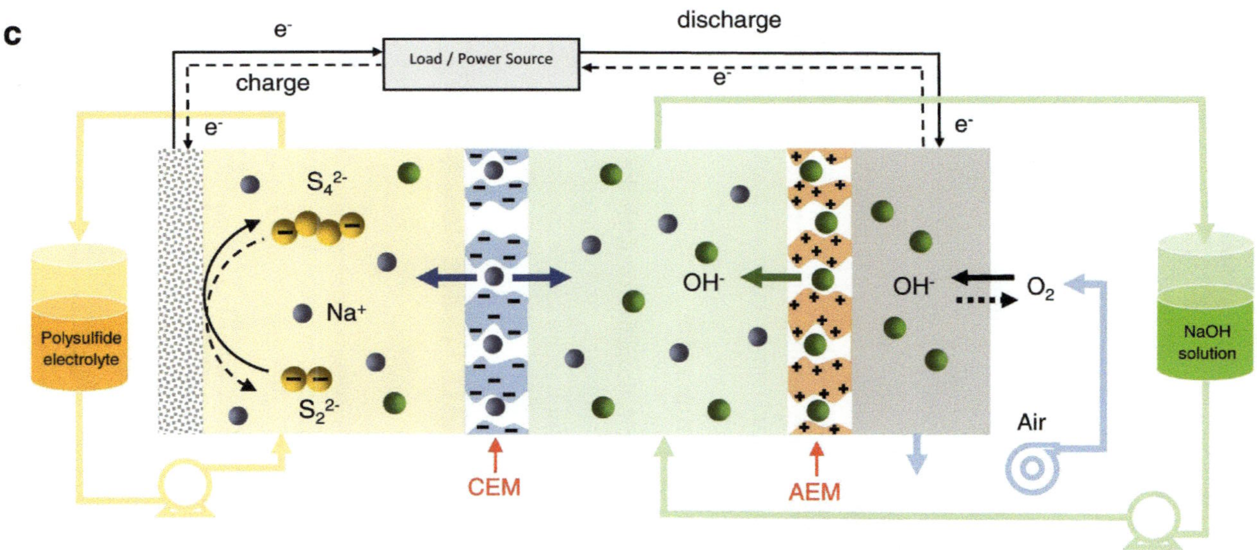

Figure 5.10 (a) PSARFB based on a single anion-exchange membrane. (b) PSARFB based on a single cation-exchange membrane. (c) PSARFB based on a double-membrane design combining anion- and cation-exchange membranes. Reprinted with permission from Ref. [44]. Copyright (2022) Springer.

recent developments [99]. Further research is needed to develop membranes with improved selectivity and conductivity to mitigate the crossover issue and enhance the performance and durability of PSARFBs.

In a recent study conducted by researchers at Imperial College, a stable alkaline PSA flow battery utilizing a modular dual-membrane architecture was demonstrated [44]. The novel design combines an AEM and a CEM in each individual cell, as shown in Fig. 5.10c. This dual-membrane configuration effectively mitigates the crossover of polysulfide ions and enables the utilization of commercially available or synthetic polymer ion-exchange membranes (IEMs).

The modular dual-membrane design offers several advantages. It reduces membrane resistance, leading to an enhanced peak power density in the flow battery system. The system was cycled 80 times at 1 mA/cm^2 and exhibited a round-trip EE of 40%, demonstrating its stability and performance. Additionally, the energy and power cost of the system was reduced compared to previous work [44].

5.4.3 Organic Redox Couple/Air

The combination of organic redox species with the oxygen redox couple could be a feasible approach, as reported by Leung et al. [104] and illustrated in Fig. 5.11.

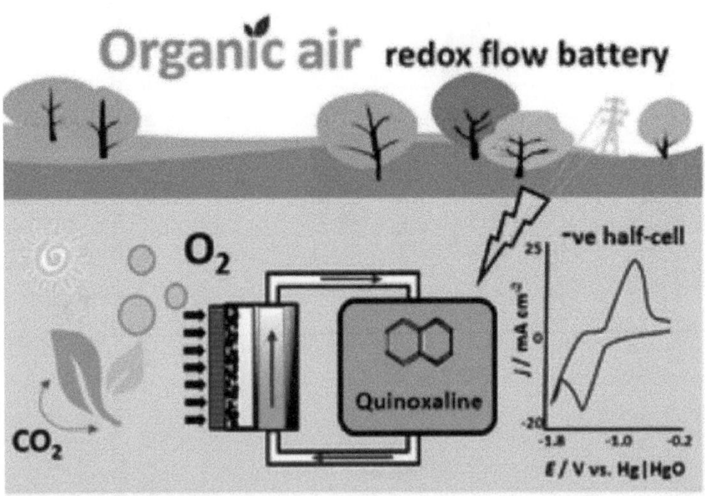

Figure 5.11 The concept of an organic/air RFB. Reprinted with permission from Ref. [104]. Copyright (2018) RSC.

In this study, the electrochemical characteristics of the quinoxaline (QX) and air electrode reactions were evaluated for the development of an organic-air flow battery (Fig. 5.11). The purpose of this evaluation was to understand the fundamental processes involved in the charging and discharging of the battery. During the charging process of the battery, QX is reduced in the negative half-cell, resulting in the formation of a divalent anion radical. This reduction reaction involves a two-electron transfer process [104]. During the discharge process, the reduced QX species undergoes oxidation, where the divalent anion radical is converted back to the original QX form. This oxidation reaction releases the stored energy in the battery, allowing it to be used for powering devices or systems. These reactions are shown in Fig. 5.12 [104].

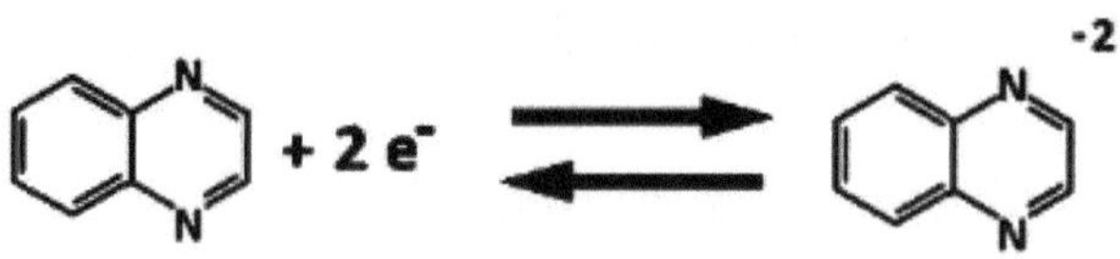

Figure 5.12 The mechanism of the QX oxidation and reduction process in an organic/air flow battery. Reproduced with permission from Ref. [104]. Copyright (2018) RSC.

Positive side:	$O_2 + 2H_2O + 4e^- \rightleftarrows 4OH^-$	$E^0 = 0.401$ V vs. SHE
Negative side:	$2QX^{2-} \rightleftarrows 2QX + 4e^-$	$E^0 = -0.78$ V vs. SHE
Overall reaction:	$2QX^{2-} + O_2 + 2H_2O \rightleftarrows 2QX + 4OH^-$	$E^0 = 1.181$ V

In the proof-of-concept for the organic-air flow battery described in the study by Leung et al., the circulation of the positive electrolyte was implemented to prevent the trapping of oxygen bubbles generated during the charging process [104]. This circulation helps avoid displacing the electrolyte, which can occur when oxygen bubbles accumulate and occupy space within the system. By continuously circulating the positive electrolyte, the bubbles are effectively removed, ensuring the smooth operation of the battery.

In practical systems, it is important to minimize the volume and flow rate of the positive electrolyte to reduce pumping losses and improve energy densities. By optimizing these parameters, the overall efficiency and performance of the battery can be enhanced.

In a recent work, Mazur et al. [3] proposed a hybrid 9,10-anthraquinone-2,7-disulfonic acid (AQDS)-oxygen flow battery that combines the reduction of the acidic AQDS-based negolyte with the OER in the organic-oxygen electrolyzer during the charging process. On the other hand, AQDS oxidation is combined with the ORR on the gas diffusion electrode of the organic-oxygen fuel cell during the discharging process. The performance of the fuel cell/ electrolyzer system (kind of RFCs) is extensively studied in their research. When compared to a VARFB, the AQDS-based RFC displays reduced ohmic resistances and minimal capacity fade because of limited AQDS permeation through the cation-exchange membrane.

5.5 Hydrogen/Liquid RFCs

5.5.1 Introduction

Hydrogen has emerged as a promising substitute energy carrier. When hydrogen is used in fuel cells to generate electricity, it produces no local pollution, with the only by-product being pure water. This characteristic makes hydrogen-based systems environmentally friendly and suitable for applications where a reduction in emissions is a priority. Compared to gasoline combustion, hydrogen can provide three times more energy per unit mass [105]. This makes it an attractive option for energy storage and transportation, enabling efficient utilization of energy resources.

Hydrogen produced through electrolysis can be efficiently utilized in fuel cells. Stationary fuel cell technologies enable the development of distributed power backup, stand-alone power plants, and cogeneration systems. These applications offer alternative options to traditional power grids, as electricity can be produced onsite, precisely when and where it is needed. This eliminates the necessity for extensive hydrogen storage and transportation infrastructure. Overall, hydrogen's characteristics, including its clean nature, high energy density, diverse production sources, and compatibility with fuel cells, position it as a promising energy carrier in the ongoing transition toward a more sustainable and decarbonized energy system [106].

Additionally, in a hydrogen-based RFC architecture, the crossover of soluble liquid active species through the separator or ion-exchange membrane is controlled, which helps reduce the loss of active materials and improves the overall efficiency of the system. This control can be achieved using selective membranes or other design features that prevent or minimize crossover [14]. By minimizing the crossover of liquid species, the hydrogen-based RFC architecture enables more efficient utilization of the electrolyte, reducing the need for frequent electrolyte replacement and lowering the associated costs. It also helps mitigate the irreversible degradation of the RFB, prolonging its lifespan and improving its long-term performance.

5.5.2 Hydrogen/Bromine

The hydrogen–bromine (H_2/Br_2) RFC has gained attention as a promising energy storage system due to its various advantages, including the following [107]:

- **Low cost:** Due to the abundance of hydrogen and bromine electrolytes as well as economically sourced carbon electrode materials, the H_2/Br_2 RFC can be made at scale [108]. For example, its capital costs can be as low as USD 220 kWh^{-1} and levelized costs of storage as low as USD 0.034 $KW\,h^{-1}$ [109].
- **High achievable energy density:** The H_2/Br_2 RFC offers a high solubility of hydrogen and bromine in their respective electrolytes, which allows for the storage of a large amount of energy per unit volume [107], achieving values up to 200 Wh/L [110].
- **Operational power density:** High operational power densities more than 1.4 W/cm^2 have been reported in laboratory-scale reactors [25].
- **Good stability:** Stability is crucial for ensuring the long-term performance and reliability of the RFC.

- **Fast electrochemical kinetics:** This characteristic enables efficient charge and discharge processes leading to high power density and rapid response times [107].
- **Shallow discharges and overcharge capability:** The H_2/Br_2 RFC can tolerate overcharging conditions without significant adverse effects. This flexibility allows for better utilization of the electrolyte and extends the battery's operational range.
- **High CE:** The low self-discharge rate contributes to this high efficiency [111].

The H_2–Br_2 involves several electrochemical reactions on the negative and positive sides. The cell potential for the H_2/Br_2 RFC, considering the electrode potentials, can be determined as 1.09 V at 298 K. This potential represents the maximum voltage that can be achieved by the RFC under ideal conditions [107]. The reactions can be summarized as follows, although it should be borne in mind that in practical systems, the actual cell potential may be slightly lower than the theoretical value due to factors such as electrode kinetics, mass transport limitations, and losses associated with internal resistance [29, 33, 74, 112, 113]:

Positive side:	$Br_2 + 2e^- \rightleftarrows 2Br^-$	$E^0 = 1.09\ \text{V vs. SHE}$
Negative side:	$H_2 \rightleftarrows 2H^+ + 2e^-$	$E^0 = 0\ \text{V vs. SHE}$
Overall reaction:	$Br_2 + H_2 \rightleftarrows 2HBr$	$E^0 = 1.09\ \text{V}$

In the field of the H_2/Br_2 RFC, the choice of positive electrode materials is crucial for achieving high performance and cost-effectiveness [107]. Carbon-based materials have been extensively studied as liquid-side electrode materials for H_2/Br_2 RFCs due to their relatively low cost, durability in corrosive environments, and reasonable activity for bromine reactions [107]. Carbon papers are commonly utilized [73, 74, 114–117], although graphite felts, carbon cloths, and carbon foams have been investigated as well [73, 114, 118]. These carbon-based materials are often modified by acid [115, 116, 118] or thermal treatments [73, 118] to introduce oxygen-containing functional groups, enhancing their activity for the bromine reaction [107]. In addition to carbon-based materials, the use of nanostructured electrode materials in H_2/Br_2 RFCs is gaining attention. For example, multiwalled carbon nanotubes (MWCNTs) synthesized directly onto a carbon paper substrate using electrodeposition/chemical vapor deposition methods have shown significantly increased active surface areas and improved performance in H_2/Br_2 RFCs [33]. Hollow-core mesoporous shell (HCMS) carbon, with its high specific surface area, has also been investigated as a bromine electrode material, demonstrating promising peak power densities in H_2/Br_2 RFCs [119].

While carbon-based materials have been widely explored, the utilization of other nanomaterials, such as MOF-derived nanoporous carbon [120], as bromine electrodes in H_2/Br_2 RFCs is not well known and requires further investigation [107]. Overall, the development of highly active and cost-effective electrode materials with large surface areas is essential for realizing the full potential of H_2/Br_2 RFCs in terms of high-power-density operation, stability, and cost efficiency. Ongoing research focuses on exploring different carbon-based materials and nanostructured positive electrode materials such as MXenes [121], to optimize their performance and contribute to the advancement of H_2/Br_2 RFC technology [107].

5.5.3 Hydrogen/Iron

Overall, the use of a redox oxidant as a replacement for oxygen in the cathode of a PEMFC offers advantages such as faster reaction kinetics and improved mass transfer. However, the specific choice of the redox oxidant (iron based) and the regeneration method should be carefully evaluated to ensure an efficient and sustainable operation of the RFC system.

Fatih et al. studied the combination of the HOR/HER with an iron-based redox couple (Fe^{3+}/Fe^{2+}) [122]. In this reaction, Fe^{2+} ions are oxidized to Fe^{3+} ions, while H^+ ions are reduced to form molecular hydrogen (H_2). This process accounts for the radiochemical regeneration of Fe^{3+} ions from Fe^{2+} ions using H_2 as a reducing agent. The use of a redox couple, in this case Fe^{3+}/Fe^{2+}, enables the conversion between different oxidation states of iron during the regeneration process. The reaction is as follows:

$$2Fe^{2+} + 2H^+ \xrightarrow{\gamma} H_2 + 2Fe^{3+} \tag{5.34}$$

In the aforementioned investigation, Fatih et al. used a concentric design for the unit cell of the fuel cell. The unit cell consisted of a microporous polyvinyl chloride (PVC) separator material called Porvic, high-loaded platinum-catalyzed wet-proofed carbon electrodes, and an electrolyte composed of equimolar $FeSO_4$ and $Fe_2(SO_4)_3$ with 1 M H_2SO_4 [122]. However, the performance results obtained in this study were poor. The fuel cell operated at voltage levels typically ranging from 0.35 to 0.50 V at a current density of 20 mA/cm^2. Additionally, the overall gravimetric (power density per unit mass) and volumetric (power density per unit volume) power densities were low [122]. Despite such limitations, the potential of the hydrogen/iron RFC was demonstrated in a recent study whereby the authors reported a new water management strategy that could enable the system to operate at high current densities for long cyclic durations [123]. Drawing on from such promising investigations, scaling up the system to operate in stacks should be the next step forward in a drive to commercialize the technology.

5.5.4 Hydrogen/Vanadium

The hydrogen–vanadium regenerative fuel cell (H_2VRFC) offers several advantages compared to conventional VRFBs. It combines the benefits of mixed liquid-gas RFBs and reduces the overall system cost by utilizing only half of the vanadium electrolyte required for a VRFB. In the H_2VRFC, a precious metal catalyst is required at the negative side for the HOR/HER [124]. This catalyst facilitates the electrochemical reaction of hydrogen and protons at the negative side [41]. To address the capacity loss in the H_2VRFC caused by the crossover of the positive liquid electrolyte solution to the negative half-cell, any crossover of vanadium electrolyte can be collected and returned to the vanadium tank [30]. This helps in regaining the lost capacity. Similarly, capacity loss due to self-discharge or side reactions in the H_2VRFC can be addressed using similar methods to those employed in conventional VRFBs [125]. The H_2VRFC was initially proposed and demonstrated by Yufit et al., achieving a peak performance of 114 mW/cm^2 at 100% SOC [30].

In the H_2VRFC (as illustrated in Fig. 5.13), the discharge process involves the reduction of V(V) to V(IV) and the oxidation of H_2 to H^+, while the reverse reactions occur during the charge process,

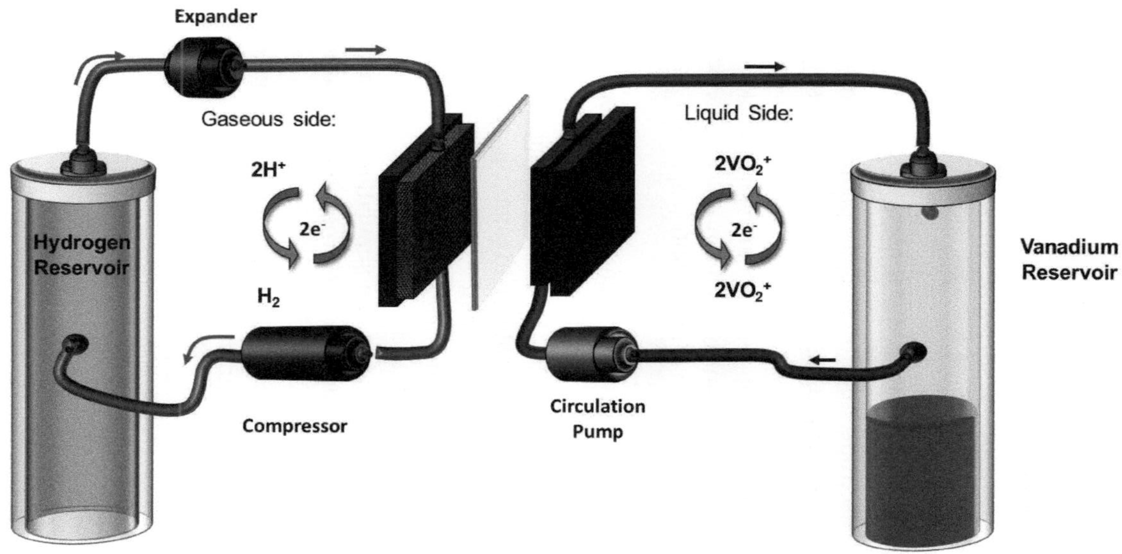

Figure 5.13 Operational principle of an H_2VRFC. Reprinted with permission from Ref. [13]. Copyright (2020) ACS.

resulting in the production of H_2 [13]. Understanding the loss mechanisms and optimizing the performance of the H_2VRFC are crucial for its efficient operation [124].

A study on the performance losses in H_2VRFC highlighted that the ion diffusion in the liquid electrolyte and ohmic losses contribute significantly to the overall polarization [30]. Additionally, the possible adsorption of vanadium ions onto the platinum catalyst was identified as a potential factor affecting performance. Positive electrode losses in the H_2VRFC can be attributed to the partial wettability of the electrode due to its hydrophobic nature and the need for heat treatment [78, 126, 127]. A recent investigation explored the use of carbon catalyst layers (CCLs) to overcome the partial wettability issues, which resulted in higher peak power densities of nearly 700 mW/cm^2 and high energy efficiencies (>90%) to be reported with long cycle lives of above 500 at operating current densities above 100 mA/cm^2 [43].

To further investigate and understand the physicochemical phenomena involved in the H_2VRFC, a first modeling approximation was proposed and validated [36]. Mathematical modeling and simulation techniques have been employed to establish relationships between operational conditions and performance, predict and rationalize experimental findings, and study performance optimization in the H_2VRFC [128–130].

By utilizing mathematical models and simulations, researchers can gain insights into the complex electrochemical processes occurring within the H_2VRFC, aiding in the design and optimization of the cell for enhanced performance and efficiency. These modeling approaches contribute to the understanding and advancement of the H_2VRFC technology [78].

For the first-generation H_2VRFC system using H_2SO_4 as the electrolyte, the reactions involved are as follows [41]:

Positive side:	$2VO_2^+ + 4H^+ + 2e^- \rightleftarrows 2VO^{2+} + 2H_2O$	$E^0 = 1.00$ V vs. SHE
Negative side:	$H_2 \rightleftarrows 2H^+ + 2e^-$	$E^0 = 0$ V vs. SHE
Overall reaction:	$2VO_2^+ + 2H^+ + H_2 \rightleftarrows 2VO^{2+} + 2H_2O$	$E^0 = 1.00$ V

However, using sulfuric acid as electrolyte has some limitations shown in the standard VRFB, such as low solubility (and stability at a temperature range from 10 to 40 °C) of VO^{2+} < 1.7 M of $VOSO_4$, limiting the capacity. Some workers improved the vanadium solubility by introducing hydrochloric acid (HCl) as the electrolyte [131]. As mentioned earlier, by changing the supporting electrolyte to full HCl, it is possible to achieve higher vanadium concentrations up to 2.5 M [131]. Rubio-Garcia et al. [42] developed a H_2VRFC system using 6 M HCl successfully. Using HCl, the potential of the cell was increased by 167 mV, compared with using H_2SO_4 as the supporting electrolyte (second-generation H_2VRFC).

Positive side:	$2VO_2Cl + 4H^+ + 2e^- \rightleftarrows 2VO^{2+} + 2H_2O + 2Cl^-$	$E^0 = 1.17\ \mathrm{V\ vs.\ SHE}$
Negative side:	$H_2 \rightleftarrows 2H^+ + 2e^-$	$E^0 = 0\ \mathrm{V\ vs.\ SHE}$
Overall reaction:	$2VO_2Cl + 2H^+ + H_2 \rightleftarrows 2VO^{2+} + 2H_2O + 2Cl^-$	$E^0 = 1.17\ \mathrm{V}$

As Pt is used in the negative side for the HOR/HER reactions in the RFC systems, Rubio-Garcia et al. found that using high concentration of HCl during the charge process, generating the V^{5+} species, led to a rapid dissolution of Pt in the electrode due to crossover issues from the positive to the negative side. Consequently, the performance of the cell could not be sustained for more than two cycles. However, replacing the metal catalyst with a chlorine-resistant one allows the second-generation RHVFC to cycle for longer [42].

5.5.5 Hydrogen/Cerium

The hydrogen/cerium (H_2/Ce) RFC system (Fig. 5.14) shows promise for high-power and high-energy-density applications due to its unusually high OCV ranging from 1.5 to 1.7 V. A proof-of-concept study on this system demonstrated a peak discharge power of 148 mW/cm^2 and an EE of 88%, indicating its potential for efficient energy conversion [132]. However, it was observed that the kinetics of the cerium electrode limit the overall performance of the H_2/Ce system. The Ce (+) electrode operates at potentials that fall outside the stability window of water, which poses a challenge. Nonetheless, extensive research on the aqueous Zn/Ce RFB has shown that proper solution chemistry and operating protocols can mitigate the negative effects of oxygen evolution, which is relevant to addressing the stability issues of the Ce electrode [133].

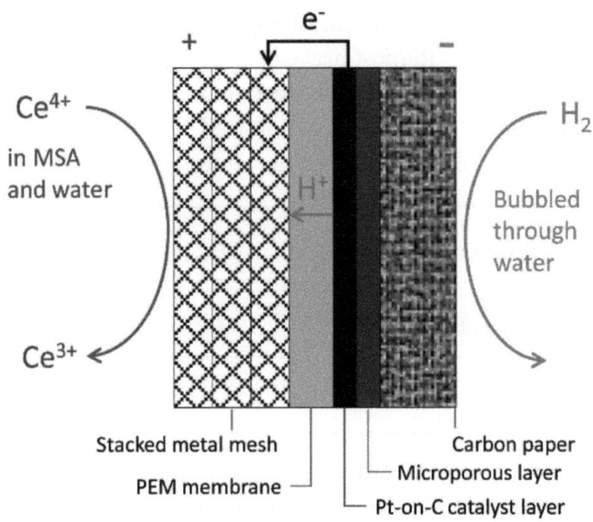

Figure 5.14 Schematic of an H_2/Ce redox-flow cell. Reprinted with permission from Ref. [35]. Copyright (2016) Elsevier.

The electrochemical reactions in the H_2/Ce RFC system during discharge and charge can be represented as follows [133, 134]:

Positive side:	$2Ce^{4+} + 2e^- \rightleftarrows 2Ce^{3+}$	$E^0 = 1.5 - 1.75$ V vs. SHE
Negative side:	$H_2 \rightleftarrows 2H^+ + 2e^-$	$E^0 = 0$ V vs. SHE
Overall reaction:	$2Ce^{4+} + H_2 \rightleftarrows 2Ce^{3+} + 2H^+$	$E^0 = 1.5 - 1.75$ V

These reactions occur in an electrochemical cell where gaseous hydrogen is fed to the negative electrode and an aqueous solution of cerium(IV) methanesulfonate is supplied to the positive electrode.

Protons (H^+) generated at the negative electrode pass through an ion-conducting membrane to the positive electrode, where they combine with cerium(IV) to form cerium(III) and release methanesulfonic acid (MSA) into the solution. It is important to note that the potential for the reduction of cerium(IV) methanesulfonate depends on the type and concentration of the solvent used in the system [35].

The high OCV offered by the H_2-Ce couple is indeed promising, but it is important to consider the challenges associated with the relatively low current density reported for aqueous cerium electrochemistry on Pt catalysts in general. Previous studies have shown that Pt catalysts exhibit moderate activity for cerium electrochemistry, resulting in limited current densities [133–136]. Furthermore, specific to the H_2/Ce cell, moderate power densities have been demonstrated [35]. The performance of the cell appears to be relatively insensitive to factors such as membrane thickness, cerium concentration, and features of the hydrogen electrode. Instead, the limitations in cell performance are attributed to mass transport and kinetics at the cerium electrode.

As disclosed in a recent patent on the H_2/Ce RFC, a maximum discharge power of 895 mW/cm^2 was observed at 60 °C, and an EE of 90% was achieved at 50 °C [137]. These results indicate that while the H_2/Ce system shows promise, there are still challenges to overcome to achieve higher current densities and power densities, particularly related to the kinetics and mass transport at the cerium electrode. In other words, considering the potential for high-energy and high-power densities, further optimization of the H_2/Ce system holds promise. By addressing the kinetics and stability challenges, an optimized H_2/Ce RFB system may enable the realization of its potential for efficient and high-performance energy storage applications [35].

5.5.6 Hydrogen/Manganese

Due to its abundance and low cost, manganese is commonly used in disposable alkaline batteries [138]. However, its application in RFBs has been limited due to the instability of Mn(III) in aqueous electrolytes, which results in the precipitation of MnO_2 through a disproportionation reaction. This undesirable reaction leads to capacity decay and hinders the use of manganese in RFBs [138]. To overcome this limitation, recent research has proposed the addition of metal additives to the Mn-electrolyte formulation [139]. This strategy aims to prevent the formation of insoluble MnO_2 and enhance the stability of the Mn redox couple in aqueous electrolytes.

As a result of these efforts, Rubio-Garcia et al. successfully demonstrated the first hydrogen/manganese hybrid RFC (RHMnFC/ H_2MnRFC) [15]. This innovative system showed promising performance characteristics and projected cost advantages compared to VRFBs. The RHMnFC (illustrated schematically in Fig. 5.15) exhibited a high round-trip EE of 82%, indicating efficient energy conversion during charge and discharge cycles. It also demonstrated a high power density of 1410 mW/cm^2 and an energy density of 33 Wh/L, highlighting its ability to deliver significant power output and store a considerable amount of energy per unit volume. In addition to its performance benefits, the RFC showed an estimated 70% cost reduction compared to VRFBs, making it an attractive option in terms of cost-effectiveness [15].

The development of the H_2MnRFC represents an exciting advancement in utilizing manganese for energy storage applications. With its high EE, power density, and projected cost reduction, this hybrid system shows promise for widespread adoption and contributes to the exploration of alternative redox materials for RFBs. The H_2MnRFC reactions are shown below [13]:

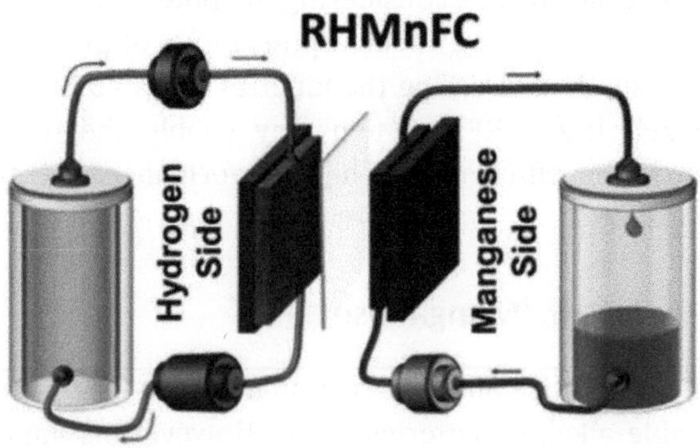

Figure 5.15 A schematic of the hydrogen/manganese regenerative fuel cell. Reprinted with permission from Ref. [13]. Copyright (2020) ACS.

Positive side:	$2Mn^{3+} + 2e^- \rightleftarrows 2Mn^{2+}$	$E^0 = 1.5$ V vs. SHE
Negative side:	$H_2 \rightleftarrows 2H^+ + 2e^-$	$E^0 = 0$ V vs. SHE
Overall reaction:	$2Mn^{3+} + H_2 \rightleftarrows 2Mn^{2+} + 2H^+$	$E^0 = 1.5$ V

The optimization of the H_2MnRFC is still an ongoing process with some investigations that reported electrode catalytic enhancements for the positive half-cell reaction in general. In this respect, electrophoretic deposition was employed to improve the electrochemical surface area of commercially sourced carbon paper electrodes by decorating these substrates with nitrogen-doped graphene [13]. A 30% increase in EE (at 100 mA/cm^2) was reported, but even better performance was obtained when a commercially sourced carbon cloth was combined with an electrospun paper (CCL) in a very recent investigation, as shown in Fig. 5.16 [45].

In the investigation depicted in Fig. 5.16, comparisons have been made between graphite felt electrodes alone and graphite felt electrodes combined with an in-house CCL [45]. The results have shown that the use of graphite felt electrodes, particularly in combination with the CCL, leads to improved KPIs. These findings align with the performance trends observed in standard VRFBs as reported in the literature [140].

For instance, the H_2MnRFC with graphite felt and CCL demonstrates more than 80% EE, peak power density exceeding 0.5 W/cm^2, and electrolyte utilization of 20 Ah/L. In comparison, the use of graphite felt alone results in approximately 75% EE, peak power density of 0.45 W/cm^2, and electrolyte utilization of 11 Ah/L [45]. These findings suggest that optimizing the electrochemical performance of CCLs could present new opportunities for the commercial utilization of H_2MnRFC systems. By enhancing the efficiency and power density while maximizing electrolyte utilization, the potential for practical implementation and commercial viability of these systems can be improved.

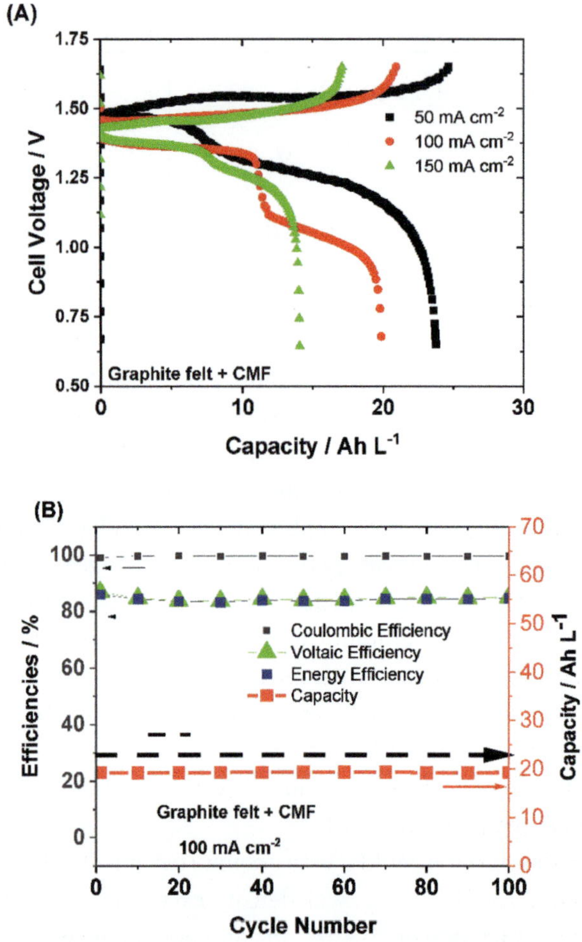

Figure 5.16 (A) Galvanostatic charge and discharge (GCD) experiments were conducted at different current densities of 50, 100, and 150 mA/cm^2. The GCD tests involve the controlled application of a constant current to charge and discharge the H_2MnRFC system. The current densities of 50, 100, and 150 mA/cm^2 were used to investigate the performance of the system under different operating conditions. (B) GCD tests of the H_2MnRFC system were performed for 100 cycles to evaluate the system's performance over an extended period. The efficiencies and electrolyte capacity utilization were monitored during these cycles. The electrolyte used in these experiments consisted of 1 M $MnSO_4$, 1 M $Ti(SO_4)_2$, and 3 M H_2SO_4. During the experiments, a hydrogen flow rate of 100 mL/min and a liquid flow rate of 50 mL/min were maintained. These flow rates ensure sufficient supply of hydrogen to the system and facilitate the movement of the electrolyte through the cell. Reprinted with permission from Ref. [45]. Copyright (2023) MDPI.

5.5.7 Hydrogen/Organic Redox Couple

The first hydrogen/benzoquinone (H_2/BQ) system was initially reported from Harvard [141]. The investigators managed to achieve a reasonable power density of 35 mW/cm^2. However, the durability in terms of long-term charge/discharge cycling was not studied. In a more recent study by Dargily et al., the performance of the H_2/BQ system was examined, and they reported approximately 200 charge/discharge cycles. However, the poor solubility of the active species, attributed to nonfunctionalized benzoquinone molecules, limited the operational current density to 20 mA/cm^2. This restriction made it impractical to conduct further comprehensive studies on the system [142]. Specifically, Rubio-Garcia et al. demonstrated the use of a functionalized benzoquinone, di-sulfonic acid (BQDS), which addressed the solubility issue [14]. The incorporation of BQDS enabled the development of a low-cost system capable of operating at a relatively high current density of 100 mA/cm^2, while maintaining high efficiency. The utilization of functionalized benzoquinone molecules, such as BQDS, in the H_2/BQ system opens up possibilities for improving the system's performance and feasibility for practical applications. By addressing solubility issues and enabling higher current densities, it becomes more viable to explore the potential of this system for energy storage applications.

Figure 5.17 shows the reduction reaction of the BQDS. During the discharge of the H_2/BQDS RFC, the reduction of BQDS to 1,2-dihydroxybenzene-3,5-disulfonic acid (BQDS–OH) occurs in the positive electrolyte, while during the charging process, the oxidation of the latter species takes place, forming BQDS. This RFC system avoids self-discharge from electrolyte crossover and achieves higher energy density because of hydrogen's compressible nature [14].

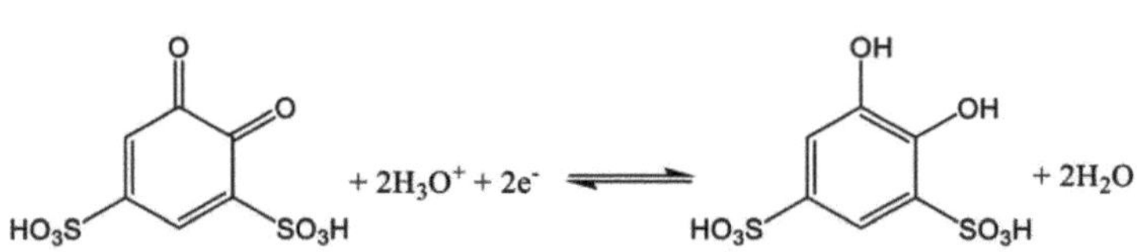

Figure 5.17 Redox reaction of the BQDS molecule. Reproduced with permission from Ref. [14]. Copyright (2020) RSC.

Positive side:	$BQDS + 2H^{+} + 2e^{-} \rightleftarrows BQDS - OH$	$E^{0} = 0.85$ V vs. SHE
Negative side:	$H_2 \rightleftarrows 2H^{+} + 2e^{-}$	$E^0 = 0$ V vs. SHE
Overall reaction:	$BQDS + H_2 \rightleftarrows BQDS - OH$	$E^{0} = 0.85$ V

According to an independent study, the specific capacity of modified quinone molecules, such as BQDS, can range from 200 to 500 Ah/kg [143]. This indicates that these molecules have a high charge storage capability per unit mass. Furthermore, the study reports that the cost of these modified quinone molecules can range from USD 10 to USD 15 per kilogram or USD 20 to USD 30 per kilowatt-hour [143]. This suggests that the cost of the active materials in the H_2/BQ system could be significantly lower compared to other RFBs [14]. In comparison, the study mentions that H_2/Br_2 RFCs have been reported to have a capital cost of USD 400 per kilowatt-hour according to the literature [113]. This suggests that the H_2/BQ system has the potential for a much lower capital cost, making it a more cost-effective alternative for energy storage applications [14].

5.6 Challenges and Prospects of RFCs

To enable the widespread adoption of RFCs, several challenges must be overcome. First, the durability of RFCs needs improvement to match the longevity of other energy storage options. The decay of RFC constituents can impact their overall durability (especially catalyst poisoning issues with H_2/Br_2 RFC systems). Additionally, the cost-effectiveness and safety of hydrogen fuel storage technology must be addressed. Reducing the storage volume required for hydrogen fuel storage is also a key consideration.

Furthermore, the maintenance cost of RFCs remains high, apart from the initial capital investment. To make RFCs competitive with other energy options, these challenges, including durability, storage technology, and maintenance costs, need to be effectively addressed.

To optimize the performance of RFCs, the cell voltage must be maximized by minimizing irreversible reactions that result in losses of the active materials. Various steps can be taken to reduce activation losses (similar to PEMFCs), which occur due to energy barriers that hinder reaction kinetics at the electrodes [144]:

- **Enhance the cell temperature:** Higher temperatures can enhance reaction kinetics and reduce activation losses.
- **Employ catalysts with improved reaction kinetics:** Catalysts can facilitate faster reaction rates, reducing activation losses.
- **Enhance electrode roughness:** Rough electrodes provide a large surface area for catalyst deposition, increasing the number of reaction sites and reducing activation losses.
- **Raise the reagent concentration:** Higher reagent concentrations increase the operating current densities, reducing activation losses.
- **Increase reagent (gas) pressure:** Elevated pressure improves reagent concentration and enhances reaction kinetics, minimizing activation losses.

To minimize losses due to charge crossover in conductor materials for ions and electrons, the electrical resistance of these materials should be minimized using the following approaches [144]:

- **Use electrodes with high conductivity:** Materials with excellent electrical conductivity can minimize resistance losses.
- **Optimize design and material selection for bipolar plates:** The design and choice of materials for bipolar plates can reduce electrical resistance.
- **Reduce membrane thickness:** Thinner membranes reduce the electrical resistance for ion transport.
- **Increase membrane humidity:** Maintaining appropriate humidity levels in the membrane improves ionic conductivity and reduces resistance.

To mitigate the negative effects on voltage during increased current density, which lowers reactant concentration at reaction sites, the following parameters can be improved [144]:

- **Enhance reagent gas supply speed:** A faster supply of reagent gases ensures an adequate concentration at reaction sites.

- **Optimize reagent-delivery channel geometry:** An efficient design of reagent-delivery channels facilitates uniform distribution and reduces concentration losses.
- **Improve diffusive characteristics of gas diffusion layers and electrodes:** Enhancing diffusion properties allows for better reactant distribution.
- **Enhance air circulation in the positive electrode (for liquid/air RFCs):** Proper air circulation prevents water accumulation and maintains optimal conditions for reaction.

It is important to consider the overall system and make informed decisions when adjusting operating parameters, as improving certain aspects may negatively impact others. Trade-offs and compromises are necessary to avoid performance degradation or reduced RFC lifespan. For example, while increasing temperature or oxygen pressure can reduce activation losses and increase power output, they may also lead to membrane dehydration or require thicker membranes to withstand higher mechanical stresses, resulting in increased ionic resistivity.

Other factors, such as using pure oxygen instead of air, can enhance performance but require additional considerations for safety and maintenance. The airflow rate and hydration of the membrane are crucial for self-humidified fuel cells, as higher airflow rates can increase power output but may limit membrane hydration due to water carry-off. Finding an optimal compromise for airflow is essential to enhance the performance of liquid/air RFCs while maintaining appropriate membrane hydration levels.

References

1. C. Minke, U. Kunz, and T. Turek, Techno-economic assessment of novel vanadium redox flow batteries with large-area cells, *J Power Sources.* **361** (2017): 105–114. https://doi.org/10.1016/J.JPOWSOUR.2017.06.066.
2. J. Grosse Austing, C. Nunes Kirchner, L. Komsiyska, and G. Wittstock, Layer-by-layer modification of Nafion membranes for increased life-time and efficiency of vanadium/air redox flow batteries, *J Memb Sci.* **510** (2016): 259–269. https://doi.org/10.1016/j.memsci.2016.03.005.

3. J. Charvát, P. Mazúr, J. Pocedič, P. Richtr, J. Mrlík, J. Kosek, J. Akrman, and L. Kubáč, New organic-air flow fuel cell and electrolyser for stationary energy storage, *J Power Sources.* **520** (2022): 230811. https://doi.org/10.1016/J.JPOWSOUR.2021.230811.
4. L. Barelli, G. Bidini, P.A. Ottaviano, D. Pelosi, M. Perla, L. Trombetti, F. Gallorini, and M. Serangeli, Electric vehicles fire protection during charge operation through Vanadium-air flow battery technology, *Heliyon.* **7** (2021): e08064. https://doi.org/10.1016/J.HELIYON.2021.E08064.
5. F. Barbir, T. Molter, and L. Dalton, Efficiency and weight trade-off analysis of regenerative fuel cells as energy storage for aerospace applications, *Int J Hydrogen Energy.* **30** (2005): 351–357. https://doi.org/10.1016/J.IJHYDENE.2004.08.004.
6. J.D. Maclay, J. Brouwer, and G. Scott Samuelsen, Dynamic analyses of regenerative fuel cell power for potential use in renewable residential applications, *Int J Hydrogen Energy.* **31** (2006): 994–1009. https://doi.org/10.1016/J.IJHYDENE.2005.10.008.
7. G.J. Suppes, Plug-in hybrid with fuel cell battery charger, *Int J Hydrogen Energy.* **30** (2005): 113–121. https://doi.org/10.1016/J.IJHYDENE.2004.04.017.
8. T. Sadhasivam, K. Dhanabalan, S.H. Roh, T.H. Kim, K.W. Park, S. Jung, M.D. Kurkuri, and H.Y. Jung, A comprehensive review on unitized regenerative fuel cells: Crucial challenges and developments, *Int J Hydrogen Energy.* **42** (2017): 4415–4433. https://doi.org/10.1016/J.IJHYDENE.2016.10.140.
9. Y. Wang, D.Y.C. Leung, J. Xuan, and H. Wang, A review on unitized regenerative fuel cell technologies, part B: Unitized regenerative alkaline fuel cell, solid oxide fuel cell, and microfluidic fuel cell, *Renew. Sust. Energ. Rev.* **75** (2017): 775–795. https://doi.org/10.1016/J.RSER.2016.11.054.
10. Z. Pu, G. Zhang, A. Hassanpour, D. Zheng, S. Wang, S. Liao, Z. Chen, and S. Sun, Regenerative fuel cells: Recent progress, challenges, perspectives and their applications for space energy system, *Appl Energy.* **283** (2021): 116376. https://doi.org/10.1016/J.APENERGY.2020.116376.
11. L. Tang, P. Leung, M.R. Mohamed, Q. Xu, S. Dai, X. Zhu, C. Flox, A.A. Shah, and Q. Liao, Capital cost evaluation of conventional and emerging redox flow batteries for grid storage applications, *Electrochim Acta.* **437** (2023): 141460. https://doi.org/10.1016/j.electacta.2022.141460.
12. L. Li, S. Kim, W. Wang, M. Vijayakumar, Z. Nie, B. Chen, J. Zhang, G. Xia, J. Hu, G. Graff, J. Liu, and Z. Yang, A stable vanadium redox-flow

battery with high energy density for large-scale energy storage, *Adv Energy Mater.* **1** (2011): 394–400. https://doi.org/10.1002/aenm.201100008.

13. B.K. Chakrabarti, J. Feng, E. Kalamaras, J. Rubio-Garcia, C. George, H. Luo, Y. Xia, V. Yufit, M.M. Titirici, C.T.J. Low, A. Kucernak, and N.P. Brandon, Hybrid redox flow cells with enhanced electrochemical performance via binderless and electrophoretically deposited nitrogen-doped graphene on carbon paper electrodes, *ACS Appl Mater Interfaces.* **12** (2020): 53869–53878. https://doi.org/10.1021/acsami.0c17616.
14. J. Rubio-Garcia, A. Kucernak, A. Parra-Puerto, R. Liu, and B. Chakrabarti, Hydrogen/functionalized benzoquinone for a high-performance regenerative fuel cell as a potential large-scale energy storage platform, *J Mater Chem A Mater.* **8** (2020): 3933–3941. https://doi.org/10.1039/C9TA12396B.
15. J. Rubio-Garcia, A. Kucernak, D. Zhao, D. Li, K. Fahy, V. Yufit, N. Brandon, and M. Gomez-Gonzalez, Hydrogen/manganese hybrid redox flow battery, *J Phys Energ.* **1** (2018): 015006. https://doi.org/10.1088/2515-7655/AAEE17.
16. W. Juda, Electrolytic process for producing halogen gases and the apparatus therefor, US3117066A, 1960. https://patents.google.com/patent/US3117066A/en.
17. W. Juda, Electrolytic process for producing a halogen from its respective acid and the apparatus therefor, US3291708A, 1964. https://patents.google.com/patent/US3291708 (accessed April 11, 2023).
18. W. Glass and G.H. Boyle, Performance of hydrogen-bromine fuel cells, *Adv Chem.* **12** (1969): 203–220. https://doi.org/10.1021/BA-1965-0047.CH015.
19. K.T. Cho, M.C. Tucker, and A.Z. Weber, A review of hydrogen/halogen flow cells, *Energy Technol.* **4** (2016): 655–678. https://doi.org/10.1002/ENTE.201500449.
20. R.S. Yeo, J. McBreen, A.C.C. Tseung, S. Srinivasan, and J. McElroy, An electrochemically regenerative hydrogen-chlorine energy storage system: Electrode kinetics and cell performance, *J Appl Electrochem.* **10** (1980): 393–404. https://doi.org/10.1007/BF00617215/METRICS.
21. G.G. Barna, S.N. Frank, T.H. Teherani, and L.D. Weedon, Lifetime studies in H_2/Br_2 fuel cells, *J Electrochem Soc.* **131** (1984): 1973–1980. https://doi.org/10.1149/1.2116003/XML.
22. E.B. Anderson, E.J. Taylor, G. Wilemski, and A. Gelb, A high performance hydrogen/chlorine fuel cell for space power applications, *J Power*

Sources. **47** (1994): 321–328. https://doi.org/10.1016/0378-7753(94)87011-X.

23. M. Thomassen, B. Børresen, G. Hagen, and R. Tunold, H_2/Cl_2 fuel cell for co-generation of electricity and HCl, *J Appl Electrochem.* **33** (2003): 9–13. https://doi.org/10.1023/A:1022942213808/METRICS.
24. M. Thomassen, B. Børresen, K. Scott, and R. Tunold, A computational simulation of a hydrogen/chlorine single fuel cell, *J Power Sources.* **157** (2006): 271–283. https://doi.org/10.1016/J.JPOWSOUR.2005.07.034.
25. V. Livshits, A. Ulus, and E. Peled, High-power H_2/Br_2 fuel cell, *Electrochem Commun.* **8** (2006): 1358–1362. https://doi.org/10.1016/J.ELECOM.2006.06.021.
26. C. Menictas and M. Skyllas-Kazacos, Performance of vanadium-oxygen redox fuel cell, *J Appl Electrochem.* **41** (2011): 1223–1232. https://doi.org/10.1007/s10800-011-0342-8.
27. B. Huskinson, J. Rugolo, S.K. Mondal, and M.J. Aziz, A high power density, high efficiency hydrogen–chlorine regenerative fuel cell with a low precious metal content catalyst, *Energy Environ Sci.* **5** (2012): 8690–8698. https://doi.org/10.1039/C2EE22274D.
28. N. Brandon, A. Kucernak, V. Yufit, *Regenerative Fuel Cells*, WO/2012/038379, 2012. https://patentscope.wipo.int/search/en/detail.jsf?docId=WO2012038379 (accessed April 11, 2023).
29. W.A. Braff, M.Z. Bazant, and C.R. Buie, Membrane-less hydrogen bromine flow battery, *Nat. Commun.* **4** (2013): 2346. https://doi.org/10.1038/ncomms3346.
30. V. Yufit, B. Hale, M. Matian, P. Mazur, and N.P. Brandon, Development of a regenerative hydrogen-vanadium fuel cell for energy storage applications, *J Electrochem Soc.* **160** (2013): A856–A861. https://doi.org/10.1149/2.086306JES/XML.
31. S. Nawar, B. Huskinson, and M. Aziz, Benzoquinone-hydroquinone couple for flow battery, *Mater Res Soc Symp Proc.* **1491** (2013): 1–6. https://doi.org/10.1557/OPL.2012.1737/metrics.
32. J. Noack, J. Tübke, and K. Pinkwart, Dithionite/air direct ion liquid fuel cell, *J Power Sources.* **286** (2015): 232–238. https://doi.org/10.1016/J.JPOWSOUR.2015.03.159.
33. V. Yarlagadda, G. Lin, P.Y. Chong, and T. Van Nguyen, High active surface area and durable multi-wall carbon nanotube-based electrodes for the bromine reactions in H_2-Br_2 fuel cells, *J Electrochem Soc.* **163** (2016): A5134–A5143. https://doi.org/10.1149/2.0181601jes.

34. J. Grosse Austing, C. Nunes Kirchner, L. Komsiyska, and G. Wittstock, Layer-by-layer modification of Nafion membranes for increased life-time and efficiency of vanadium/air redox flow batteries, *J Memb Sci.* **510** (2016): 259–269. https://doi.org/10.1016/J.MEMSCI.2016.03.005.

35. M.C. Tucker, A. Weiss, and A.Z. Weber, Improvement and analysis of the hydrogen-cerium redox flow cell, *J Power Sources.* **327** (2016): 591–598. https://doi.org/10.1016/j.jpowsour.2016.07.105.

36. C.A.P. Muñoz, H.H. Dewage, V. Yufit, and N.P. Brandon, A unit cell model of a regenerative hydrogen-vanadium fuel cell, *J Electrochem Soc.* **164** (2017): F1717–F1732. https://doi.org/10.1149/2.1431714jes.

37. N. Brandon, V. Yufit, A. Kucernak, J. Rubio-Garcia, *Regenerative Fuel Cells,* 2017. https://patentscope.wipo.int/search/en/detail.jsf?docId=WO2017103578 (accessed April 11, 2023).

38. K.M. Tenny, V.S. Lakhanpal, R.P. Dowd, V. Yarlagadda, and T. Van Nguyen, Impact of multi-walled carbon nanotube fabrication on carbon cloth electrodes for hydrogen-vanadium reversible fuel cells, *J Electrochem Soc.* **164** (2017): A2534–A2538. https://doi.org/10.1149/2.1151712JES/XML.

39. Z. Li, M.S. Pan, L. Su, P.C. Tsai, A.F. Badel, J.M. Valle, S.L. Eiler, K. Xiang, F.R. Brushett, and Y.M. Chiang, Air-breathing aqueous sulfur flow battery for ultralow-cost long-duration electrical storage, *Joule.* 1 (2017): 306–327. https://doi.org/10.1016/J.JOULE.2017.08.007.

40. K. Saadi, P. Nanikashvili, Z. Tatus-Portnoy, S. Hardisty, V. Shokhen, M. Zysler, and D. Zitoun, Crossover-tolerant coated platinum catalysts in hydrogen/bromine redox flow battery, *J Power Sources.* **422** (2019): 84–91. https://doi.org/10.1016/J.JPOWSOUR.2019.03.043.

41. B. Chakrabarti, V. Yufit, A. Kavei, Y. Xia, G. Stevenson, E. Kalamaras, H. Luo, J. Feng, F. Tariq, O. Taiwo, M.M. Titirici, and N. Brandon, Charge/discharge and cycling performance of flexible carbon paper electrodes in a regenerative hydrogen/vanadium fuel cell, *Int J Hydrogen Energy.* **44** (2019): 30093–30107. https://doi.org/10.1016/J.IJHYDENE.2019.09.151.

42. J. Rubio-Garcia, J. Cui, A. Parra-Puerto, and A. Kucernak, Hydrogen/vanadium hybrid redox flow battery with enhanced electrolyte concentration, *Energy Storage Mater.* **31** (2020): 1–10. https://doi.org/10.1016/j.ensm.2020.05.031.

43. B.K. Chakrabarti, E. Kalamaras, M. Ouyang, X. Liu, G. Remy, P.F. Wilson, M.A. Williams, J. Rubio-Garcia, V. Yufit, G. Bree, Y.S. Hajimolana, A. Singh, F. Tariq, C.T.J. Low, B. Wu, C. George, and N.P. Brandon, Trichome-like

carbon-metal fabrics made of carbon microfibers, carbon nanotubes, and Fe-based nanoparticles as electrodes for regenerative hydrogen/vanadium flow cells, *ACS Appl Nano Mater.* **4** (2021): 10754–10763. https://doi.org/10.1021/acsanm.1c02195.

44. Y. Xia, M. Ouyang, V. Yufit, R. Tan, A. Regoutz, A. Wang, W. Mao, B. Chakrabarti, A. Kavei, Q. Song, A.R. Kucernak, and N.P. Brandon, A cost-effective alkaline polysulfide-air redox flow battery enabled by a dual-membrane cell architecture, *Nat Commun.* **13** (2022): 2388. https://doi.org/10.1038/s41467-022-30044-w.
45. J. Rubio-Garcia, A. Kucernak, B.K. Chakrabarti, D. Zhao, D. Li, Y. Tang, M. Ouyang, C.T.J. Low, and N. Brandon, High performance H_2–Mn regenerative fuel cells through an improved positive electrode morphology, *Batteries* **9** (2023): 108. https://doi.org/10.3390/BATTERIES9020108.
46. Y.A. Hugo, *Membrane Development for Large-Scale Hydrogen-Bromine Flow Batteries*, Technische Universiteit Eindhoven, 2020.
47. R. Seeber, C. Zanardi, and G. Inzelt, Links between electrochemical thermodynamics and kinetics, *ChemTexts.* 1 (2015): 1–16. https://doi.org/10.1007/s40828-015-0018-9.
48. J.K. Wlodarczyk, M. Küttinger, A.K. Friedrich, and J.O. Schumacher, Exploring the thermodynamics of the bromine electrode in concentrated solutions for improved parametrisation of hydrogen–bromine flow battery models, *J Power Sources.* **508** (2021): 230202. https://doi.org/10.1016/J.JPOWSOUR.2021.230202.
49. W. Sheng, H.A. Gasteiger, and Y. Shao-Horn, Hydrogen oxidation and evolution reaction kinetics on platinum: Acid vs alkaline electrolytes, *J Electrochem Soc.* **157** (2010): B1529. https://doi.org/10.1149/1.3483106.
50. T. Reier, H.N. Nong, D. Teschner, R. Schlögl, and P. Strasser, Electrocatalytic oxygen evolution reaction in acidic environments: Reaction mechanisms and catalysts, *Adv Energy Mater.* **7** (2017): 1601275. https://doi.org/10.1002/AENM.201601275.
51. J.K. Nørskov, J. Rossmeisl, A. Logadottir, L. Lindqvist, J.R. Kitchin, T. Bligaard, and H. Jónsson, Origin of the overpotential for oxygen reduction at a fuel-cell cathode, *J Phys Chem B.* **108** (2004): 17886–17892. https://doi.org/10.1021/jp047349j.
52. J. Durst, C. Simon, F. Hasche, and H.A. Gasteiger, Hydrogen oxidation and evolution reaction kinetics on carbon-supported Pt, Ir, Rh, and Pd electrocatalysts in acidic media, *J Electrochem Soc.* **162** (2014): F190–F203. https://doi.org/10.1149/2.0981501jes.

53. C. Jackson, X. Lin, P.B.J. Levecque, and A.R.J. Kucernak, Toward understanding the utilization of oxygen reduction electrocatalysts under high mass transport conditions and high overpotentials, *ACS Catal.* **12** (2022): 200–211. https://doi.org/10.1021/acscatal.1c03908.

54. P.J. Rheinlander, J. Herranz, J. Durst, and H.A. Gasteiger, Kinetics of the hydrogen oxidation/evolution reaction on polycrystalline platinum in alkaline electrolyte reaction order with respect to hydrogen pressure, *J Electrochem Soc.* **161** (2014): F1448–F1457. https://doi.org/10.1149/2.0501414jes.

55. E.J.F. Dickinson and A.J. Wain, The Butler–Volmer equation in electrochemical theory: Origins, value, and practical application, *J Electroanal Chem.* **872** (2020): 114145. https://doi.org/10.1016/J.JELECHEM.2020.114145.

56. C. Jackson, L.F.J.M. Raymakers, M.J.J. Mulder, and A.R.J. Kucernak, Assessing electrocatalyst hydrogen activity and CO tolerance: Comparison of performance obtained using the high mass transport 'floating electrode' technique and in electrochemical hydrogen pumps, *Appl Catal B.* **268** (2020): 118734. https://doi.org/10.1016/J.APCATB.2020.118734.

57. C.M. Zalitis, J. Sharman, E. Wright, and A.R. Kucernak, Properties of the hydrogen oxidation reaction on Pt/C catalysts at optimised high mass transport conditions and its relevance to the anode reaction in PEFCs and cathode reactions in electrolysers, *Electrochim Acta.* **176** (2015): 763–776. https://doi.org/10.1016/j.electacta.2015.06.146.

58. A.R. Kucernak and C. Zalitis, General models for the electrochemical hydrogen oxidation and hydrogen evolution reactions: Theoretical derivation and experimental results under near mass-transport free conditions, *J Phys Chem C.* **120** (2016): 10721–10745. https://doi.org/10.1021/acs.jpcc.6b00011.

59. H. Wang, S.Y. Sayed, E.J. Luber, B.C. Olsen, S.M. Shirurkar, S. Venkatakrishnan, U.M. Tefashe, A.K. Farquhar, E.S. Smotkin, R.L. McCreery, and J.M. Buriak, Redox flow batteries: How to determine electrochemical kinetic parameters, *ACS Nano.* **14** (2020): 2575–2584. https://doi.org/10.1021/acsnano.0c01281.

60. H.M.A. Amin, Y. Uchida, E. Kätelhön, and R.G. Compton, Determination of standard electrochemical rate constants from semi-circular sweep voltammetry: A combined theoretical and experimental study, *J Electroanal Chem.* **880** (2021): 114891. https://doi.org/10.1016/J.JELECHEM.2020.114891.

61. K.J. Lee, C.T. Gruninger, K.M. Lodaya, S. Qadeer, B.E. Griffith, and J.L. Dempsey, Analysis of multi-electron, multi-step homogeneous catalysis by rotating disc electrode voltammetry: Theory, application, and obstacles, *Analyst.* **145** (2020): 1258–1278. https://doi.org/10.1039/C9AN02192B.

62. J. Heinze, Cyclic voltammetry—"Electrochemical spectroscopy". New analytical methods (25), *Angew Chem Int Ed English.* **23** (1984): 831–847. https://doi.org/10.1002/ANIE.198408313.

63. W. He and T. Van Nguyen, Edge effects on reference electrode measurements in PEM fuel cells, *J Electrochem Soc.* **151** (2004): A185. https://doi.org/10.1149/1.1634272/XML.

64. K. Mitsuda and T. Murahashi, Polarization study of a fuel cell with four reference electrodes, *J Electrochem Soc.* **137** (1990): 3079–3085. https://doi.org/10.1149/1.2086163/XML.

65. A.A. Kulikovsky and P. Berg, Positioning of a reference electrode in a PEM fuel cell, *J Electrochem Soc.* **162** (2015): F843–F848. https://doi.org/10.1149/2.0231508JES/XML.

66. G. Hinds and E. Brightman, In situ mapping of electrode potential in a PEM fuel cell, *Electrochem Commun.* **17** (2012): 26–29. https://doi.org/10.1016/J.ELECOM.2012.01.007.

67. H.H. Dewage, V. Yufit, and N.P. Brandon, Study of loss mechanisms using half-cell measurements in a regenerative hydrogen vanadium fuel cell, *J Electrochem Soc.* **163** (2016): A5236–A5243. https://doi.org/10.1149/2.0301601jes.

68. Z. Wang, Y. Li, J. Ren, J. Sun, T. Wang, B. Liu, X. Fan, and T. Zhao, A response time-based method to operando decouple the polarizations in redox flow batteries, *Cell Rep Phys Sci.* **4** (2023): 101395. https://doi.org/10.1016/j.xcrp.2023.101395.

69. M.R. Gerhardt, L.M. Pant, J.C. Bui, A.R. Crothers, V.M. Ehlinger, J.C. Fornaciari, J. Liu, and A.Z. Weber, Method: Practices and pitfalls in voltage breakdown analysis of electrochemical energy-conversion systems, *J Electrochem Soc.* **168** (2021): 074503. https://doi.org/10.1149/1945-7111/ABF061.

70. Y. Yao, J. Lei, Y. Shi, F. Ai, and Y.C. Lu, Assessment methods and performance metrics for redox flow batteries, *Nature Energy* **6** (2021): 582–588. https://doi.org/10.1038/s41560-020-00772-8.

71. D.G. Kwabi, Y. Ji, and M.J. Aziz, Electrolyte lifetime in aqueous organic redox flow batteries: A critical review, *Chem Rev.* **120** (2020): 6467–6489. https://doi.org/10.1021/acs.chemrev.9b00599.

72. J. Shin, C. Kim, B. Jeong, N. Vaz, and H. Ju, New operating strategy for all-vanadium redox flow batteries to mitigate electrolyte imbalance, *J Power Sources*. **526** (2022): 231144. https://doi.org/10.1016/j.jpowsour.2022.231144.

73. M.C. Tucker, K.T. Cho, A.Z. Weber, G. Lin, and T. Van Nguyen, Optimization of electrode characteristics for the Br_2/H_2 redox flow cell, *J Appl Electrochem*. **45** (2015): 11–19. https://doi.org/10.1007/s10800-014-0772-1.

74. K.T. Cho, P. Ridgway, A.Z. Weber, S. Haussener, V. Battaglia, and V. Srinivasan, High performance hydrogen/bromine redox flow battery for grid-scale energy storage, *J Electrochem Soc*. **159** (2012): A1806–A1815. https://doi.org/10.1149/2.018211jes.

75. J. Wlodarczyk, *A 1D Model of a Hydrogen-Bromine Redox Flow Battery*, Master's Thesis, Zurich University of Applied Sciences, 2018.

76. A. H. Burr and J. B. Cheatham, *Mechanical Analysis and Design*, 2nd edition, Prentice Hall, Englewood Cliffs, NJ 07632, 1995.

77. Y. Shi, Z. Wei, H. Liu, and J. Zhao, Dynamic modeling of long-term operations of vanadium/air redox flow battery with different membranes, *J Energy Storage*. **50** (2022): 104171. https://doi.org/10.1016/J.EST.2022.104171.

78. B.K. Chakrabarti, E. Kalamaras, A.K. Singh, A. Bertei, J. Rubio-Garcia, V. Yufit, K.M. Tenny, B. Wu, F. Tariq, Y.S. Hajimolana, N.P. Brandon, C.T. John Low, E.P.L. Roberts, Y.-M. Chiang, and F.R. Brushett, Modelling of redox flow battery electrode processes at a range of length scales: A review, *Sustain Energy Fuels*. **4** (2020). https://doi.org/10.1039/d0se00667j.

79. M. Skyllas-Kazacos, Review: Highlights of UNSW all-vanadium redox battery development: 1983 to present, *J Electrochem Soc*. **169** (2022): 070513. https://doi.org/10.1149/1945-7111/AC7BAB.

80. C.L. Chen, H.K. Yeoh, and M.H. Chakrabarti, One dimensional mathematical modelling of the all-vanadium and vanadium/oxygen redox flow batteries, *ECS Trans*. **66** (2015): 1–23. https://doi.org/10.1149/06610.0001ecst.

81. M. Risbud, C. Menictas, M. Skyllas-Kazacos, and J. Noack, Vanadium oxygen fuel cell utilising high concentration electrolyte, *Batteries* **5**, (2019): 24. https://doi.org/10.3390/BATTERIES5010024.

82. J. Grosse Austing, C. Nunes Kirchner, L. Komsiyska, and G. Wittstock, Investigation of crossover processes in a unitized bidirectional vanadium/air redox flow battery, *J Power Sources*. **306** (2016): 692–701. https://doi.org/10.1016/J.JPOWSOUR.2015.12.052.

83. Y. Shi, Z. Wei, H. Liu, and J. Zhao, Dynamic modeling of long-term operations of vanadium/air redox flow battery with different membranes, *J Energy Storage.* **50** (2022): 104171. https://doi.org/10.1016/J.EST.2022.104171.

84. Y. Shi, C. Eze, B. Xiong, W. He, H. Zhang, T.M. Lim, A. Ukil, and J. Zhao, Recent development of membrane for vanadium redox flow battery applications: A review, *Appl Energy.* **238** (2019): 202–224. https://doi.org/10.1016/j.apenergy.2018.12.087.

85. O. David, K. Percin, T. Luo, Y. Gendel, and M. Wessling, Proton-exchange membranes based on sulfonated poly(ether ether ketone)/polyaniline blends for all- and air-vanadium redox flow battery applications, *J Energy Storage.* **1** (2015): 65–71. https://doi.org/10.1016/j.est.2015.01.001.

86. Y. Shi and J. Zhao, A dynamic model incorporating the effects of the ion diffusion and side reactions for the vanadium/air redox flow battery, in: *ASME 2018 12th International Conference on Energy Sustainability*, American Society of Mechanical Engineers, 2018. https://doi.org/10.1115/ES2018-7120.

87. S. Sreenath, P.S. Nayanthara, C.M. Pawar, M.C. Noufal, and R.K. Nagarale, Phenolic triamine dangling poly(VDF-co-HFP) anion exchange membrane for all aqueous organic redox flow battery, *J Energy Storage.* **40** (2021): 102689. https://doi.org/10.1016/j.est.2021.102689.

88. H. Cheng, J. Zhao, and M.K. Rowinski, Study on two wall temperature peaks of supercritical fluid mixed convective heat transfer in circular tubes, *Int J Heat Mass Transf.* **113** (2017): 257–267. https://doi.org/10.1016/j.ijheatmasstransfer.2017.05.078.

89. C.H.L. Tempelman, J.F. Jacobs, R.M. Balzer, and V. Degirmenci, Membranes for all vanadium redox flow batteries, *J Energy Storage.* **32** (2020): 101754. https://doi.org/10.1016/j.est.2020.101754.

90. S. Zhang, W. Guo, F. Yang, P. Zheng, R. Qiao, and Z. Li, Recent progress in polysulfide redox-flow batteries, *Batter Supercaps.* **2** (2019): 627–637. https://doi.org/10.1002/BATT.201900056.

91. S. Licht, An energetic medium for electrochemical storage utilizing the high aqueous solubility of potassium polysulfide, *J Electrochem Soc.* **134** (1987): 2137–2141. https://doi.org/10.1149/1.2100838.

92. C. Zhao, G.L. Xu, Z. Yu, L. Zhang, I. Hwang, Y.X. Mo, Y. Ren, L. Cheng, C.J. Sun, Y. Ren, X. Zuo, J.T. Li, S.G. Sun, K. Amine, and T. Zhao, A high-energy and long-cycling lithium–sulfur pouch cell via a macroporous catalytic cathode with double-end binding sites, *Nature Nanotechnology* (2020): 166–173. https://doi.org/10.1038/s41565-020-00797-w.

93. Z. Li and Y.-C. Lu, Polysulfide-based redox flow batteries with long life and low levelized cost enabled by charge-reinforced ion-selective membranes, *Nat Energy*. **6** (2021): 517–528. https://doi.org/10.1038/s41560-021-00804-x.

94. Z. Li, G. Weng, Q. Zou, G. Cong, and Y.-C. Lu, A high-energy and low-cost polysulfide/iodide redox flow battery, *Nano Energy*. **30** (2016): 283–292. https://doi.org/10.1016/j.nanoen.2016.09.043.

95. L. Su, A.F. Badel, C. Cao, J.J. Hinricher, and F.R. Brushett, Toward an inexpensive aqueous polysulfide–polyiodide redox flow battery, *Ind Eng Chem Res*. **56** (2017): 9783–9792. https://doi.org/10.1021/acs.iecr.7b01476.

96. S.H. Ge, B.L. Yi, and H.M. Zhang, Study of a high power density sodium polysulfide/bromine energy storage cell, *J Appl Electrochem*. **34** (2004): 181–185. https://doi.org/10.1023/B:JACH.0000009936.82613.ad.

97. P. Zhao, H. Zhang, H. Zhou, and B. Yi, Nickel foam and carbon felt applications for sodium polysulfide/bromine redox flow battery electrodes, *Electrochim Acta*. **51** (2005): 1091–1098. https://doi.org/10.1016/j.electacta.2005.06.008.

98. X. Wei, G.-G. Xia, B. Kirby, E. Thomsen, B. Li, Z. Nie, G.G. Graff, J. Liu, V. Sprenkle, and W. Wang, An aqueous redox flow battery based on neutral alkali metal ferri/ferrocyanide and polysulfide electrolytes, *J Electrochem Soc*. **163** (2016): A5150–A5153. https://doi.org/10.1149/2.0221601jes.

99. I.A. Khan, A.S. Alzahrani, S. Ali, M. Mansha, M.N. Tahir, M. Khan, H.A. Qayyum, and S.A. Khan, Development of membranes and separators to inhibit cross-shuttling of sulfur in polysulfide-based redox flow batteries: A review, *Chem. Rec*. (2023): e202300171. https://doi.org/10.1002/tcr.202300171.

100. W.F. Giggenbach, Equilibriums involving polysulfide ions in aqueous sulfide solutions up to 240.deg., *Inorg Chem*. **13** (1974): 1724–1730. https://doi.org/10.1021/ic50137a038.

101. S. Licht, G. Hodes, and J. Manassen, Numerical analysis of aqueous polysulfide solutions and its application to cadmium chalcogenide/polysulfide photoelectrochemical solar cells, *Inorg Chem*. **25** (1986): 2486–2489. https://doi.org/10.1021/ic00235a003.

102. S.R. Narayan, A. Manohar, and S. Mukerjee, Bi-functional oxygen electrodes: Challenges and prospects, *Interface Magazine*. **24** (2015): 65–69. https://doi.org/10.1149/2.F06152if.

103. M.J. Baran, M.N. Braten, S. Sahu, A. Baskin, S.M. Meckler, L. Li, L. Maserati, M.E. Carrington, Y.-M. Chiang, D. Prendergast, and B.A. Helms, Design rules for membranes from polymers of intrinsic microporosity for crossover-free aqueous electrochemical devices, *Joule.* **3** (2019): 2968–2985. https://doi.org/10.1016/j.joule.2019.08.025.

104. P. Leung, D. Aili, Q. Xu, A. Rodchanarowan, and A.A. Shah, Rechargeable organic–air redox flow batteries, *Sustain Energy Fuels.* **2** (2018): 2252–2259. https://doi.org/10.1039/C8SE00205C.

105. G. Nicoletti, N. Arcuri, G. Nicoletti, and R. Bruno, A technical and environmental comparison between hydrogen and some fossil fuels, *Energy Convers Manag.* **89** (2015): 205–213. https://doi.org/10.1016/j.enconman.2014.09.057.

106. M. Yue, H. Lambert, E. Pahon, R. Roche, S. Jemei, and D. Hissel, Hydrogen energy systems: A critical review of technologies, applications, trends and challenges, *Renewable and Sustainable Energy Reviews.* **146** (2021): 111180. https://doi.org/10.1016/j.rser.2021.111180.

107. D.P. Trudgeon and X. Li, Enhanced surface area carbon cathodes for the hydrogen–bromine redox flow battery, *Batteries* **8** (2022): 276. https://doi.org/10.3390/BATTERIES8120276.

108. M.L. Perry and A.Z. Weber, Advanced redox-flow batteries: A perspective, *J Electrochem Soc.* **163** (2016): A5064–A5067. https://doi.org/10.1149/2.0101601jes.

109. Y.A. Hugo, W. Kout, G. Dalessi, A. Forner-Cuenca, Z. Borneman, and K. Nijmeijer, Techno-economic analysis of a kilo-watt scale hydrogen-bromine flow battery system for sustainable energy storage, *Processes.* **8** (2020): 1492. https://doi.org/10.3390/pr8111492.

110. M. Küttinger, J.K. Wlodarczyk, D. Daubner, P. Fischer, and J. Tübke, High energy density electrolytes for H_2/Br_2 redox flow batteries, their polybromide composition and influence on battery cycling limits, *RSC Adv.* **11** (2021): 5218–5229. https://doi.org/10.1039/D0RA10721B.

111. Y. Munaiah, S. Dheenadayalan, P. Ragupathy, and V.K. Pillai, High performance carbon nanotube-based electrodes for zinc bromine redox flow batteries, *ECS J Solid State Sci Technol.* **2** (2013): M3182–M3186. https://doi.org/10.1149/2.024310jss.

112. Y. Popat, D. Trudgeon, C. Zhang, F.C. Walsh, P. Connor, and X. Li, Carbon materials as positive electrodes in bromine-based flow batteries, *ChemPlusChem.* **87** (2022): 16. https://doi.org/10.1002/cplu.202100441.

113. G. Lin, P.Y. Chong, V. Yarlagadda, T.V. Nguyen, R.J. Wycisk, P.N. Pintauro, M. Bates, S. Mukerjee, M.C. Tucker, and A.Z. Weber, Advanced hydrogen-bromine flow batteries with improved efficiency, durability and cost, *J Electrochem Soc.* **163** (2016) A5049–A5056. https://doi.org/10.1149/2.0071601jes.

114. Y. Popat, D.P. Trudgeon, X. Li, P. Connor, A. Asokan, and M.E. Suss, Electrochemical testing of carbon materials as bromine electrodes for the hydrogen-bromine redox flow battery, *Batteries.* **8** (2022): 166. https://doi.org/10.3390/batteries8100166.

115. K.T. Cho, P. Albertus, V. Battaglia, A. Kojic, V. Srinivasan, and A.Z. Weber, Optimization and analysis of high-power hydrogen/bromine-flow batteries for grid-scale energy storage, *Energy Technology.* **1** (2013): 596–608. https://doi.org/10.1002/ente.201300108.

116. K.T. Cho, M.C. Tucker, M. Ding, P. Ridgway, V.S. Battaglia, V. Srinivasan, and A.Z. Weber, Cyclic performance analysis of hydrogen/bromine flow batteries for grid-scale energy storage, *ChemPlusChem.* **80** (2015): 402–411. https://doi.org/10.1002/cplu.201402043.

117. M.E. Suss, K. Conforti, L. Gilson, C.R. Buie, and M.Z. Bazant, Membraneless flow battery leveraging flow-through heterogeneous porous media for improved power density and reduced crossover, *RSC Adv.* **6** (2016): 100209–100213. https://doi.org/10.1039/C6RA22608F.

118. L. Zhang, Z.-G. Shao, X. Wang, H. Yu, S. Liu, and B. Yi, The characterization of graphite felt electrode with surface modification for H_2/Br_2 fuel cell, *J Power Sources.* **242** (2013): 15–22. https://doi.org/10.1016/j.jpowsour.2013.05.049.

119. M.C. Karaeyvaz, B. Duman, and B. Fıçıcılar, An alternative HCMS carbon catalyst in bromine reduction reaction for hydrogen-bromine flow batteries, *Int J Hydrogen Energy.* **46** (2021): 29512–29522. https://doi.org/10.1016/j.ijhydene.2020.11.055.

120. M.S. Javed, N. Shaheen, S. Hussain, J. Li, S.S.A. Shah, Y. Abbas, M.A. Ahmad, R. Raza, and W. Mai, An ultra-high energy density flexible asymmetric supercapacitor based on hierarchical fabric decorated with 2D bimetallic oxide nanosheets and MOF-derived porous carbon polyhedra, *J Mater Chem A Mater.* **7** (2019): 946–957. https://doi.org/10.1039/C8TA08816K.

121. M.S. Javed, A. Mateen, S. Ali, X. Zhang, I. Hussain, M. Imran, S.S.A. Shah, and W. Han, The emergence of 2D MXenes based Zn-ion batteries: Recent development and prospects, *Small.* 18 (2022): 2201989. https://doi.org/10.1002/smll.202201989.

122. K. Fatih, D.P. Wilkinson, F. Moraw, A. Ilicic, and F. Girard, Advancements in the direct hydrogen redox fuel cell, *Electrochem Solid State Lett.* **11** (2008): B11. https://doi.org/10.1149/1.2812788.

123. L. Yan, Y. Shao, and W. Wang, A hydrogen iron flow battery with high current density and long cyclability enabled through circular water management, *Energy Environ. Mater.* **6** (2023): e12478. https://doi.org/10.1002/eem2.12478.

124. C.A. Pino-Muñoz, B.K. Chakrabarti, V. Yufit, and N.P. Brandon, Characterization of a regenerative hydrogen-vanadium fuel cell using an experimentally validated unit cell model, *J Electrochem Soc.* **166** (2019): A3511–A3524. https://doi.org/10.1149/2.0211914JES/XML.

125. R.P. Dowd, A. Verma, Y. Li, D. Powers, R. Wycisk, P.N. Pintauro, and T. Van Nguyen, A hydrogen-vanadium reversible fuel cell crossover study, *J Electrochem Soc.* **164** (2017): F1608–F1614. https://doi.org/10.1149/2.11001714jes.

126. M.H. Chakrabarti, N.P. Brandon, S.A. Hajimolana, F. Tariq, V. Yufit, M.A. Hashim, M.A. Hussain, C.T.J. Low, and P.V. Aravind, Application of carbon materials in redox flow batteries, *J Power Sources.* **253** (2014): 150–166. https://doi.org/10.1016/j.jpowsour.2013.12.038.

127. A.M. Pezeshki, J.T. Clement, G.M. Veith, T.A. Zawodzinski, and M.M. Mench, High performance electrodes in vanadium redox flow batteries through oxygen-enriched thermal activation, *J Power Sources.* **294** (2015): 333–338. https://doi.org/10.1016/j.jpowsour.2015.05.118.

128. L.F. Arenas, C. Ponce de León, and F.C. Walsh, Engineering aspects of the design, construction and performance of modular redox flow batteries for energy storage, *J Energy Storage.* **11** (2017): 119–153. https://doi.org/10.1016/J.EST.2017.02.007.

129. C.L. Chen, H.K. Yeoh, and M.H. Chakrabarti, An enhancement to Vynnycky's model for the all-vanadium redox flow battery, *Electrochim Acta.* **120** (2014): 16–179. https://doi.org/10.1016/j.electacta.2013.12.074.

130. Q. Zheng, X. Li, Y. Cheng, G. Ning, F. Xing, and H. Zhang, Development and perspective in vanadium flow battery modeling, *Appl Energy.* **132** (2014): 254–266. https://doi.org/10.1016/j.apenergy.2014.06.077.

131. L. Li, S. Kim, W. Wang, M. Vijayakumar, Z. Nie, B. Chen, J. Zhang, G. Xia, J. Hu, G. Graff, J. Liu, and Z. Yang, A stable vanadium redox-flow battery with high energy density for large-scale energy storage, *Adv Energy Mater.* **1** (2011): 394–400. https://doi.org/10.1002/aenm.201100008.

132. H. Hewa Dewage, B. Wu, A. Tsoi, V. Yufit, G. Offer, and N. Brandon, A novel regenerative hydrogen cerium fuel cell for energy storage applications, *J Mater Chem A Mater.* **3** (2015): 9446–9450. https://doi.org/10.1039/C5TA00571J.

133. F.C. Walsh, C. Poncedeléon, L. Berlouis, G. Nikiforidis, L.F. Arenas-Martínez, D. Hodgson, and D. Hall, The development of Zn–Ce hybrid redox flow batteries for energy storage and their continuing challenges, *ChemPlusChem.* **80** (2015): 288–311. https://doi.org/10.1002/CPLU.201402103.

134. P.K. Leung, C. Ponce De León, C.T.J. Low, and F.C. Walsh, Ce(III)/Ce(IV) in methanesulfonic acid as the positive half cell of a redox flow battery, *Electrochim Acta.* **56** (2011): 2145–2153. https://doi.org/10.1016/j.electacta.2010.12.038.

135. Z. Xie, Q. Liu, Z. Chang, and X. Zhang, The developments and challenges of cerium half-cell in zinc–cerium redox flow battery for energy storage, *Electrochim Acta.* **90** (2013): 695–704. https://doi.org/10.1016/J.ELECTACTA.2012.12.066.

136. G. Nikiforidis, L. Berlouis, D. Hall, and D. Hodgson, Impact of electrolyte composition on the performance of the zinc–cerium redox flow battery system, *J Power Sources.* **243** (2013): 691–698. https://doi.org/10.1016/j.jpowsour.2013.06.045.

137. M.C. Tucker and A.Z. Weber, Optimization of the cerium-hydrogen redox flow cell, 15/582,225, 2019. https://www.osti.gov/biblio/1576292 (accessed September 6, 2023).

138. K. Kordesh and M. Weissenbacher, Rechargeable alkaline manganese dioxide/zinc batteries, *J Power Sources.* **51** (1994): 61–78. https://doi.org/10.1016/0378-7753(94)01955-X.

139. D. Reynard, S. Maye, P. Peljo, V. Chanda, H.H. Girault, and S. Gentil, Vanadium–manganese redox flow battery: Study of MnIII disproportionation in the presence of other metallic ions, *Chem A Eur J.* **26** (2020): 7250–7257. https://doi.org/10.1002/chem.202000340.

140. H.R. Jiang, J. Sun, L. Wei, M.C. Wu, W. Shyy, and T.S. Zhao, A high power density and long cycle life vanadium redox flow battery, *Energy Storage Mater.* **24** (2020): 529–540. https://doi.org/10.1016/j.ensm.2019.07.005.

141. B. Huskinson, S. Nawar, M.R. Gerhardt, and M.J. Aziz, Novel quinone-based couples for flow batteries, *ECS Trans.* **53** (2013): 101. https://doi.org/10.1149/05307.0101ecst.

142. N. Christudas Dargily, R. Thimmappa, Z. Manzoor Bhat, M.C. Devendrachari, A.R. Kottaichamy, M. Gautam, S.P. Shafi, and M.O. Thotiyl, A rechargeable hydrogen battery, *J Phys Chem Lett.* **9** (2018): 2492–2497. https://doi.org/10.1021/acs.jpclett.8b00858.

143. B. Yang, L. Hoober-Burkhardt, S. Krishnamoorthy, A. Murali, G.K.S. Prakash, and S.R. Narayanan, High-performance aqueous organic flow battery with quinone-based redox couples at both electrodes, *J Electrochem Soc.* **163** (2016): A1442. https://doi.org/10.1149/2.1371607jes.

144. A. Alaswad, A. Omran, J.R. Sodre, T. Wilberforce, G. Pignatelli, M. Dassisti, A. Baroutaji, and A.G. Olabi, Technical and commercial challenges of proton-exchange membrane (PEM) fuel cells, *Energies (Basel).* **14** (2020): 144. https://doi.org/10.3390/en14010144.

Index